Unmanned System Technologies

Springer's Unmanned System Technologies (UST) book series publishes the latest developments in unmanned vehicles and platforms in a timely manner, with the highest of quality, and written and edited by leaders in the field. The aim is to provide an effective platform to global researchers in the field to exchange their research findings and ideas. The series covers all the main branches of unmanned systems and technologies, both theoretical and applied, including but not limited to:

- Unmanned aerial vehicles, unmanned ground vehicles and unmanned ships, and all unmanned systems related research in:
- Robotics Design
- Artificial Intelligence
- Guidance, Navigation and Control
- Signal Processing
- Circuit and Systems
- Mechatronics
- Big Data
- Intelligent Computing and Communication
- Advanced Materials and Engineering

The publication types of the series are monographs, professional books, graduate textbooks, and edited volumes.

Mingfang Du

Autonomous Vehicle Technology

Global Exploration and Chinese Practice

人民交通出版社股份有限公司
China Communications Press Co.,Ltd.

Springer

Mingfang Du
Tsinghua University
Beijing, China

ISSN 2523-3734 ISSN 2523-3742 (electronic)
Unmanned System Technologies
ISBN 978-981-19-4142-9 ISBN 978-981-19-4143-6 (eBook)
https://doi.org/10.1007/978-981-19-4143-6

Jointly published with China Communications Press Co., Ltd.
The print edition is not for sale in China (Mainland). Customers from China (Mainland) please order the print book from: China Communications Press Co., Ltd.

This Springer imprint is published by the registered company Springer Nature Singapore Pte Ltd.
The registered company address is: 152 Beach Road, #21-01/04 Gateway East, Singapore 189721, Singapore

Preface

In recent years, with the development of the economy and society, automobile automation, networking, and intelligent technology have made great progress. Autonomous vehicle technology is gradually becoming more and more practical and has been applied in some specific fields. It is influencing the global economy and quietly changing human productive activity and lifestyle. Autonomous vehicle is a type of intelligent vehicle, which can also be regarded as a type of fast-wheeled autonomous mobile robot. It uses on-board sensors to sense the surrounding environment of the vehicle and controls the vehicle's movement according to the information on the road, vehicle position, and obstacles, so that the vehicle can move safely, reliably a,nd autonomously. Autonomous vehicle integrates automatic control, artificial intelligence, navigation and positioning, machinery, vehicle, and other multidisciplinary technology. It is the product of the development of vehicle engineering, control science, computer science, and artificial intelligence theory and technology.

Autonomous vehicle has a bright future and great social significance. At present, the US, Germany, the UK, France, Japan, China, and other countries have all made the development of intelligent vehicles a part of national strategies and are investing more and more in the development and testing of autonomous vehicles. The situation is closely related to the arrival of the third historical climax of artificial intelligence. It is also related to autonomous vehicle's technical characteristics, industrial advantages, and impact on urban development. In terms of technology, autonomous vehicle is a combination of artificial intelligence, big data, Internet of Things, cloud computing, mobile Internet, advanced communication, intelligent control, cognitive computing, navigation, transportation, machinery, and other high and new technologies. It is presently the best application form of artificial intelligence. In terms of industry, autonomous vehicle is the core unit connecting automotive electronics and Internet of Vehicles industry chain and can drive the transformation and upgrading of the entire automotive industry. It may even benefit artificial intelligence, electronic information, network communication, automation, machinery, intelligent transportation, and other related industries, powering the transformation and upgrading of related fields. From the perspective of long-term development, autonomous vehicle will bring about the reconstruction of automobile-related industry chain and the extension of value chain.

From the perspective of urban development, autonomous vehicle and the Internet of Vehicles are gradually infiltrating into the urban intelligent transportation system and undertaking the historical mission of the ultimate realization of the future urban green transportation system. This book systematically discusses the core technology and industrialization methods of autonomous vehicle. Artificial intelligence city will be a new historical form for the further development of a new smart city. Autonomous vehicle and its industrial chain lay the cornerstone of the development of artificial intelligence city. It is the entrance and pillar of the construction and development of artificial intelligence city. The development of autonomous vehicle and autonomous system technology will add bright colors to the development of Digital China, smart society, and smart city in the new era. It will also contribute to the realization of "transportation power, manufacturing power, and network power".

The innovations of this book include the following: fast and robust road perception algorithm and system based on image pattern recognition, especially traffic sign recognition and complex road recognition, which are still technical problems in the field; Intelligent autonomous control of unmanned vehicle and intelligent decision-making based on machine learning; The Internet of Vehicles system based on 5G-V2X technology; Planning and design of unmanned vehicle test field and characteristic town; and Artificial intelligence city constructed by unmanned vehicle. The features of this book are as follows: complete content system, from the core technology of autonomous vehicle to the Internet of Vehicles, to industrialization and artificial intelligence city; at the micro level, in order to solve the existing technical problems as the background, the key technologies are deeply studied, and the theoretical foundation is profound; at the macro level, combining with the industrialization of unmanned vehicles and the application of intelligent transportation and smart city in specific scenarios, it pays attention to practicality.

This book consists of 10 chapters, which covers the development overview, general technology, vehicle radar system, visual lane detection and identification technology, navigation and positioning technology, motion control and planning decision-making technology, autonomous vehicle energy power system, vehicle networking, industrialization of autonomous vehicle, and the transition from autonomous vehicle to artificial intelligence cities. This book systematically and comprehensively expounds the core technology and industrialization focus of today's autonomous vehicle, which can be used as a reference book for the R&D and industrialization of autonomous vehicle. The content of this book is a comprehensive study of the research achievements of Prof. Mingfang Du in the doctoral research stage and the practical project experience in recent years. Except for Chap. 7 which was written by Wang Yawen, the rest of the book was written by Mingfang Du. The innovation of this book includes the fast robust road sensing algorithm and system based on image pattern recognition, especially traffic sign recognition and complex road recognition which are still technical problems in the field; intelligent autonomous control of vehicles and intelligent decision-making based on machine learning; the vehicle network system based on 5G-V2X technology; the planning and design of the test ground and characteristic town of autonomous vehicle; and artificial intelligence city, which is based on autonomous vehicle. The content of the book is comprehensive and detailed at the same time, starting from solving the existing technical problems and going into the industrialization of autonomous vehicle in specific scenarios, the application of intelligent transportation and smart city. In the process of writing this book, the author has been guided and assisted by Stanford University, Tsinghua University, Peking University, Beijing Institute of Technology, and experts from Academy of Military Sciences. It has also been assisted by the Standard and Quota Department of Ministry of Housing and Urban-Rural Development, the Development Center of Science, Technology and Industrialization of Ministry of Housing and Urban-Rural Development, Fujian Network Information Office, Xuzhou Bureau of Industry and Information Technology, and China Construction Technology Group. In 2011, due to the need for doctoral research, I began to learn the theory and technology of autonomous control of unmanned system and participate in the research and development of autonomous vehicle and autonomous mobile robots. I especially want to thank my supervisor Prof. Wang Junzheng of Beijing Institute of Technology, Academician Li Deyi and Chen Jie, and Prof. Xia Yuanqing who gave me guidance in the direction of research. Thanks for the guidance and help of Prof. Zeng Rong and Prof. Zhu Yan of Tsinghua University. Thanks for the responsible and hard work of editor Liu Bo and Director Xia Wei of People's Communications Press.

Due to the limited knowledge of the author, there may be mistakes in the book. Please feel free to contact me with corrections and advice at dumingfang@mail.tsinghua.edu.cn.

Beijing, China
August 2019

Mingfang Du

About This Book

This book covers the development overview of autonomous vehicle, general technology, vehicle radar system, visual lane detection and identification technology, navigation and positioning technology, motion control and planning decision-making technology, autonomous vehicle energy power system, Internet of Vehicles, autonomous vehicle industrialization and the transition from autonomous vehicle to artificial intelligence city.

This book provides reference for institutes and enterprises on research, development, and industrial application of autonomous vehicle. It can also be used as undergraduate and postgraduate teaching material for related majors such as artificial intelligence, automation, computer, and transportation.

Contents

About the Author

Mingfang Du, doctor of Engineering (major in control science and Engineering), associate professor, and graduate supervisor.

She is an Associate Professor of the national information research center of Tsinghua University and the head of smart city. In 2015, she graduated from the State Key Laboratory of complex system intelligent control and decision-making, School of automation, Beijing Institute of Technology, majoring in control science and engineering, and received a doctoral degree. During the period of doctoral study, she visited Germany to study German intelligent automation. She has a master's degree in control theory and control engineering and a bachelor's degree in industrial automation. She has won the title of young and middle-aged backbone teacher in Beijing and the second prize of Beijing Education and teaching achievement. She is an expert member of the information technology application Standardization Technical Committee of the Ministry of housing and urban-rural development of the People's Republic of China, an expert member of the digital twin technology application working committee of China Information Association, an expert member of the energy intelligence special committee of China Society of automation and China Electrotechnical Society, and an expert member of the building and urban information model professional committee and building robot Professional Committee of China Engineering Construction Standardization Association, Expert member of IEC/SEG expert group. As the first author, she has published more than 100 papers in international and domestic academic journals and conferences and published 10 monographs and teaching materials. She presided over or participated in more than 20 scientific research projects. She participated in the formulation or review of more than 10 national and industrial standards, such as the technical standard of urban information model basic platform, the data processing standard of urban information model, and the technical specification of the construction site supervision information system. She applied for 11 patents. She has five years of experience in the R&D and implementation of large-scale automation engineering projects of Honeywell. As a core R&D Engineer, she participated in the R&D of OPC interface software, industrial communication, and a series of middleware products and solved the integration problems of complex and heterogeneous industrial control systems.

Chapter 1
Overview of Autonomous Vehicle

1.1 Overview of Research and Development of Autonomous Vehicle Around the World

1.1.1 Basic Concepts

Autonomous vehicle, or self-piloting automobile, driverless car, computer-driven car, or wheeled mobile robot, is a type of intelligent car autonomously controlled by computer system. In essence, autonomous vehicle is a type of fast wheeled autonomous mobile robot. It uses on-board sensors to sense the surrounding environment of the vehicle, and controls the vehicle's movement according to the information of road, vehicle position and obstacles, so that the vehicle can move safely, reliably and autonomously. The basic mechanism of autonomous vehicle operation in the city can be illustrated in Fig. 1.1.

The laser locator on car roof radiates 64 laser beams, which bounce back after hitting objects around the car, giving the distance between. Another system at the bottom measures the accelerated speed, angular speed in three directions, then calculates the location of car with GPS or Beidou Data. All these data are sent to the onboard computer together with the images captured by the car cameras. The software processes these data at a very high speed, and makes judgments.

The society of Automotive Engineers of America classifies autonomous driving technology (as shown in Fig. 1.2). The classification is internationally recognized.

L0 belongs to traditional driving; L1 and L2 belong to assisted driving; L3-L5 belong to automatic driving. L5 technology is also known as unmanned driving. Therefore, in the order of automatic level, starting from the lowest, it should be assisted driving, automatic driving, unmanned driving.

Currently, driving assistance technology has been deployed in mass production vehicles, commonly known as advanced driver assistant systems (ADAS). ADAS uses all kinds of sensors installed on the car to sense the surrounding environment at any time during the driving, collect data, identify and track stationary and moving

M. Du, *Autonomous Vehicle Technology*, Unmanned System Technologies,
https://doi.org/10.1007/978-981-19-4143-6_1

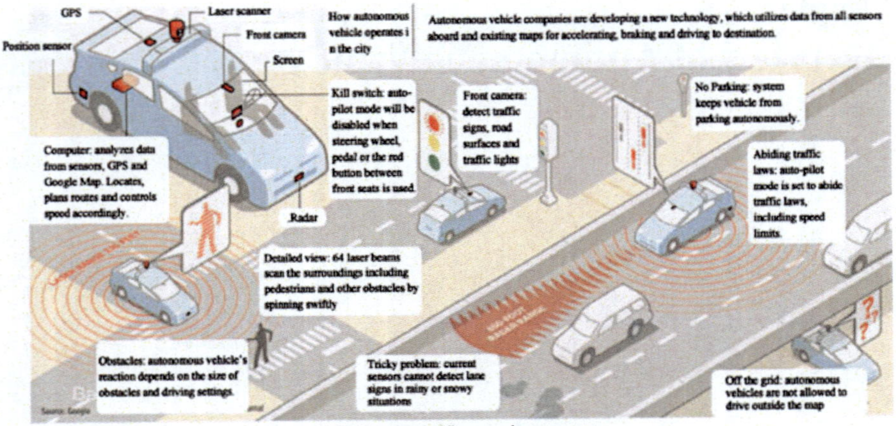

a) Urban traffic environment

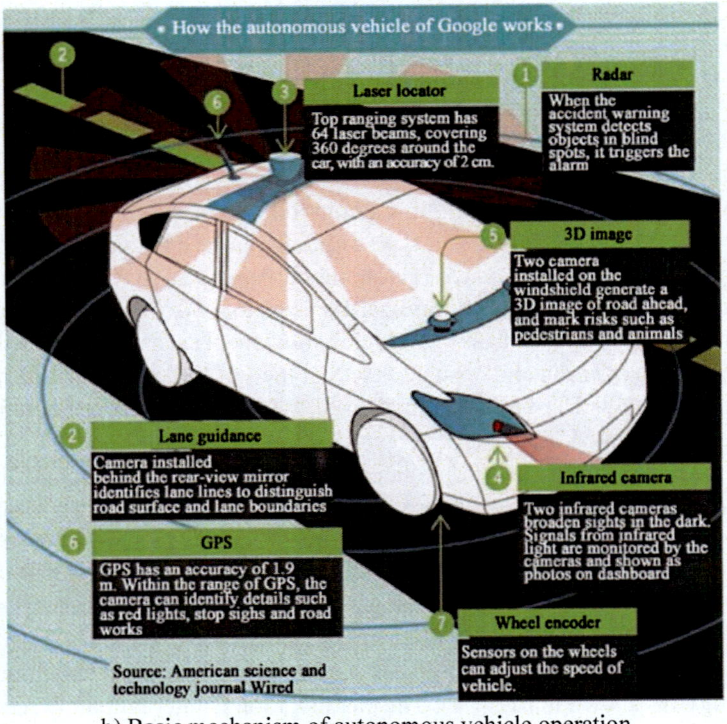

b) Basic mechanism of autonomous vehicle operation

Fig. 1.1 The basic mechanism of autonomous vehicle operation in the city

<table>
<tr><th colspan="2">Level</th><th rowspan="2">Name (SAE)</th><th rowspan="2">Definition (SAE)</th><th colspan="4">Subject</th></tr>
<tr><th>NHTSA</th><th>SAE</th><th>Execution of steering and acceleration/ deceleration</th><th>Monitoring of driving environment</th><th>Fallback performance of dynamic driving tasks</th><th>System capability</th></tr>
<tr><td>0</td><td>0</td><td>No Automation</td><td>the full -time performance by the human driver of all aspects of the dynamic driving task, even when enhanced by warning or intervention systems</td><td>human driver</td><td rowspan="3">human driver</td><td rowspan="4">human driver</td><td>n/a</td></tr>
<tr><td>1</td><td>1</td><td>Driver Assistance</td><td>the driving mode -specific execution by a driver assistance system of either steering or acceleration/deceleration using information about the driving environment and with the expectation that the human driver perform all remaining aspects of the dynamic driving task</td><td>human driver and system</td><td rowspan="4">some driving modes</td></tr>
<tr><td>2</td><td>2</td><td>Partial Automation</td><td>the driving mode-specific execution by one or more driver assistance systems of both steering and acceleration/deceleration using information about the driving environment and with the expectation that the human driver perform all remaining aspects of the dynamic driving task</td><td rowspan="4">system</td></tr>
<tr><td>3</td><td>3</td><td>Conditional Automation</td><td>the driving mode -specific performance by an automated driving system of all aspects of the dynamic driving task with the expectation that the human driver will respond appropriately to a request to intervene</td><td rowspan="3">system</td></tr>
<tr><td rowspan="2">4</td><td>4</td><td>High Automation</td><td>the driving mode -specific performance by an automated driving system of all aspects of the dynamic driving task, even if a human driver does not respond appropriately to a request to intervene</td><td rowspan="2">system</td></tr>
<tr><td>5</td><td>Full Automation</td><td>the full -time performance by an automated driving system of all aspects of the dynamic driving task under all roadway and environmental conditions that can be managed by a human driver</td><td>all driving modes</td></tr>
</table>

Fig. 1.2 Autonomous driving technology classification

objects, carry out systematic calculation and analysis with map data, so as to make the driver aware of the possible danger in advance, effectively increasing the comfort and safety level of driving. The primary ADAS is mainly passive alarm. When the vehicle detects potential danger, it reminds the driver to pay attention to abnormal vehicles or road conditions. For the latest ADAS technology, active intervention has become more common. Driving assistance is used in L1 and L2 of automatic driving technologies. At L1, the vehicle intervenes in the control of braking or steering, sharing the driver's duty. The features include adaptive cruise control (ACC), lane keep assist (LKA), automatic emergency braking (AEB), etc. At L2, the vehicle takes over multiple lateral and longitudinal controls, and the driving operation is completed by the system. However, the driver's attention should still be kept on the road, so as to take over the vehicle at any time. Unlike L1, lateral and longitudinal systems need to be integrated. Although compared with L1, the main part of driving operation in L2 is shifted from human to system, the principal subject responsible for peripheral monitoring is still human. Whether the principal subject responsible for peripheral monitoring is human or autonomous system, is the key of the transition from advanced driving assistance system (ADAS) to autonomous driving.

Unmanned driving technology is the highest level of autonomous driving technology, which means the driving is completely automatic. There may be only one start/stop button in an autonomous vehicle. Click the start button and inform the system of the destination, then the vehicle will take you there. The choice of the route and speed is made by the autonomous vehicle. With current technologies, there is still a lot of work to be done to achieve completely automatic driving, including high-precision electronic map positioning, environment-aware pattern recognition and cognitive algorithm, security of software and system, reliability and reconfigurability of system architecture. Further improvement of any one of these research tasks is quite difficult. According to the current technology development tendency, it is expected that after 2020, autonomous vehicles designed for specific scenarios will be massively produced. However, it will take at least 15 years to achieve all-weather and all-region unmanned driving.

At present, the leading foreign companies in the field of autonomous vehicle are Google, Ford, Tesla, NVIDIA, Apple, Uber, Honda, Qualcomm, Samsung, etc., while Chinese companies include Baidu, BAIC, SAIC, BYD, Geely, Chang'an, FAW and so on. A Chinese autonomous vehicle was tested on the Beijing-Tianjin Expressway in 2012.

1.1.2 Research and Development of Autonomous Vehicle Abroad

The study of automatic driving can be traced back to the beginning of the last century. In 1912, Cadillac's automatic start system meant that drivers no longer had to crank their cars manually. In 1939, Oldsmobile introduced the first automatic transmission system. In 1951, Chrysler introduced the first hydraulic steering system. In 1958, with Chrysler's cruise control system, drivers did not have to pay attention to speed. In 1970, Chrysler Imperial was equipped with anti-lock braking system. In 1997, some Toyota cars were equipped with radar-based adaptive cruise control, which can automatically maintain a safe distance from the vehicle ahead. In 2002, Toyota launched NightView, an in-car monitor that can display near-infrared images of the road ahead, highlighting obstacles. In 2003, Mercedes launched the Pre-Safe system, which uses sensors to predict imminent impact and take various preventive measures. In 2004, Infiniti (a Japanese luxury car brand) launched its first Out-of-Lane warning system to remind drivers when they leave the lane. In 2005, Volvo launched the first blind spot warning system, which gives an alarm when a vehicle enters the driver's blind spot. In 2006, Lexus (Toyota's high-end brand) launched a camera-sonar assisted parallel parking system. In 2007, Carnegie Mellon University's Tartan Team won the Department of Defense's automatic vehicle award. In 2008, Mercedes introduced Attention Assist to warn drivers when they show signs of fatigue. In 2009, Volvo launched a pedestrian monitoring system. In 2010, Audi driverless automatic car TTS drove 12.42 km to reach Pikes Peak of Rock Mountains; Google's fleet of seven driverless cars started to trial on California roads; Mercedes concept car F800 style showed Racing Assistant, a low speed adaptive cruise control system. In 2011, a driverless car created by National University of Defense Technology of China drove 177 miles from Changsha to Wuhan. In 2012, Alan Taub of General Motors predicted that the automatic driving system would become standard configuration of cars that year. In 2015, Google's driverless car has completed more than 1.3 million miles in automatic mode. Today, research of automatic driving technology is still the forefront and focus of the world's major car companies.

In recent years, the Defense Advanced Research Projects Agency of America (DARPA) has played an important role in the research of autonomous vehicles. DARPA launched the annual desert robot challenge "DARPA Grand Challenge" in 2004, and changed its name to "DARPA Urban Challenge" in 2007. The milestones in the research and development of autonomous vehicle include: (1) Demo III military robot jointly developed by US Army Research Laboratory, relevant government departments and industrial partners, which is the pioneer of semi-autonomous mobile robot research (2) The NavLab series of autonomous vehicles developed by Carnegie Mellon University (CMU), including NavLab-1, NavLab-5 and NavLab-11 (3) military autonomous vehicle projects led by DARPA, including Unmanned Ground Combat Vehicle (UGCV), Perception Off-road Tactical Mobile Robots (TMR), Organic Air Vehicle (OAV), etc. (4) "Boss" autonomous vehicle developed by CMU, which won the championship in "DARPA Urban Challenge" in 2007 (5) "Junior" developed by Stanford University, which won the second place in "DARPA Urban Challenge" in 2007 (6) The "Leonie" autonomous vehicle developed by German Technical University of Braunschweig, which demonstrated successful operation on the inner ring road of Braunschweig in 2010 (7) The driverless car for urban environment developed by Google (8) In July 2010, four autonomous vehicles developed by the University of Parma in Italy set out from Parma, Italy, and arrived in Shanghai, China safely after driving for 100 days, passing through 9 countries, completing the first international travel in the history of autonomous vehicle.

The latest reports on autonomous vehicles in practice are as follows: in early 2014, the Navia autonomous vehicle developed by French technology company Induct became the first autonomous vehicle model sold in the world, with a maximum speed of 20 km/h, which is suitable for airports, university campuses, gymnasiums, arenas and other places. In October 2013, Nissan was officially authorized to test its automatic driving technology in the household market, which was expected to enter the market in 2020. In May 2012, the Nevada motor vehicle administration issued the first autonomous vehicle license plate for Google, allowing its vehicles to be tested on public roads. At present, the world's mainstream car manufacturers, including General Motors, Ford, Mercedes Benz, Volkswagen, Audi, Nissan, Toyota, BMW, Volvo and Cadillac, are all developing driverless driving technology. As of 2015, 16 states in the United States have launched driverless driving legislation. There were only 12 states as of 2014, 9 states as of 2013, and 6 states as of 2012. Nevada authorized the operation of autonomous vehicle in 2011, which is the first state in the United States to do so. After that, California, Florida, Michigan, North Dakota, Tennessee and Washington D.C. passed laws on driverless driving. The governor of Arizona signed an executive order for autonomous vehicle.

Future combat systems (FCS) is one of the most important fields where the research of autonomous vehicle can be put into application. According to the file released recently by the U.S. Department of Defense, *Unmanned Systems Integrated Roadmap FY2011–2036*, in the U.S. integrated modern war plan from 2011 to 2036, unmanned systems are divided into three parts: unmanned aircraft systems (UAS), unmanned ground systems (UGS) and unmanned maritime systems (UMS). The planned unmanned system has the following capabilities: environmental awareness, battlefield awareness, chemical/biological/radiological/nuclear detection, explosive disposal, accurate targeting, and accurate strike. Unmanned ground vehicle (UGV) is an important part of the U.S. unmanned war system and will become an important weapon in ground battlefield in the twenty-first century. The U.S. Department of Defense has invested billions in unmanned autonomous robots, and said it was in the initial stage of autonomous systems research. At present, the main applications of unmanned ground vehicles are reconnaissance, explosive disposal and anti-tank missile launching. These vehicles such as Talon, an American bomb removal robot, and Guardium, an Israeli multipurpose unmanned reconnaissance vehicle, have successfully served in Afghanistan and Iraq.

1.1.3 Research and Development of Autonomous Vehicle in China

In order to follow the research on autonomous vehicle in the United States, China began to study autonomous vehicle on land since the early 90s. At that time, it was carried out under the framework of intelligent mobile robot technology under the background of military platform. Meanwhile, the National 863 Plan intelligent mobile robot project was set up to carry out the development of remote driving anti-nuclear/chemical reconnaissance vehicle. During the Eighth Five-Year Plan period, under the leadership of the General Equipment Department (the former National Defense Science and Technology Commission), a team composed of Beijing Institute of Technology, National University of Defense Technology, Nanjing University of Science and Technology, Tsinghua University, Zhejiang University and Harbin Institute of Technology developed China's first autonomous land vehicle prototype ATB-1 (Autonomous Test Bed-1), namely 7B.8 military intelligent robot platform, which is currently parked in Beijing Institute of Technology. During the Ninth Five-Year Plan period, the second generation of autonomous land vehicle ATB-2 was developed, and then the third. Now the fourth generation is under development. The research of military autonomous land vehicle has accumulated key technologies and cultivated the talent team for the research of China's autonomous vehicle. He Hangen, Yang Jingyu, He Kezhong, Fu Mengyin, Lu Jilian, Liu Jilin, Zhu Miaoliang and other professors have made great contributions to this. Since 2009, the National Natural Science Foundation of China (NSFC) has set up a major research program of "cognitive computing of audio visual information", which takes the research and development of autonomous vehicle verification platform with natural environment perception and intelligent behavior decision-making ability as its scientific goal. By 2014, the "China Smart Car Future Challenge" has been carried out for six consecutive years, which has greatly promoted the research of autonomous vehicle in China, thanks to the strong promotion of Academician Li Deyi, Zheng Nanning and other experts.

Domestic automobile brands have been widely exploring in the field of driverless and automatic driving. In June 2016, the first "National Intelligent Connected Vehicle (Shanghai) Pilot Demonstration Zone" approved by the Ministry of Industry and Information Technology was officially put into operation in Jiading, Shanghai. The closed test area (phase I) can provide a comprehensive test site for autonomous vehicle. At present, the state is actively promoting the compilation of automatic driving related technical specifications. Although China entered the field of autonomous vehicle relatively late, analysts believe that China may become a major market for autonomous vehicle, due to a more favorable policy and consumption environment. According to Boston Consulting, 12 million driverless cars will be sold worldwide by 2035, more than a quarter of which will be sold in China. It is predicted that in 10 years' time, driverless cars will be popularized in China, accounting for 80% of the automobile market.

Figure 1.3 shows the progress of China's top ten auto companies that develop driverless technology.

The representative research works of autonomous vehicle in China are as follows:

1. Baidu Driverless Car

In 2013, Baidu launched its driverless car project. On December 10, 2015, Baidu driverless car conducted its first road test. On December 14, 2015, automatic driving

Company	Progress
Chang'an	Trial production vehicle on demonstration and to be put to industrial application in 2020
BAIC	Trial driving a nd riding vehicle on demonstration
BYD	Working with Baidu and no plan for commercial application
Lifan	Low- speed autonomous vehicle released in June 2016
Great Wall	Demonstrated assisted driving technology with high-speed driverless autonomous vehicle to be released in 2020
Dongfeng	Working with Huawei aiming to achieve driverless driving
Geely	Volvo has achieved mass production of autonomous vehicles. Geely is to release autonomous vehicles in 2 to 3 years.
GAC	Working with Chinese Academy of Scien ces and demonstrated autonomous concept car
SAIC	Demonstrated technology progress and to achieve all road condition autonomous driving in 10 years
FAW	Demonstrated driverless driving technology with 50% of models to be highly autonomous by 2025

Fig. 1.3 China's top ten auto companies developing driverless technology

division of Baidu was established, and Baidu driverless car appeared in front of the public for the first time.

2. Tsinghua University THMR Series Autonomous Vehicles

In the early 90s, Tsinghua University began to study the technology of autonomous vehicles. In the late 90s, the THMR series of autonomous vehicles developed by Tsinghua University came out. In 2003, Tsinghua University successfully developed THMR-V (Tsinghua Mobile Robot-V) autonomous vehicle, which could complete line-tracking drive on structured roads with clear lane lines, with a maximum speed over 100 km/h. In 2008, Tsinghua University intelligent car team, led by Academician Li Deyi of the Chinese Academy of engineering, was formed.

3. Hongqi HQ3 Autonomous Vehicle of National University of Defense Technology

National University of Defense Technology developed the CITAVT autonomous car, then the Hongqi HQ3 autonomous car. On July 14, 2011, Hongqi HQ3 completed its first high-speed driving experiment of 286 km from Changsha to Wuhan, creating a new record of autonomous driving of Chinese autonomous vehicle under complex traffic conditions. The average speed of autonomous driving in the whole journey was 87 km/h, and the distance of manual intervention under special circumstances was 2.24 km, only 0.78% of the total mileage.

4. Beijing Institute of Technology

The research on autonomous vehicle technology of Beijing Institute of Technology started in 1990, which was one of the first institutions to carry out relevant research in

China. The first autonomous vehicle in China is still parked in Xishan Test Ground of Beijing Institute of Technology. In November 2013, Ray fleet of Beijing Institute of Technology won the championship in the fifth "China Smart Car Future Challenge" which took place in Changshu, Jiangsu Province. In addition, in terms of unmanned off-road vehicle and formula racing, the Beili Jinggong autonomous vehicle was successfully developed in 2006, which was reformed on the basis of Dongfeng Mengshi 1.5-ton high mobility off-road vehicle. Driverless College Formula Car was released in January 2016, which was the first driverless college formula car in the world.

Other representative projects of autonomous vehicles include the Intelligent Pioneer autonomous vehicle jointly developed by Hefei Institute of Intelligent Machinery of the Chinese Academy of Sciences, Chery Automobile; the Springrobot experimental vehicle developed by Xi'an Jiaotong University; the Fierce Lion intelligent vehicle jointly developed by Military Transportation College and Great Wall Motor; and the JLUIV series experimental vehicles developed by Jilin University. Generally speaking, the research and development of autonomous vehicle in China has gone through the initial stage of technological exploration, where universities and research institutes were the main participants. At present, it is entering the stage of practical R&D for commercial purposes. Traditional automobile companies and related high-tech enterprises have become the main force, and the government is also participating.

1.1.4 Latest Trends of Commercialization of Autonomous Vehicle

The leading edge of the commercialization of autonomous vehicle came from Google. In February 2016, the National Highway Traffic Safety Administration of the United States confirmed that the artificial intelligence system used by Google's driverless car can be regarded as "driver". It cleared the way for Google's driverless car to be allowed to go on the road and finally commercialized. The latest test scenario is shown in Fig. 1.4.

From a global perspective, the latest achievements of autonomous vehicle R&D are as follows. The driverless cars developed by Google has traveled more than three hundred thousand kilometers. Google's second generation driverless car, as shown in Fig. 1.5, is based on the white Lexus RX 450H hybrid SUV, while the previous first generation driverless car was based on the Toyota Prius hybrid.

With support of the government, the two-seater small autonomous car developed by British company Oxbotica as shown in Fig. 1.6 was tested on the road of Milton Keynes, England, and scheduled to officially go on the road 2020. Since 2016, the British government has been accelerating the development of autonomous vehicle technology and the revision of relevant laws and regulations that allow autonomous vehicle to travel on the road.

Fig. 1.4 Autonomous vehicle of Google in test (early 2018)

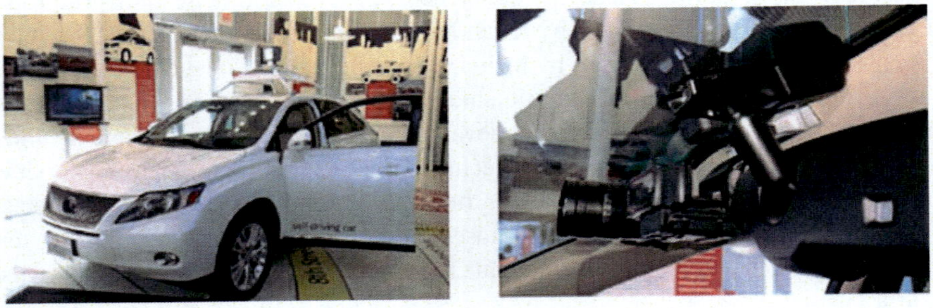

Fig. 1.5 Google's second generation driverless car

Fig. 1.6 Two-seater small autonomous car of Oxbotica

In January 2018, U.S. R&D center of Baidu in Silicon Valley opened up to the media showing the latest automatic driving technology of Apollo open platform. The autonomous cars with Apollo 2.0, carrying dozens of journalists and partners, took a public test drive on the open road in California. They completed traffic light recognition, U-turn and other complex actions, as shown in Fig. 1.7.

Fig. 1.7 Test of Baidu autonomous vehicle in California

1.2 Development of Autonomous Vehicle Test

Autonomous vehicle test is an emerging field with the development of autonomous driving technology. It is also an important link between the development and commercialization of autonomous vehicle. The test of autonomous vehicle can be divided into two types: field test and road test. An automatic driving test field is a field that reproduces all kinds of road conditions and situations for autonomous vehicle. It is used to verify and test the performance of the software algorithm of autonomous vehicle. The test road is a truthfully concentration of actual roads, including highway, urban road, rural road and tough roads that can cause strong bumps.

1.2.1 Field Test

Mcity Test Facility (as shown in Fig. 1.8) is the world's first driverless test field built to test autonomous vehicle and V2V/V2I networking technology, with a total road length of 4 miles. Mcity, built by the University of Michigan's Mobility Transformation Center (MTC), is located in Ann Arbor, Michigan. It cost 10 million dollars (jointly funded by the University of Michigan and the Michigan Department of Transportation), and covers 129,000 m^2. At present, it is cooperating with 15 auto companies and parts suppliers such as Ford, GM, Honda, Nissan, Toyota and Delphi through capital injection. Mcity simulates urban and suburban environments. This virtual city will build the facades of 40 buildings, right-angled intersections, roundabouts, bridges, tunnels, gravel roads, building barriers and many other obstacles. Mcity includes two major test areas: the high-speed test area used to simulate expressway environments, and the low-speed test area used to simulate urban and suburban areas. The goal of this facility is to build an online and automated mobile system on the streets of southern Michigan by 2021. It plans to deploy more than 20,000 cars, trucks and buses in southern Michigan as an experiment to assess consumer behavior and explore market opportunities.

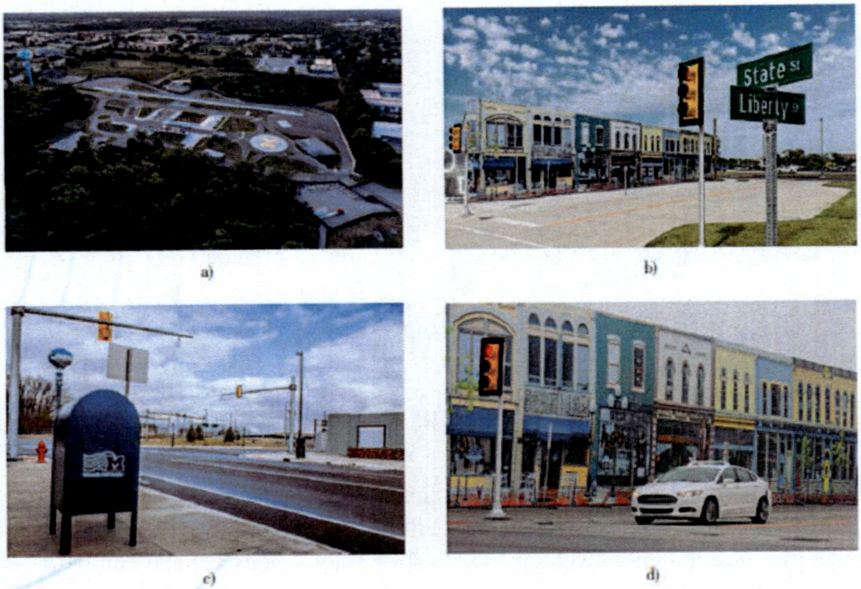

Fig. 1.8 Mcity test facility **a** Aerial view of Mcity; **b Test field; c** Road; **d** Ford autonomous vehicle test

According to incomplete statistics, there are 19 large-scale automobile test fields in China, including Hainan Automobile Test Field and FAW Technology Center Test Field. Among them, the closed test field of the National Intelligent Network Connected Vehicle (Shanghai) Pilot Demonstration Zone is one of the first and most important test centers for autonomous vehicle in China. The Innovation Demonstration Zone of Intelligent Vehicle and Transportation Industry in Beijing Yizhuang Pilot Zone is planned to be under construction from 2016 to 2020. It will carry out six application demonstrations including green car, intelligent road network, intelligent driving, convenient parking, happy vehicle lifestyle and intelligent management. By the end of 2018, a closed test field with a total length of 10 km has been built, which contains more than 30 types of urban road elements, and 100 km of open/semi-open municipal road reconstruction, covering more than 180 scenarios. The features of the Zone will continue to expand. By the end of 2020, the closed test field would contain more than 100 types of city road elements and 200 km open/semi-open municipal road reconstruction, covering 300 or more scenarios, supporting tests of 1000 fully autonomous vehicles under various conditions. In Wuhan, French company Renault opened China's first autonomous vehicle test field, which has been officially put into use. The test field jointly planned by Geely and Volvo is located in Hangzhou Bay New Zone, Ningbo City, Zhejiang Province. In the future, it will work with the R&D center of Geely to develop autonomous vehicle. Baidu signed a cooperation agreement with the Government of Wuhu City, Anhui Province, claiming they would jointly build an autonomous vehicle operation area in Wuhu. Baidu also reached an agreement with Wuzhen tourist area in Zhejiang Province, and they will promote L4 level driverless vehicle operations on the roads of Wuzhen.

1.2.2 Road Test

California is the first place in the world to allow autonomous vehicle to be tested on the road. In February 2012, California became the first state in the world that can independently formulate laws for autonomous vehicle. At present, most enterprises choose to test autonomous vehicle in California, and the test environment is more sophisticated. The California Department of Motor Vehicle (DMV) is the agency for major companies to apply for tests on driverless technology. It also designs regulations on these tests. As of September 2017, a total of 39 companies worldwide were allowed to test their driverless cars in California. As of January 2018, more than 40 companies in the United States had obtained license plates for autonomous vehicle. Four states in the United States have passed laws that allow driverless cars on the road.

In January 2015, the UK allowed driverless cars on the road, and considered readjusting traffic rules. In early 2015, driverless cars were put into trial in four towns in England. Bristol in the southwest and Coventry in the middle have launched their own pilot projects for driverless cars. Greenwich and Milton Keynes, two small towns on the outskirts of London, have also joined the trial.

In December 2007, Beijing issued the *Guiding Opinions of Beijing Municipality on Accelerating the Road Test of Autonomous vehicle (for Trial Implementation)* and the *Detailed Rules for the Management Implementation of Road Test of Autonomous vehicle in Beijing Municipality (for Trial Implementation)*, which are the first government guidance documents issued in China. The purpose is to provide a legal basis for autonomous vehicle, and encourage researchers to conduct scientific tests. Then in January 2018, Beijing Yizhuang Pilot Zone set fixed road sections and time windows to allow autonomous vehicle to test on the road. Some road sections contain vehicle infrastructure cooperation mode, and high-precision map mode.

In September 2017, the traffic management department of Taiwan, China drew up the road test regulations for autonomous vehicle, which were implemented at the end of the year. In October 2017, some companies in Taiwan completed the development of autonomous vehicle and carried out simulation environment test. The authorities require that vehicles must be successfully tested in the closed field before they can apply to advance to designated lanes, such as bus lanes, and then to mixed traffic. The conditions which a local government will consider for the application of autonomous vehicle test include vehicle requirements, qualification of test personnel, insurance, liability for compensation, etc. The validity period of the license plate for the test is six months.

1.3 Bottleneck of Autonomous Vehicle Commercialization

According to American standards, there are five levels of autonomous driving. At present, it is still at Level 3 and 4 for experiment. It is estimated that in the next decade or so, the autonomous driving industry will witness great development, and the market scale may reach 100 billion. However, at present, there are still technical bottlenecks in autonomous vehicle, including road environment perception and behavior planning decision-making. These are the main factors which restrict the large-scale commercialization of autonomous vehicle.

Although a lot of research and trial production work has been carried out on autonomous vehicle all over the world since the 1960s, the environment perception of autonomous vehicle and the autonomous navigation technology driven by perception information have only been verified in contests, which are limited to the prescribed tasks. In practice, there are very few successful applications, and these applications are limited to specific road sections. Examples include the explosive disposal and reconnaissance robots launched by the United States in Iraq and Afghanistan. The immaturity of the environment perception system is still the main bottleneck hindering the improvement of the overall performance of autonomous vehicle. There are still a lot of unsolvable scientific and technical problems, which are also the biggest obstacles for wider practice of the autonomous vehicle. In recent years, the failure statistics of autonomous vehicle competitions show that most of the failures come from the environmental perception part. For example, in the 2007 DARPA Urban Challenge, the champion car Boss, developed by Carnegie Mellon University, summed up its own shortcomings afterwards and proposed the following key problems: in the dust road environment, the dust was "suspected" as an obstacle in the detection algorithm, which led to increased behavior decision-making time and seriously affected performance. In the China Smart Car Future Challenge of 2009, the champion car developed by Tianjin Military Transportation Institute summed up the issues: it was unable to identify fake obstacles such as birds and leaves that randomly flew into the vehicle's field of vision, causing unnecessary braking for numerous times. As of 2013, the major unsolved problems of Google driverless car still came from environment perception: when dealing with reflection of shiny surface (such as rainy road), the sensor appears clueless. Also, vehicles don't know how to cross intersections, since they lack relevant cognitive ability.

The main factor that affects the perception performance is that most of the existing algorithms are only suitable for optimal visual conditions, or only for certain kinds of interference. It is difficult to develop robust, fast and unified visual detection and recognition algorithms that can overcome all kinds of random heterogeneous disturbances. Therefore, a robust visual perception system ready for complex and uncertain environment is yet to come.

1.4 Content Arrangement of This Book

This book systematically discusses the technology of autonomous vehicle, and integrates the author's practical research and engineering experience in the field for years. This book consists of 10 chapters:

This chapter is an overview of the development of autonomous vehicle around the world, which includes typical products, test methods, commercial test progress, commercialization bottlenecks.

Chapter 2 is an overall technical analysis, introducing the architecture of autonomous vehicle, the function of each component module and the world's leading autonomous driving operation control hardware unit technology.

Chapter 3 is about the vehicle borne radar system, which introduces the lidar, millimeter wave radar, ultrasonic radar and the lane detection technologies used in autonomous vehicle.

Chapter 4 is about lane visual detection and recognition technology. It introduces the basis of lane image processing and understanding, mechanism of lane line visual detection, lane line feature extraction and positioning, lane line tracking and lane self-correction, lane image segmentation, navigation sign detection based on prior knowledge reasoning and color-shape combination feature, navigation sign recognition based on SURF feature optimization matching and deep learning.

Chapter 5 is about navigation and positioning technology, including high-precision map of road network, vehicle inertial navigation and positioning, GPS/Beidou navigation and positioning system, etc.

Chapter 6 is about motion control and planning decision-making technology. It discusses the automatic and autonomous control technology of autonomous vehicle, as well as action planning, behavior planning and intelligent decision-making based on reinforcement learning.

Chapter 7 is about the energy and power system of autonomous vehicle, which comprehensively discusses the current mainstream energy and power system technology of autonomous vehicle from the perspectives of fuel, battery and hybrid.

Chapter 8 is about the Internet of Vehicles, which discusses intelligent connected vehicles, C-V2X, 5G-V2X and relevant national policies and standards.

Chapter 9 is about the industrialization of autonomous vehicle. It introduces the research and development of special practical autonomous vehicles for various application scenarios, the design of autonomous vehicle test field and the planning and design of driverless town, which provides guidance for practice.

Chapter 10 is about the transition from autonomous vehicle to artificial intelligence city. It discusses the future development trend of autonomous vehicle, the new business form of future intelligent transportation with autonomous driving technology, as well as smart city and artificial intelligence city. Finally, the artificial intelligence city model is proposed from the perspective of complex giant system.

Chapter 2
General Technology

2.1 Technical Framework

The three core tasks of autonomous vehicle are cognition, decision-making and control. In the process of driving, the vehicle perceives the surrounding environment and road conditions by itself, completes the processing and fusion process based on perceptual information to form an understanding of the overall situation, then makes decisions on how to deal with it through various algorithms. Finally, the decision information is transmitted to each control system to form executive commands and complete the driving action.

In order to complete the above tasks, the overall technical system of autonomous vehicle is mainly composed of the following modules: robot operating system (ROS), map, navigation and positioning (GPS + IMU), optical radar (lidar), radar, vision (camera), planning, control, communications, cloud platform, Internet of Vehicles, simulation.

The technical components of autonomous vehicle are shown in Fig. 2.1.

Macroscopically, they can be divided into two categories: vehicle end and cloud end, as shown in Fig. 2.2.

Each category includes different technical modules.

I. Vehicle end

Vehicle end has upper and lower layers.

(1) The upper layer includes Software and algorithm:

Sensor module: how to better obtain environmental data;

Perception module: how to better understand the environment, including positioning, object recognition, object tracking;

Control module: including control of brake and throttle, direction, posture, etc.;

Decision module: how to make better decision after understanding the environment, including path planning, behavior prediction, obstacle avoidance, etc.

M. Du, *Autonomous Vehicle Technology*, Unmanned System Technologies,
https://doi.org/10.1007/978-981-19-4143-6_2

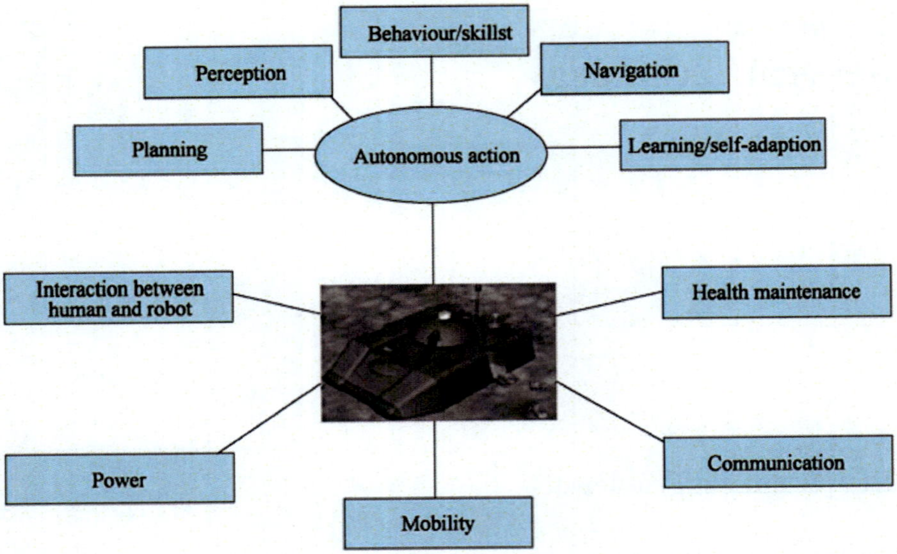

Fig. 2.1 Technical components of autonomous vehicle

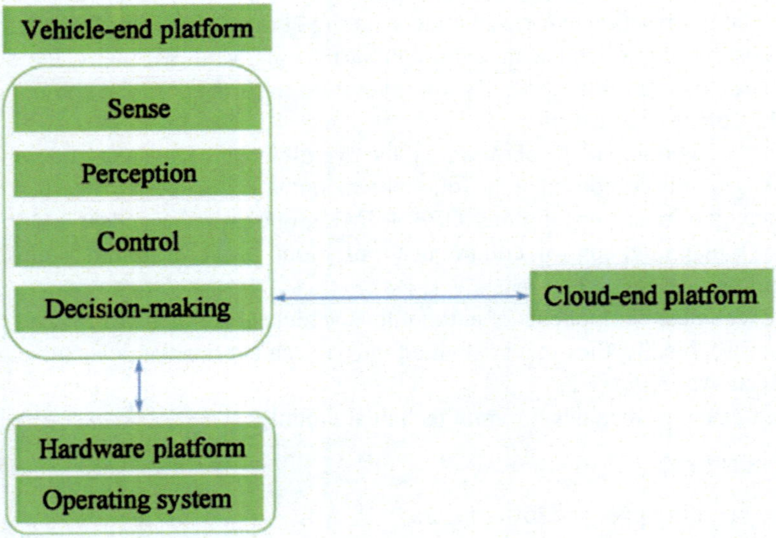

Fig. 2.2 Overall structure of autonomous vehicle

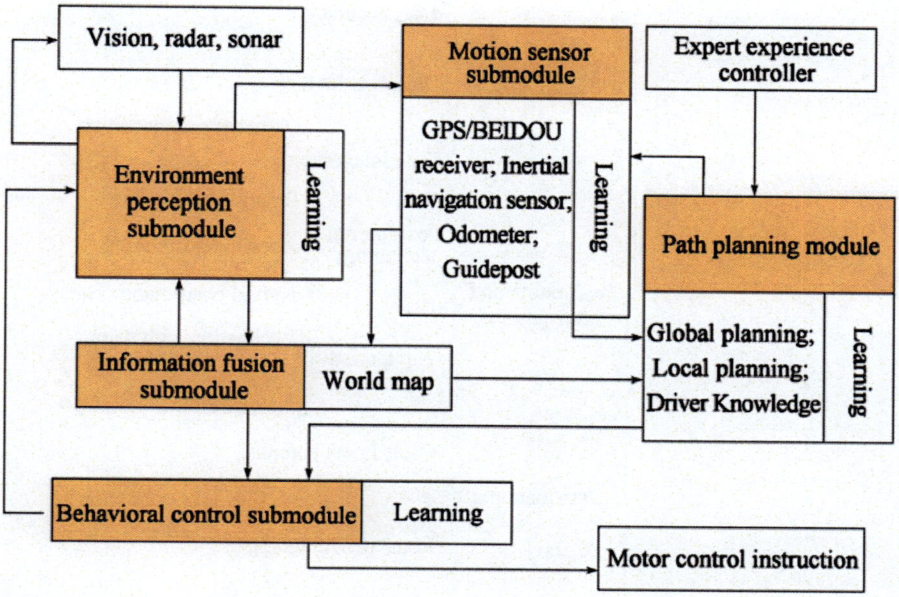

Fig. 2.3 Function modules of autonomous vehicle and their relationships

(2) The lower layer includes operating system and hardware platform

I. Cloud end (driverless cloud platform)

The cloud end includes high-precision map, model training, simulation calculation and data storage modules. Figure 2.3 shows the function modules and their relationships contained in an autonomous vehicle with learning ability.

The key technologies involved in autonomous vehicle are shown in Fig. 2.4.

The technical architecture of autonomous vehicle proposed by NXP is shown in Fig. 2.5.

The detailed technical architecture and technical composition modules of typical autonomous vehicle are shown in Fig. 2.6.

2.2 Main Module Functions

2.2.1 Perception Module

The environment perception task of autonomous vehicle is very complex, which can be summarized into four aspects.

(1) Road shape estimation and cognition;
(2) Static, traffic participants and other dynamic obstacles detection and identification;

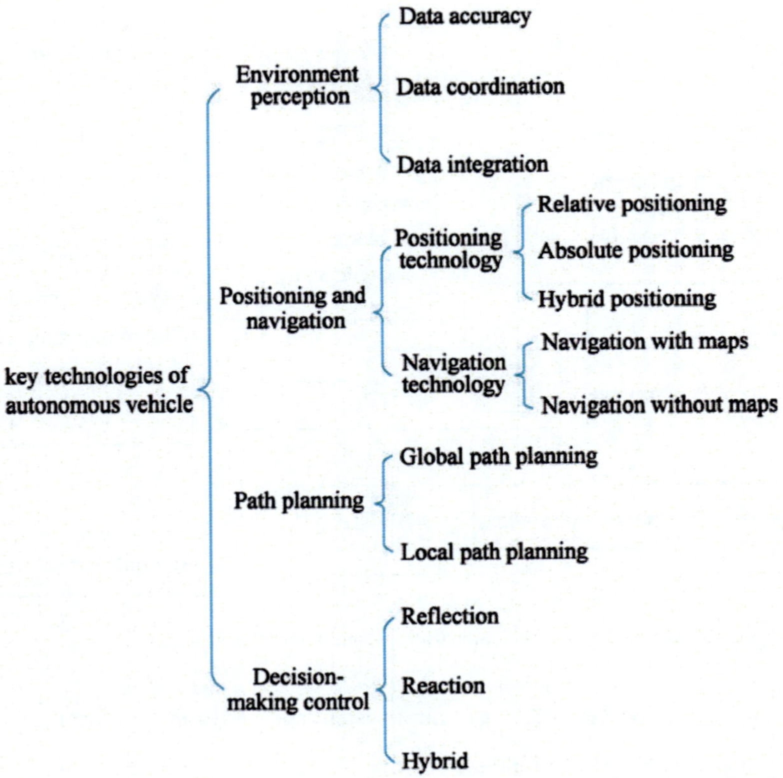

Fig. 2.4 Key technologies of autonomous vehicle

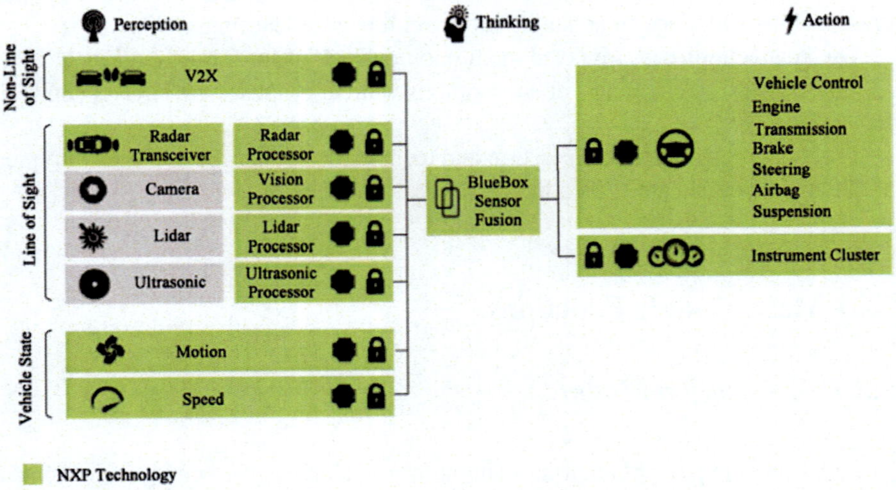

Fig. 2.5 Technical architecture of autonomous vehicle proposed by NXP

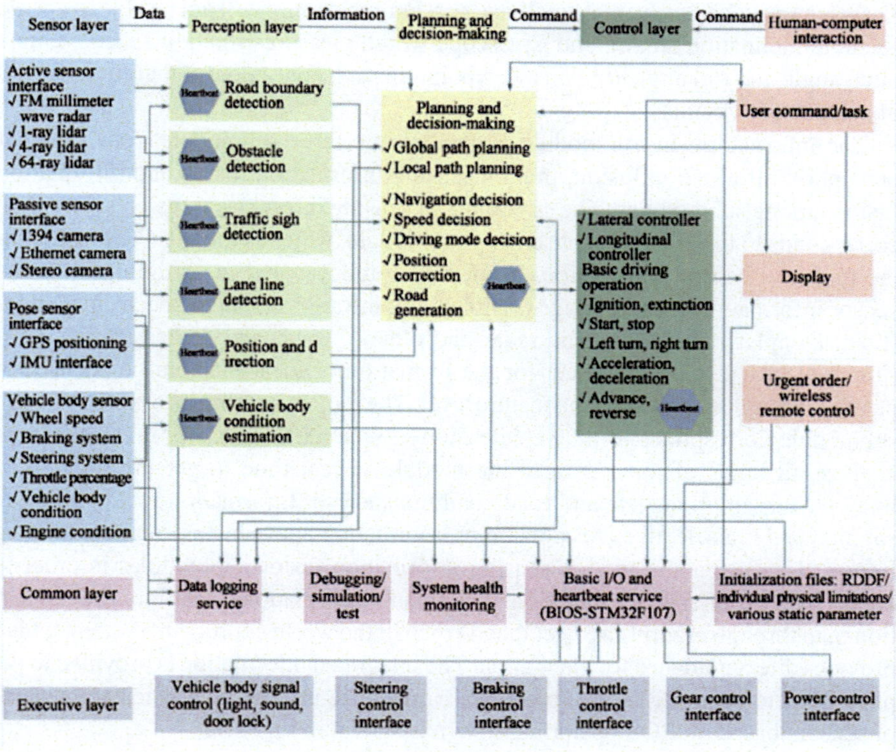

Fig. 2.6 Detailed technical architecture of autonomous vehicle

(3) Vehicle state estimation and motion compensation;
(4) Map aided or no map positioning.

The perception module is responsible for providing the real world model for the path planning module, which is composed of three submodules: road environment perception, vehicle motion sensing and information fusion.

The environment perception submodule uses optical binocular stereo camera, optical monocular camera, infrared camera, lidar, microwave radar, ultrasonic radar, sonar and other sensors to obtain the three-dimensional environment information around the vehicle. On the premise of meeting the system requirements, the key task of the submodule is to reduce the input information redundancy as much as possible, and to design and build the architecture optimal minimum vision and radar combined sensing system.

The vehicle motion sensing submodule mainly uses inertial navigation sensor, odometer, GPS receiver to provide vehicle azimuth, attitude angle, motion speed and other information for global navigation. Its objective is to measure the real-time acceleration and viewpoint of the vehicle. The measured data are acceleration and angular velocity along three axes. The main sensing device is the accelerometer (when there is motion acceleration, an inaccurate inclination angle will be measured.

That is to say, when the vehicle inclines, accelerometer only will not be able to get the accurate inclination angle), and gyroscope which measures rotation speed, vehicle pitch angle and roll angle (obtains viewpoint through continuous integration instead of rotation viewpoint).

The information fusion module is divided into three submodules: environment perception information fusion, motion sensing information fusion and information fusion among submodules. The environment sensing information fusion submodule corresponds to the environment sensing submodule, which is mainly responsible for the fusion of sensor information. It effectively reduces and fuses the multi-modal sensor information such as image (visible light, infrared and other spectrums), radar (lidar, Doppler radar, ultrasonic radar and others), sonar and so on through algorithms. It provides original data for the formation of local map, and evidence for passing clearance analysis at the feature level. The motion sensing information fusion submodule corresponds to the vehicle motion sensing submodule. It can fuse and analyze all kinds of input data in the module in real time to get information of body posture, motion state and road conditions ahead. Information fusion between submodules is a selective cross fusion of information between environment sensing information fusion and motion sensing information fusion submodules in order to obtain data beneficial to navigation. The fusion information is put into the navigation map through algorithm to get the 3D map in the world coordinate system, which provides direct guidance for navigation. The integrated information is provided to the planning and behavior control module as input. The information fusion framework of sensors based on ROS operating system is shown in Fig. 2.7.

2.2.2 Path Planning Module

Based on the theory of artificial intelligence, the path planning module performs the task of vehicle behavior and motion planning, reasoning and decision-making. The whole module centers on trajectory prediction algorithm and is divided into two submodules: high-level behavior planning and low-level motion planning. Its architecture is shown in Fig. 2.8.

Based on the road network database, the high-level behavior planning submodule is responsible for executing the commands issued by the task database, and formulating intelligent strategies for lane change, priority and safety decision-making. The structure of behavior planning is based on identifying a series of driving contexts, and relies on concise environmental features. The three typical contexts are: lane, intersection and area. The corresponding independent behavior types are: lane driving, intersection handling and achieving a zone pose. Area refers to unstructured or unconstrained driving environment, such as a parking lot and toll station. From the perspective of behavior sequence, behavior structure can be divided into two layers: one is target selection layer, the other is behavior execution layer. There are three types of behaviors in the behavior execution layer of structured road: lane driving, intersection handling and achieving a zone pose. Each type can be decomposed

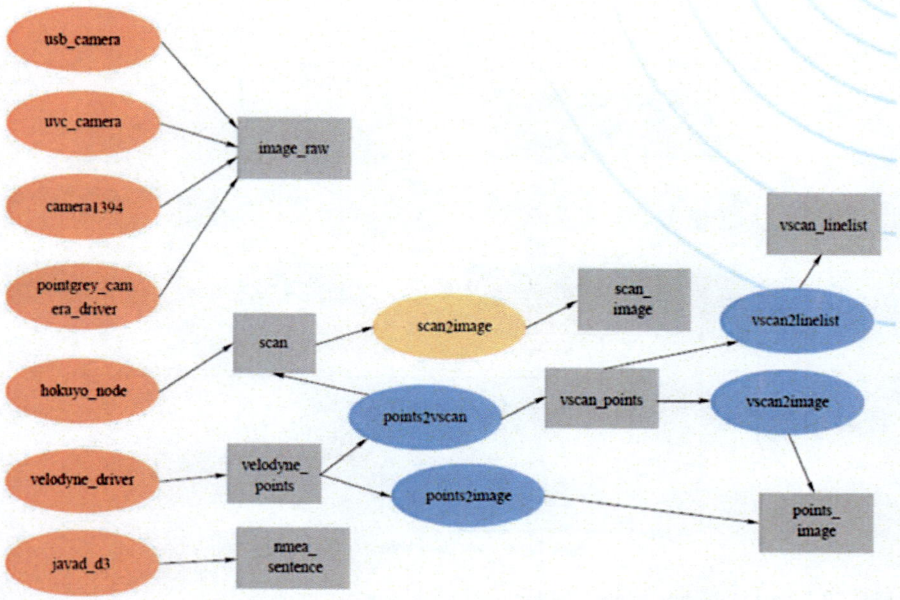

Fig. 2.7 Sensor information fusion based on ROS operating system

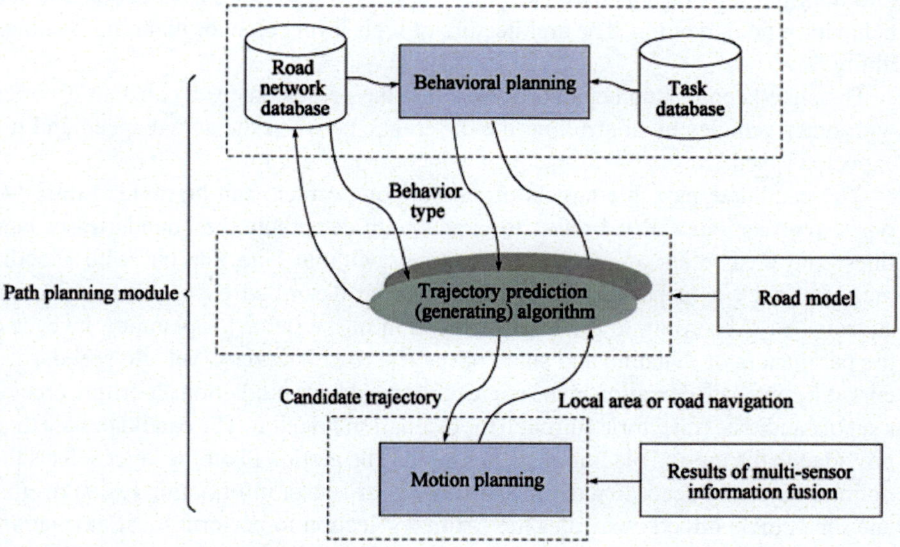

Fig. 2.8 Path planning module

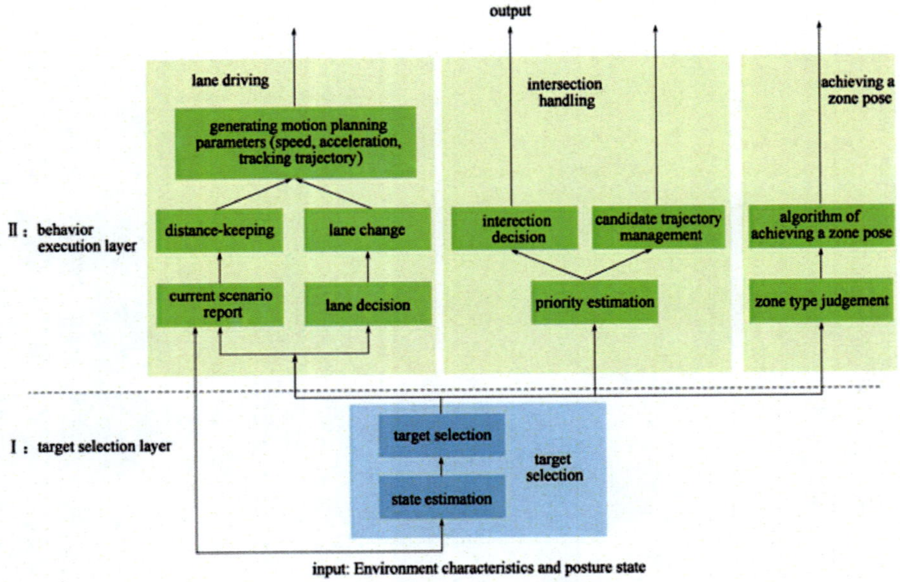

Fig. 2.9 The architecture of modular high-level behavior planning

into multiple sub behaviors with sequence based on behavior classification and sub behavior modularization. The architecture of high-level behavior planning is shown in Fig. 2.9.

Distance-keeping sub behavior means that the vehicle keeps a constant distance with other vehicles by controlling the difference between the actual speed and the expected speed.

The candidate passable targets of autonomous vehicle can be divided into two types: road and area. Road refers to targets with detectable the boundaries or lane lines; and area refers to special shape targets without lane line but with specific meaning, such as a parking area, which can be described by regional inflection coordinates. According to the target indication of the behavior planning layer and the parameters of dynamic and static obstacles, road boundary, vehicle speed, road curvature and path deviation in the context, the trajectory prediction algorithm creates a set of candidate trajectories through the evaluation function. The candidates are then provided to the motion planning layer as input. The motion planning layer selects the optimal trajectory according to the real-time multi-sensor information fusion results, and the vehicle directly executes the optimal selection to perform local navigation. Driver behavior knowledge can be integrated into both behavior planning layer and motion planning layer.

The most typical candidate trajectory generation method is model-predictive algorithm, which was proposed by Howard and Kelly in 2007. The algorithm provides a method to generate parametric control variables under state constraints. Three key concepts are involved: expected trajectory, actual trajectory and candidate trajectory.

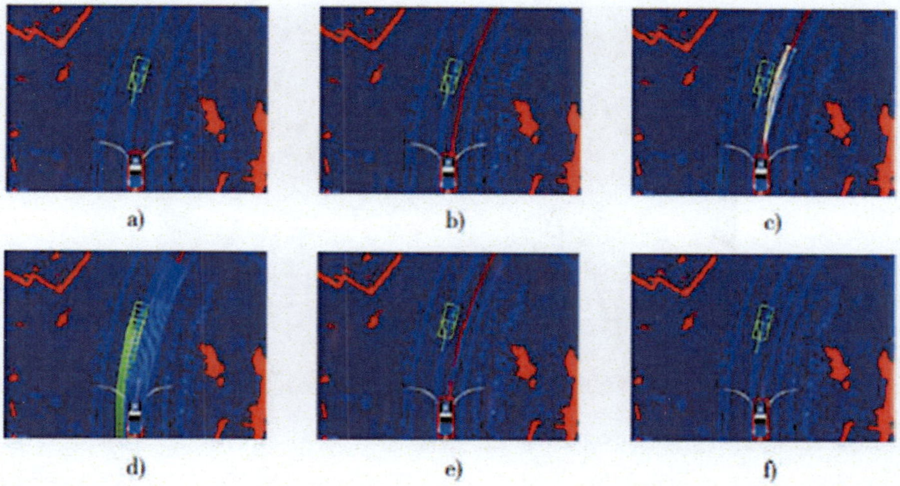

Fig. 2.10 Typical results of a two-lane scenario

By implementing the trajectory self-optimization algorithm, the vehicle can successfully avoid obstacles and drive along the middle line of the lane. The typical results of a two-lane scenario are shown in Fig. 2.10. In Fig. 2.10a, the middle blue line is the lane line. The white short arc is the minimum allowed turning radius, and the green rectangle is another vehicle. In Fig. 2.10b, the red line is the extracted expected trajectory. In Fig. 2.10c, the white line is the candidate trajectory. Figure 2.10d shows the estimation of passable area considering dynamic obstacles. Figure 2.10e shows the estimation of passable area considering static obstacles, and Fig. 2.10f shows the actual selected trajectory.

2.2.3 Motion Control Module

The motion control module is the executive part of the planning layer, which compares the actual motion posture with the expected motion posture in real time, and uses intelligent control algorithm to control the steering wheel, accelerator, brake and other actuators. The most typical reasoning method is fuzzy reasoning, and the most commonly used control algorithm is PID algorithm.

There are three kinds of motion controller: single chip microcomputer or microprocessor as the core; special chip as the core processor; DSP and FPGA as the core processor based on PC bus. The standard layout of control devices in the trunk of an autonomous vehicle includes: industrial control computer (GPU calculation card, CAN card), UPS power/regulated power supply, switch, low voltage power distributor, GNSS/IMU module, vehicle control unit (usually dSPACE MicroAutobox), heat dissipation and cooling mechanism, etc. If specific sensors are used,

Fig. 2.11 Autopilot operation control unit of Tesla

there will be independent industrial computer, fuser, interface and power module. The Autopilot operation control unit of Tesla is shown in Fig. 2.11.

2.3 Hardware Platform and Computing Control Hardware

2.3.1 Hardware Platform

The hardware platform of autonomous vehicle is shown in Fig. 2.12.

The control platform is the core component of autonomous vehicle, which controls various control systems of the vehicle, including anti-lock braking system (ABS), acceleration slip regulation (ASR), electronic stability program (ESP), sensotronic

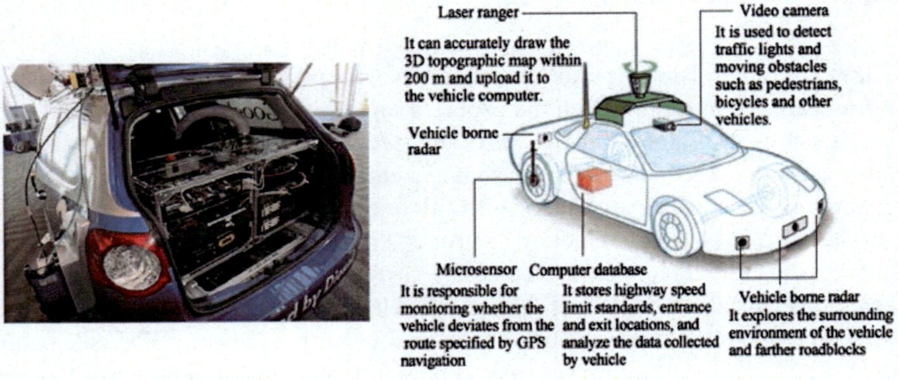

Fig. 2.12 Hardware platform of autonomous vehicle **a** Actual product; **b** Composition

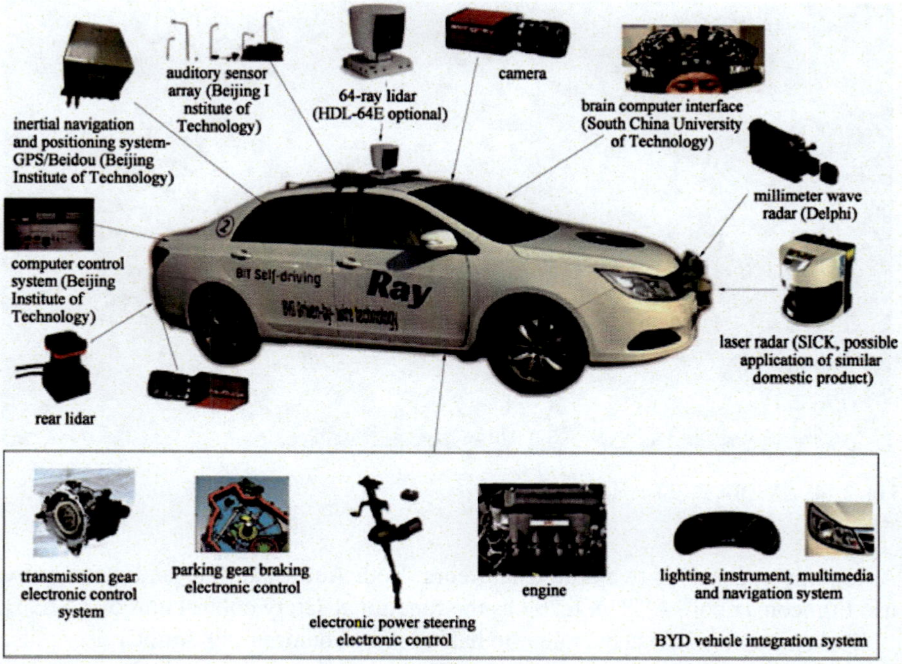

Fig. 2.13 Hardware composition of Ray autonomous vehicle

brake control system (SBC), electronic brake-force distribution (EBD), brake assist system (BAS), supplement restraint system (SRS) and collision avoidance radar system, electronic automatic transmission (EAT), constantly variable transmission (CVT), cruise control system, etc.

The hardware composition of Ray autonomous vehicle of Beijing Institute of Technology is shown in Fig. 2.13 as an example.

2.3.2 Computing Control Hardware

At present, the most advanced driverless operation control unit and domain controller in the world include NVIDIA Drive PX2, Tesla AutoPilot operation unit, NXP BlueBox, Qualcomm 9150 C-V2X chipset, Audi zFAS, etc.

Drive PX2 AutoChauffeur development board—NVIDIA PX2 Drive series is the latest generation of driverless series products, as shown in Fig. 2.14.

The lowest configuration, AutoCruiser in this series, has a single Parker SOC, and has high-speed cruise capability. The middle-end configuration, AutoChauffeur, is at L3, and its operation part is a Pascal architecture independent operation unit with dual Parker SOC and dual MXM3.1 interface. The higher level product, Full

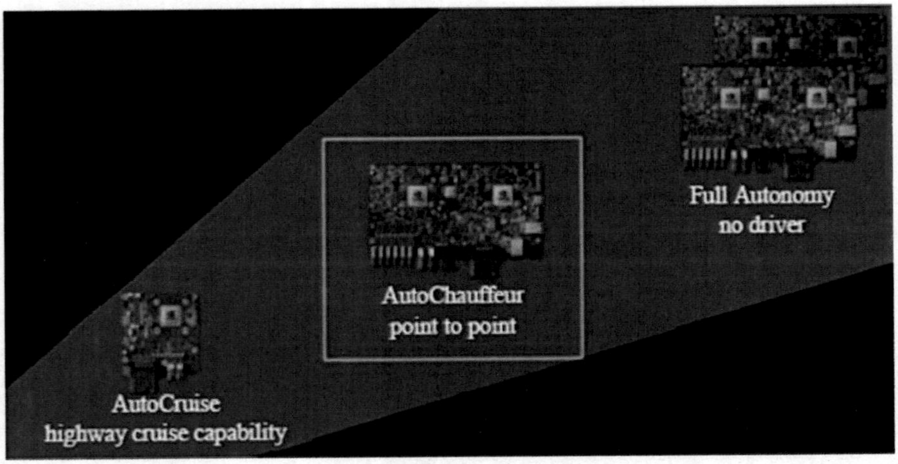

Fig. 2.14 NVIDIA PX2 Drive series

Autonomy consists of two AutoChauffeurs. Both AutoChauffeur and AutoCruiser use Infineon Tricore AUTIX tc297 as the functional safety control unit of ASIL-D.

Tesla autopilot 2.0 is equivalent to half of AutoChauffeur, or AutoCruiser plus a Pascal independent operation unit, with GNSS receiver chip added. The main chip models included in the board are as follows:

1. NVIDIA "Parker" P94W97.01P TA795SA-A2, Parker SOC master, built-in 256 CUDA units, 4-core A5 64-bit ARM and 2-core Denver 64-bit ARM.
2. Four Samsung DRAM K4F8E3S4HBMHCJ.
3. The MXM card of NVIDIA GP106-505-KC, 4 GB GDDR graphic memory, 4 pads reserved and up to 8 GB. Equivalent to a GP106 series operation card without the display part, with 1280 CUDA operation units.
4. INFINEON TriCore AUTRIX TC297TX-128 ASIL-D MCU.
5. UBlox NEO-M8L GNSS receiving module.
6. Toshiba eMMC and Spansion MOR Flash.
7. Marvell 88EA1512 AVB/Ethernet transceiver.
8. Marvell 88EA6321 7-port AVB switching chip.
9. Maxim MAX9260 GMSL display output.
10. TI DS90UB964 LVDS camera input.

The CPU is 4xa57 + 2xdenver + Aurix, which has enough computing power to handle HMI, bus communication, result output sensor fusion, GNSS data processing and vehicle motion control related to automatic driving, with about half of redundancy.

GPU is 256 + 1280 CUDA units, with six cameras, four for looking around and SFM, two forward-looking for lane line, pedestrian/target detection, traffic sign detection and free space detection. According to the codes and resource occupation

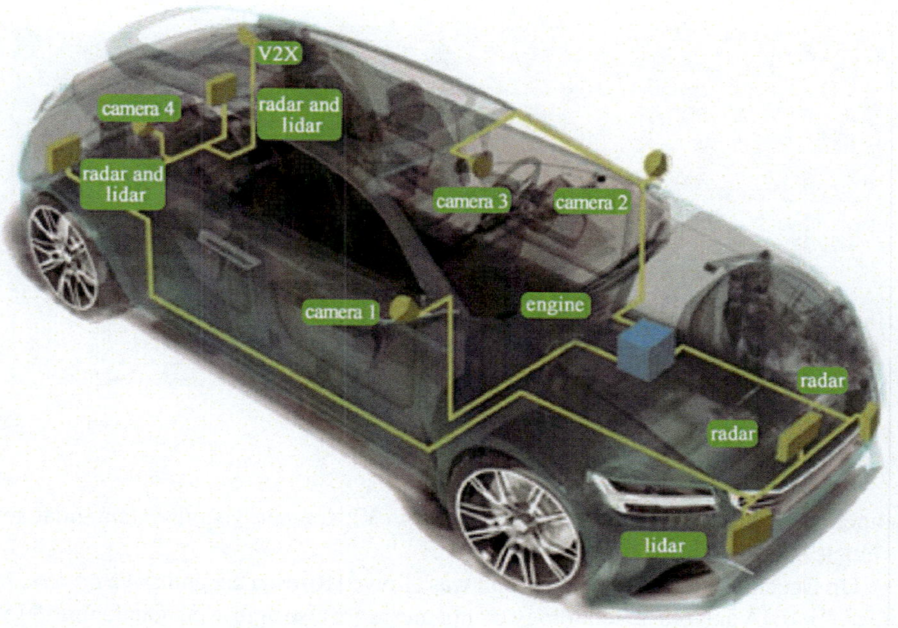

Fig. 2.15 BlueBox computing platform of NXP

provided by DriveWorks, its ability is not enough to meet the requirements of simultaneous operations. Lane line, target and traffic sign detection takes about 90% of GPU ability.

The BlueBox computing platform of NXP, a Dutch semiconductor supplier, as shown in Fig. 2.15, is mainly used to help OEMs produce and test autonomous vehicles. NXP chips (BlueBox, millimeter wave radar, V2X and safety) will support Baidu Apollo open automatic driving platform, which aims to become the new Android system in the autonomous driving industry. BlueBox is a central computer, which can process all the data collected from radar, camera, lidar, visual sensor and V2X system installed on the car at the same time. It will promote the launch of L4 driverless car in 2020.

BlueBox is equipped with an NXP S32V automotive vision processor and an LS2088A embedded computer processor. S32V chip contains different graphics processing engines, special high-performance graphics processing accelerators, high-performance ARM cores, advanced APEX graphics processing and sensor fusion. S32V belongs to the category of safety controller. Its functions include sensor/actuator management and fault detection. It can analyze the driving environment, evaluate risk factors, and then lead the behavior of the car. The LS2088A embedded processor, which is responsible for high-performance computing, is shown in Fig. 2.14. It is composed of eight 64-bit ARM Cortex-a72 cores, with 2 GHz special accelerator, high-performance communication interface and DDR4 memory controller. Its delay is extremely low. In addition to the two core processors of S32V and LS2088A, BlueBox also carries other chips that implement different sensor node

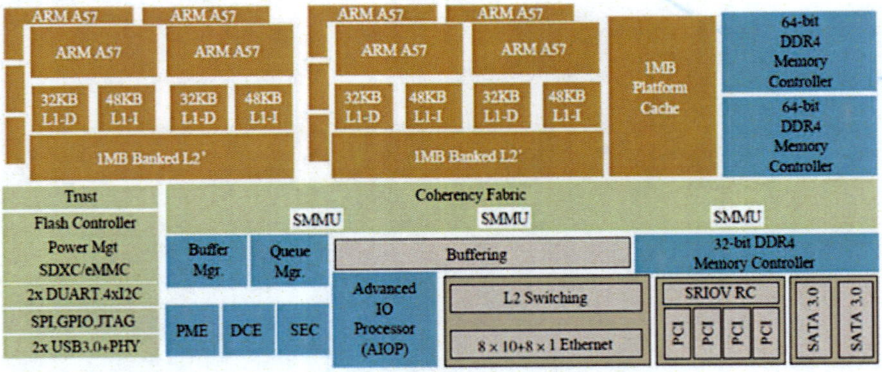

Fig. 2.16 LS2088A embedded processor

functions, which can process information from V2X, radar, vision system, lidar and vehicle status.

On December 12, 2017, Qualcomm was allowed to test a car and three drivers. Its rival NVIDIA and other technology companies were also approved. Qualcomm 9150 C-V2X chipset can provide 360-degree non line of sight perception for vehicles, and can communicate with other vehicles, traffic lights and other infrastructure. The test of vehicles equipped with Qualcomm 9150 C-V2X chipset is shown in Fig. 2.17.

2.4 Robot Operating System (ROS)

Robot operating system (ROS) can be selected as the operating system of autonomous vehicle.

ROS was originally a robot project of Stanford University, later developed by Willow Garage Company, and now maintained by OSRF (Open Source Robotics Foundation, Inc.). ROS is a set of computer operating system architecture designed for robot software development. It is an open source meta level operating system (post operating system), which provides services similar to an operating system, including hardware abstract description, underlying driver management, implementation of common functions, message delivering between programs and program package management. It also provides some tools and libraries for acquiring, building, writing and executing multi-machine integrated programs. ROS uses the concept of node to represent an application. Different nodes are connected by pre-defined topics, services and actions. The architecture of ROS is shown in Fig. 2.18.

ROS almost includes all the most advanced algorithms in the field of robotics. Typical algorithms are as follow.

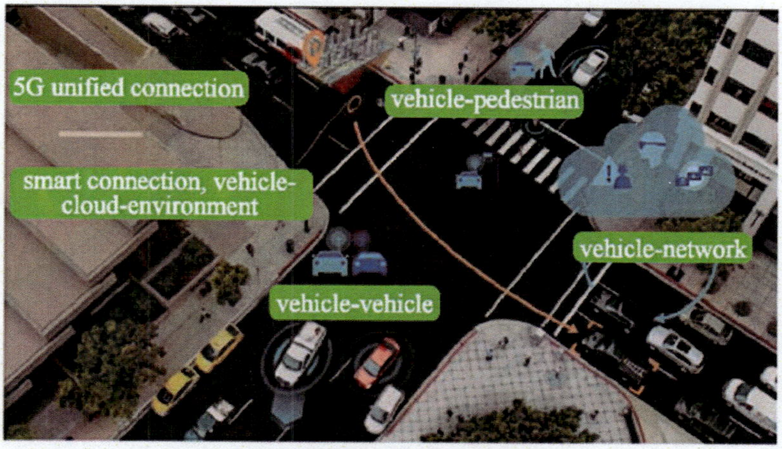

a) Autonomous vehicle test

b) Qualcomm 9150 C-V2X chipset

Fig. 2.17 Test of vehicles equipped with Qualcomm 9150 C-V2X chipset

Localization: the localization algorithm of autonomous vehicle based on Extended Kalman Filter (EKF) and Unscented Kalman Filter (UKF) can fuse the positioning information of various sensors to obtain more accurate positioning results (Fig. 2.19).

Navigation: Based on Dijkstra, A* algorithm (global planner) and Dynamic Window Method (local planner), the path planning module of mobile robot can achieve the navigation of autonomous vehicle on two-dimensional map (Fig. 2.20).

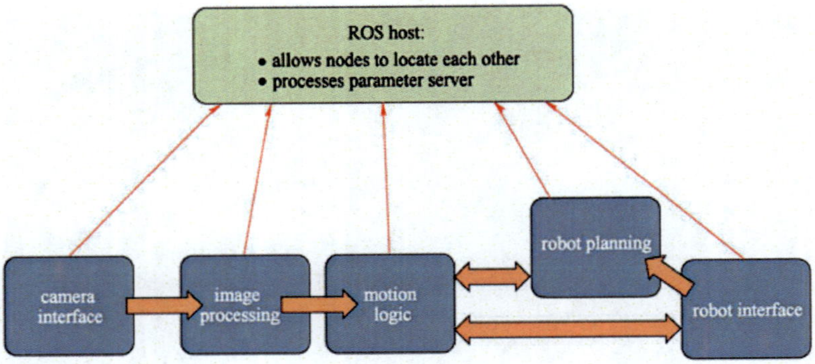

Fig. 2.18 ROS architecture

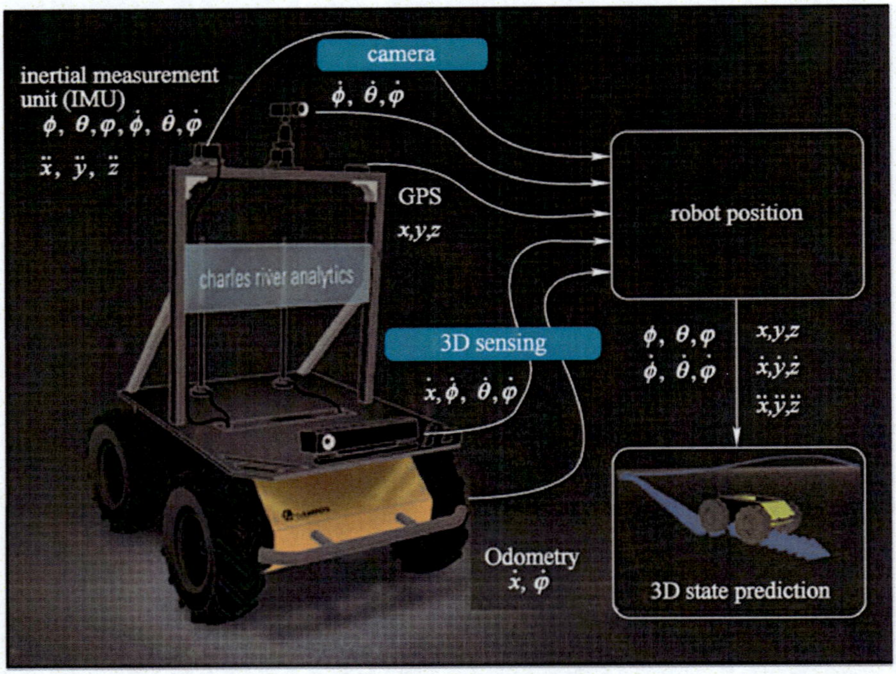

Fig. 2.19 Positioning of autonomous vehicle based on ROS

Gmapping: it can carry out laser-based SLAM and quickly build 2D maps in closed areas (Fig. 2.21).

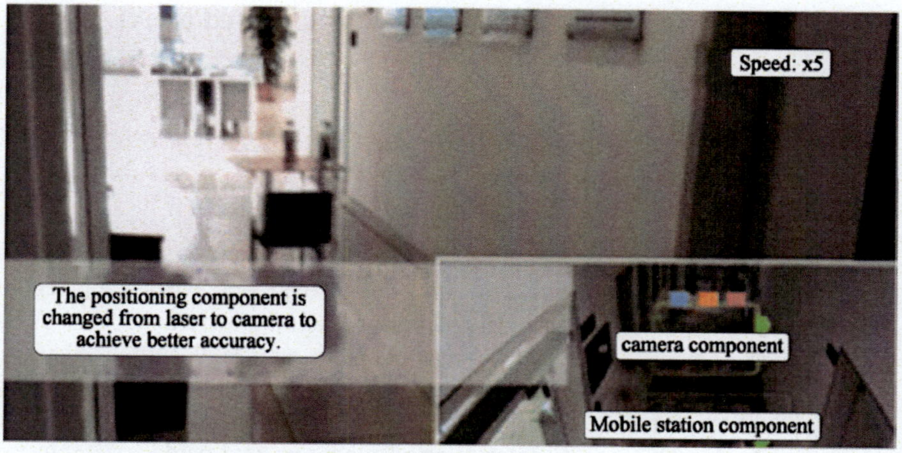

Fig. 2.20 Navigation of autonomous vehicle based on ROS

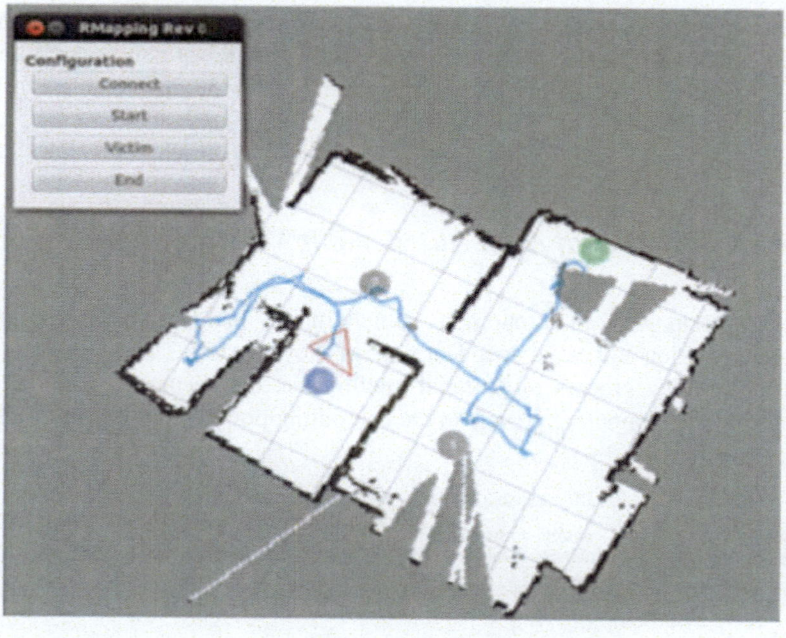

Fig. 2.21 Two-dimensional map construction based on ROS

2.5 The Ecosystem of Autonomous Vehicle Based on AI + 5G

2.5.1 The Architecture of Autonomous Vehicle Ecosystem Based on AI + 5G

The operation of autonomous vehicle produces massive data. The demand for fast and intelligent processing of these data will continue to promote the development of special computing processing unit technology for autonomous vehicle. AI + 5G is influencing and reconstructing the whole autonomous vehicle electronics industry. 5G communication technology is expected to be commercialized in the next two years, while AI will penetrate into all aspects of autonomous vehicle electronic applications under the two-way interaction of algorithm and hardware. From the perspective of hardware innovation, the realization of 5G requires the vehicle terminal products to have faster processing speed, better display quality, more efficient signal processing and longer battery life. In addition to the evolution of algorithm, the implementation of AI also needs comprehensive upgrades of chip, sensor, camera, radar, control unit and so on. The architecture of autonomous vehicle ecosystem supported and led by AI + 5G is shown in Fig. 2.22.

2.5.2 Application of AI in Autonomous Vehicle

The application of AI in autonomous vehicle mainly includes speech recognition, image recognition and cloud deep learning, as shown in Fig. 2.23.

In the training process of artificial intelligence, the accuracy of model fitting depends on the computing power, so the computing power requirement of artificial

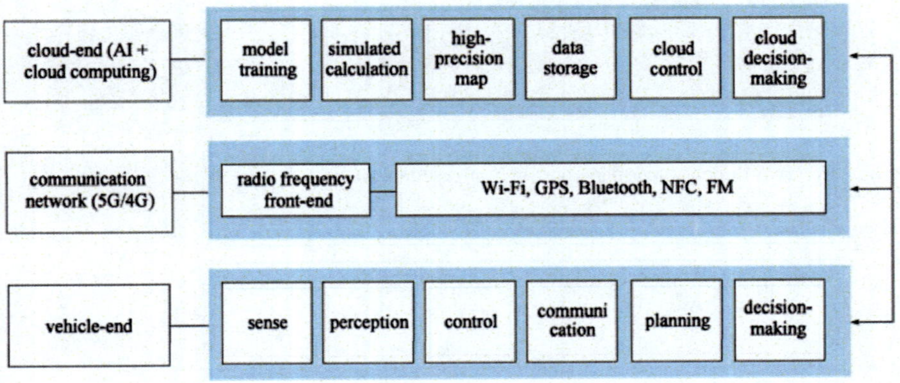

Fig. 2.22 Architecture of autonomous vehicle ecosystem supported and led by AI + 5G

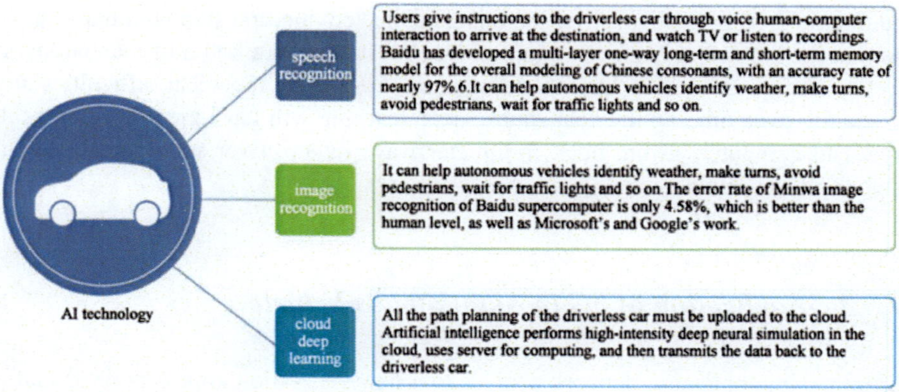

Fig. 2.23 Application of AI in autonomous vehicle

intelligence on the terminal is showing explosive growth. In terms of the amount of computation, parallel computing can improve the computing speed. In terms of calculation accuracy, the stronger the floating-point operation ability is, the higher the calculation accuracy is. Artificial intelligence chip is an important part of the AI market. In the era of AI, chips based on GPU, FPGA and ASIC will be widely used. GPU with parallel computing capability better meets the needs of deep learning, and will become the main tool for intelligent computing in the future with large new demand. In addition, intelligence also drives the complexity of core chips to increase rapidly. In addition to a CPU, Apple's newly released A11 bionic chip also integrates GPU, performance controller, neural network unit, ISP and other modules. It adopts 10 nm FinFET process and integrates 4.3 billion transistors in a single chip. Huawei's Kirin 970 also uses 10 nm process. The chip integrates CPU, GPU, ISP/DSP and NPU. A single chip integrates more than 5 billion transistors.

The advantage of deep learning is that it can simulate the brain recognition mechanism and better recognize, judge and classify unstructured data (such as image and speech), so that the algorithm can learn. This process is just like the human brain. It only requires engineers to continuously train the machine through similar scenes, so that it can learn to make its own judgment. In this way, even in a brand new scene, the car knows how to deal with it, which is more conducive to adaptation and expansion. Similarly, in terms of recognizing traffic lights, rule-base autonomous vehicle needs to mark all the traffic lights on the high-precision map and keep the machine looking at the same direction. However, the deep learning algorithm can recognize the color of traffic lights directly from the camera image, so the vehicle can understand the traffic lights and the driving situation of the whole intersection, so as to decide whether to move forward. The advantages of deep learning in the environment perception system of autonomous vehicle have been fully discovered, so that more and more companies claim that their technology is based on deep learning. In fact, that is not usually true. The reason is that the technical difficulties cannot be completely overcome in a short time. For example, the amount of data generated by automatic driving is huge. After the data is collected, the first step is to mark them so that the algorithm engine can be trained. Yet it may take 800 h of manual labor to tag them, even for big Internet companies. After these technical difficulties are gradually overcome, in the near future, deep learning will have great development space in computer vision, lidar, millimeter wave radar, driver state perception of autonomous vehicle.

2.5.3 Application of 5G in Autonomous Vehicle

Mobile communication network evolves to a new generation almost every decade. From 1 to 2G, the mobile communication technology has completed the transformation from analog to digital. On the basis of voice services, it has expanded to support low-speed data services. From 2 to 3G, the data transmission capacity has been significantly improved, and the peak rate can reach 2Mbps to tens of Mbps, supporting mobile multimedia services such as video call. The transmission capacity of 4G is an order of magnitude higher than that of 3G, with a peak rate of 100Mbps to 1Gbps. Compared with 4G technology, 5G will provide a peak bandwidth of more than 10Gbps with a new network architecture, and the user experience rate can be stabilized at 1–2 Gbps. 5G refers to the fifth generation mobile communication network. The definition of 5G by the International Telecommunication Union (ITU) is: 5G network is the next generation cellular wireless communication network which can provide 20 Gbps rate, 1 ms delay, 1 million connections per square kilometer and 99.999% network stability.

5G is the basis of automatic driving, and it will also play a key role in supporting the application of intelligent connected vehicle. 4G mainly supports the communication between people, while 5G not only connects people, but also connects vehicles, people and roads, realizing the connection between cars and people, between cars and roads, and between cars.

5G has the advantage of high connection density, strong stability and high security. Delay can be one of the most troublesome problems in communication, especially in automatic driving. 5G network not only has low latency, but also can allocate network processing capability according to the priority of data, so as to ensure the transmission of control signal of autonomous vehicle and maintain a faster response speed. Broadband is another important factor. As an alternative to human driver, each vehicle must be able to maintain information exchange with the network data center at any time, provide real-time vehicle location information and related information, and also interact with other vehicles and infrastructure for real-time data to complete precise positioning. If every car needs to be connected to the Internet at the same time, there will be a lot of data traffic. Therefore, high bandwidth will become one of the necessary features for the Internet of Vehicles and automatic driving. In addition, 5G has two more advantages. On the one hand, 5G technology can allocate the network according to the priority of data, so as to ensure the control signal transmission of the autonomous vehicle and maintain a faster response speed. On the other hand, 5G technology will allow devices close to each other to communicate directly, which can greatly reduce the overall network pressure and average delay.

At present, the global consensus on 5G spectrum is gradually formed, which basically focuses on 3.4–3.8 Ghz, 26 Ghz, 28 Ghz, 39 Ghz and other frequency bands. However, there are differences in the planning of different frequency bands. For low frequency bands, 3.5 Ghz is the preferred frequency band for 5G deployment in various regions. For high frequency bands, such as 26 and 39 Ghz, the plans of China and Europe are basically the same, but the United States, Japan, South Korea and other countries hold different opinions.

It is expected that 5G will be commercially available in 2020. At present, large-scale 5G test and deployment is being carried out all over the world. In China, 5G technology has entered the second phase of testing. The industry generally expects 5G to be commercialized on a large scale around 2020. Compared with 4G, 5G has higher transmission rate in the application scenario of autonomous vehicle networking, which allows information to be uploaded and downloaded quickly. Multiple access terminals can accommodate more data sources. Low delay is conducive to ensuring promptness, synchronization and security. After the completion of 5G channel construction, the data will be transmitted to the cloud in a timely and effective manner. By using intelligent cores such as cloud computing and AI algorithm, the data will be summarized and analyzed to complete the decision-making and control with global vision, and truly achieve transparent computing and holographic interaction.

Chapter 3
Vehicle Borne Radar System

3.1 Overview of Vehicle Borne Radar System

Autonomous vehicle sensor system generally includes four kinds of radars: lidar, millimeter wave radar, ultrasonic radar and infrared radar. Lidar and millimeter wave radar are basic and necessary onboard sensor devices, while ultrasonic radar and infrared radar are optional according to the situation. Tables 3.1 and 3.2 show the comparisons between advantages/disadvantages and working abilities of vehicle borne radar and camera.

3.2 Lidar

3.2.1 Technical Principles

Lidar technology, which first appeared in the 1960s, was originally used by astronauts to map the surface of the moon or archaeologists to draw maps. Recently its application has been expanding. It has been used in the field of automatic driving since the start of 21st Century and become one of the core technologies of ADAS. It can be applied to adaptive cruise control (ACC), forward collision warning (FCW) and automatic emergency braking (AEB). Sensors and lidar used in ADAS are shown in Fig. 3.1.

Lidar (light detection and ranging) is a radar system which detects the position, velocity and other characteristics of the target by emitting laser beams. It is a system integrating laser, global positioning system (GPS) and inertial navigation system (INS) technologies. It is used to obtain data and generate accurate ground digital elevation model (DEM). The components of lidar are shown in Fig. 3.2.

The basic working principle of lidar is: it transmits the detection signal (laser beam) to the target, and then compare the received signal (target echo) reflected from the target with the transmitted signal. After proper processing, the relevant

M. Du, *Autonomous Vehicle Technology*, Unmanned System Technologies,
https://doi.org/10.1007/978-981-19-4143-6_3

Table 3.1 Advantage/disadvantage comparison of onboard radar and camera

Radars	Ultrasonic radar	Monocular/stereo camera	Infrared radar	Lidar	Millimeter wave radar
Advantage	Simple structure, low price and small size	The cost is moderate; can detect the distance and size of obstacles; can identify pedestrians and bicycles	Low cost; not affected at night	High ranging accuracy; strong directivity; fast response; no ground clutter interference	Not affected by weather and light; can detect objects at a long distance (more than 100 m)
Disadvantage	Will be affected by weather and temperature changes; the maximum measuring distance is only a few meters	Affected by the range of vision like the human eye	Subject to weather conditions; only close range objects can be detected; bad at identifying pedestrians	Cost is high; cannot work 24 h a day	Cost is high; not likely to work 24 h a day

Table 3.2 Working ability comparison of onboard radar and camera

Working ability	Ultrasonic radar	Camera	Infrared radar	Lidar	Millimeter wave radar
Long range detection capability	Weak	Strong	Mediocre	Strong	Strong
Night working ability	Strong	Weak	Strong	Strong	Strong
All weather working ability	Weak	Weak	Weak	Weak	Strong
Affected by climate	Slightly	Greatly	Greatly	Greatly	Slightly
Working ability in smog environment	Mediocre	Strong	Weak	Strong	Strong
Working ability in rain and snow environment	Strong	Mediocre	Weak	Mediocre	Strong
Weak temperature stability	Weak	Strong	Mediocre	Strong	Strong
Speed measurement capability	Mediocre	Weak	Weak	Weak	Strong

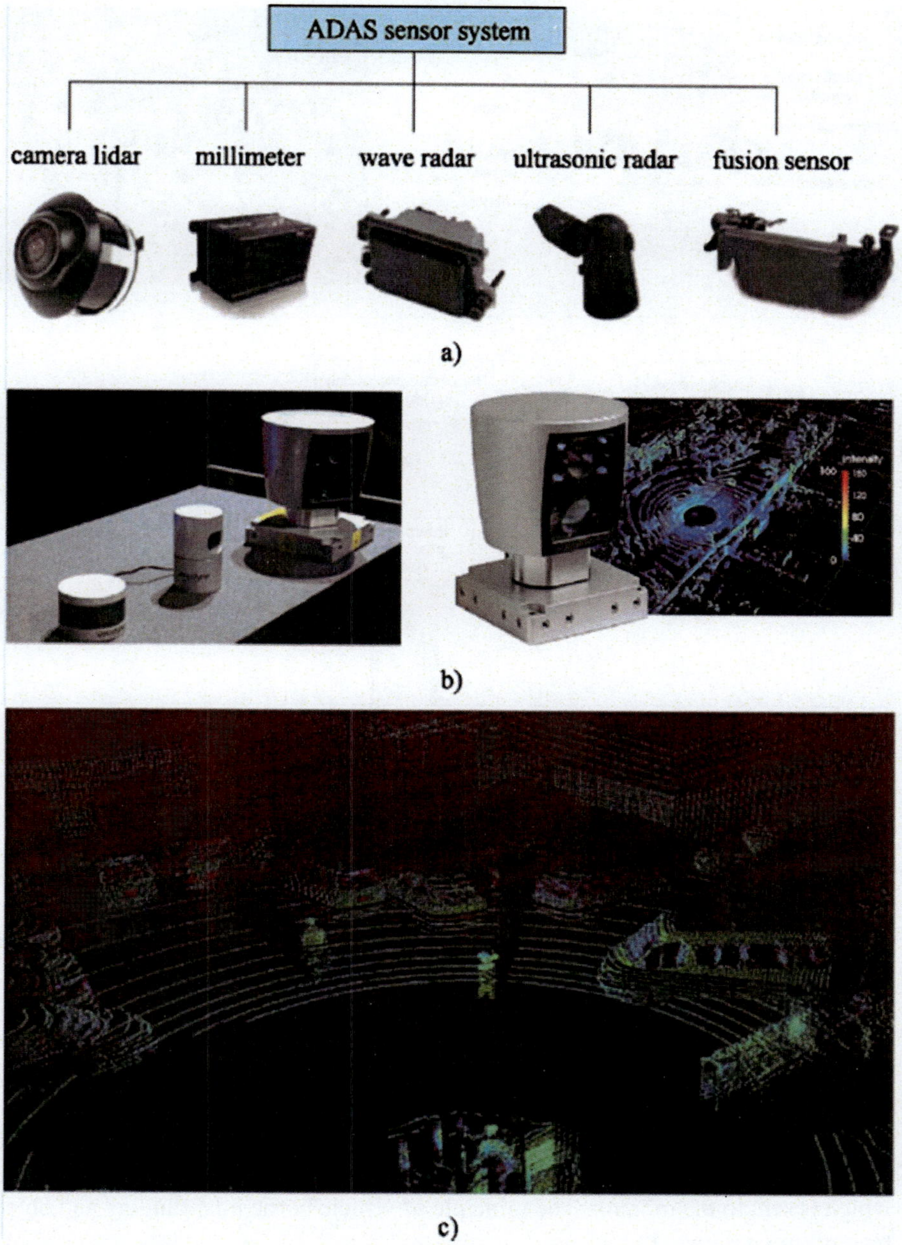

Fig. 3.1 ADAS sensor system and vehicle borne lidar

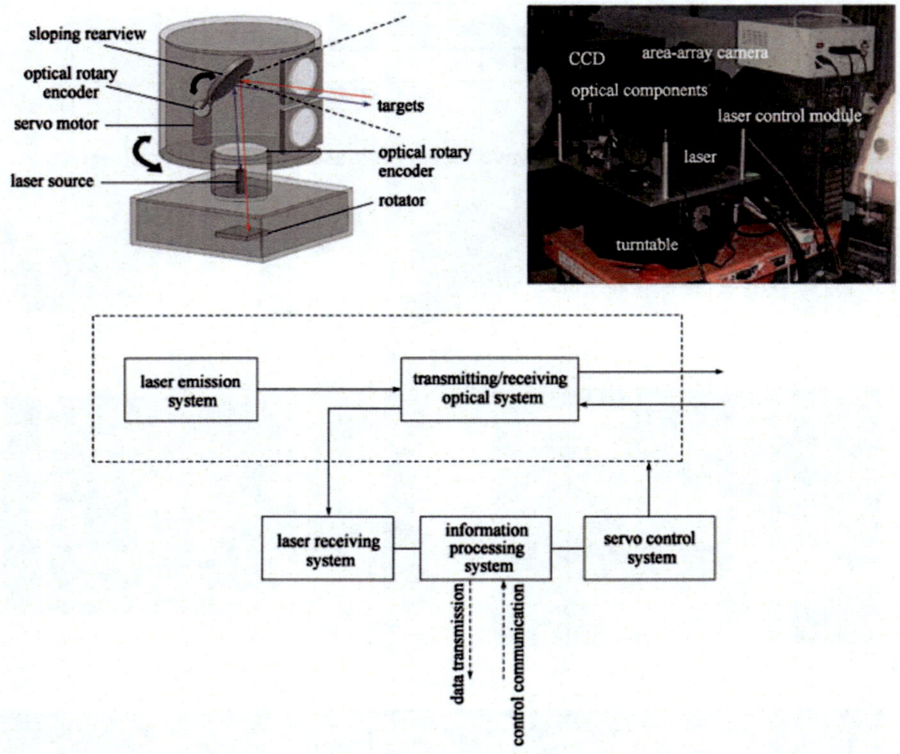

Fig. 3.2 Components of lidar

information of the target can be obtained, such as the target's distance, azimuth, altitude, speed, attitude, even shape and other parameters, which can be used to detect, track and identify the targets including aircrafts and missiles. The sensor emits tens or hundreds of thousands of laser pulses per second. When the light pulse is emitted, the stopwatch is started. Then when the light pulse (reflected from the first person/object) returns, the timer stops. By measuring the flight time of the light pulse, the distance between the sensor and the person/object is calculated. The lidar is composed of laser transmitter, optical receiver, turntable and information processing system. The laser converts the electric pulse into the optical pulse, and the optical receiver converts the optical pulse reflected from the target into the electric pulse, which is sent to the monitor. The principle of vehicle borne lidar ranging and object recognition is shown in Fig. 3.3.

The commonly used performance indexes of lidar are as follows:

1. Maximum radiated power;
2. Horizontal field of view (FOV);
3. Vertical FOV;
4. Wavelength of optical source;
5. Maximum distance to be measured;

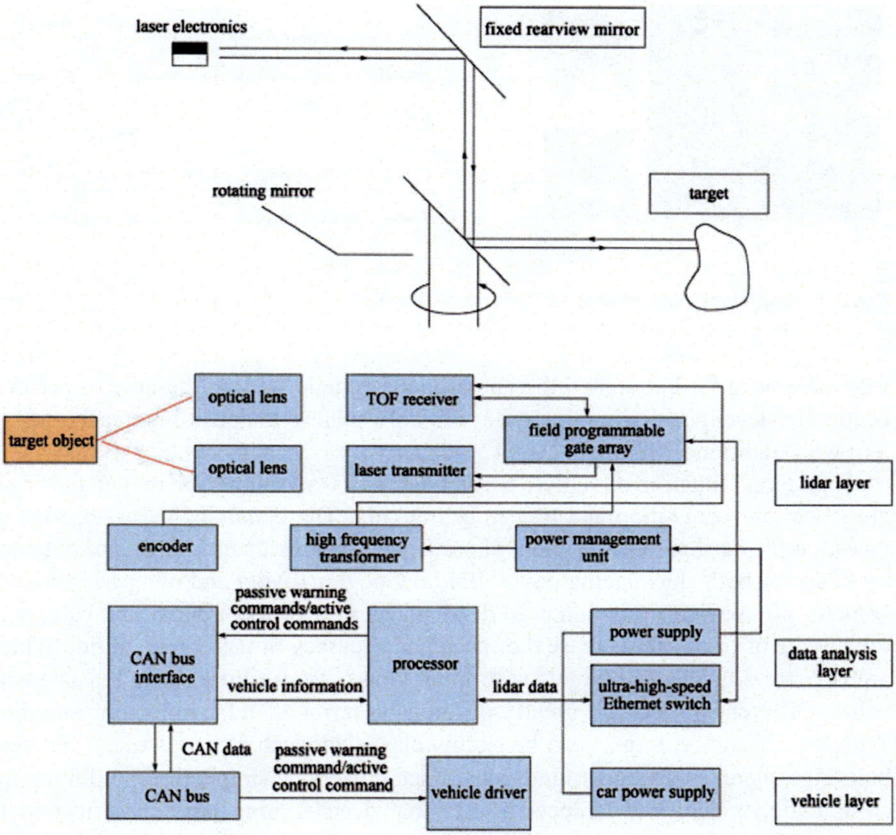

Fig. 3.3 Principle of vehicle borne lidar ranging and object recognition

6. Measurement time/frame rate;
7. Depth resolution;
8. Angular resolution;
9. Range measurement accuracy.

3.2.2 Functions

The primary function of vehicle borne lidar is positioning. With the loop detection of lidar SLAM or matching with the global map, we can achieve centimeter-level positioning, which is actually the case in the industry. Baidu, Google and mainstream manufacturers all use this method. GM has produced laser maps of all highways in the United States in its super cruise. Baidu autonomous vehicle's positioning method is shown in Fig. 3.4.

The disadvantage of this positioning method is that the lidar map, or Prior Lidar Map, must be made in advance. If the lidar map is absent (for sparsely populated areas,

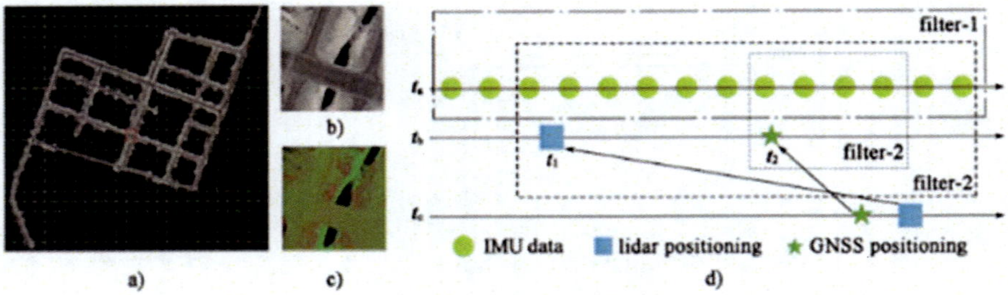

Fig. 3.4 Positioning method of Baidu autonomous vehicle

vast suburbs and rural areas), the autonomous vehicle will not be able to perform centimeter-level positioning. Instead it can only achieve the positioning accuracy of 3 m with traditional GPS.

The second function of vehicle borne lidar is to work with camera data for target classification recognition and trajectory tracking. The details of the road, such as lanes, curbs, barriers, virtual and real roads, and pedestrian crossings, are obtained by lidars on both sides facing down. Baidu, Google, Toyota and other autonomous vehicles all use lidar to obtain road details. The data fusion of lidar and camera in the middle of front can increase the speed and accuracy of target recognition. There are two ways to identify targets with lidar. One is to use lidar reflection intensity value. Different objects are greatly different in terms of lidar reflection intensity. With the difference, targets can be simply classified, such as pedestrians, vehicles, buildings, plants, roads, grasslands, etc. Since it is only a simple threshold filter, the computation is small and the speed is faster than deep learning target classification. In January 2018, Renesas announced that it would cooperate with the start-up company Dibotics to embed the Augmented Lidar software into its chip and use this method for classification and recognition. The second method is to transform the Lidar point cloud into dense image data with tensor structure, and then recognize it with Faster RCNN. The latter requires powerful computing resources and the processing speed is relatively slow. Faster RCNN is recognized as the best method in the field of image recognition. Another advantage of lidar is that it is good at predicting and tracking the trajectory of moving objects, which is the basis of behavioral decision-making. Just like human drivers, it can predict the next movement of pedestrians or other vehicles, and make decisions based on this prediction. Lidar is born with the ability of trajectory yielding. Velodyne 16-ray lidar can accurately predict the movement of the ball after the quarterback throws it in the Super Bowl in the United States. It is much faster to calculate and predict the trajectory with lidar than with optical flow method, and the consumption of computing resources is also much lower.

At present, there are four major methods to detect lane lines by lidar. The first one is based on lidar echo width. The second one is to filter out the invalid information according to the gray image formed by lidar reflection intensity information or the combination of intensity information and elevation information. The third one is the combination of lidar SLAM and high-precision map, which can detect the lane line and locate the vehicle. The fourth one is to detect the road edge based on its height information or physical reflection information, then calculate the lane line position according to the distance and road width. This method cannot be used for some roads of which the height difference between the edge and road surface is less than 3 cm. The latter three methods need at least 16-ray lidar. The former can work with 4-ray or 1-ray lidar. At present, the second method is the most widely used one. The feature of this method is that the lidar should be close to the road or facing the road as much as possible to obtain more information of reflection intensity. Toyota and Google both install lidar on the front bumper of the car. GM has 9 mm wave radars and 1 camera near the bumper, so the lidar can only be placed on the roof due to possible lack of installation space.

3.2.3 Advantages and Disadvantages

Compared with microwave radar, the working frequency of lidar is much higher, which brings many advantages.

1. High resolution

Lidar can obtain high resolution of angle, range and velocity. In general, the angular resolution is not less than 0.1 mrad, that is to say, it can distinguish two targets with a distance of 0.3 m, at a distance of 3 km (which is impossible for microwave radar). It can also track multiple targets at the same time. The range resolution can reach 0.1 m, and the velocity resolution can reach within 10 m/s. The high resolution of range and velocity means that a clear image of target can be obtained by using range Doppler imaging technology.

2. Good concealment and strong anti-active jamming ability

The laser travels in a straight line with good directivity with a narrow beam, and can only be received in its path. Therefore, it is very difficult to be intercepted. Moreover, the aperture of the lidar transmitting system (transmitting telescope) is small, and the receiving area is narrow. The probability of intentional jamming signals getting transmitted into the receiver is very low. In addition, unlike microwave radar which is easily affected by the widespread electromagnetic waves in nature, there are few signal sources that can interfere with lidar. Therefore, lidar has a strong ability to resist active jamming and is suitable for working in the increasingly complex information environment.

3. Decent low altitude detection performance

Due to the influence of various ground object echoes, there is a certain blind area (undetectable area) in the low altitude for microwave radar. For lidar, only the target will cause reflection, and there is no influence of ground echo, so it can work at zero altitude. Its low altitude detection performance is much better than that of microwave radar.

4. Small volume and light weight

Generally, the volume of ordinary microwave radar is huge. The weight of the whole system reaches tons, and the aperture of optical antenna is several meters or even tens of meters. The lidar is much more portable. The aperture of the transmitting telescope is generally at a centimeter level, and the minimum weight of the whole system can be just tens of kilograms. It is very easy to set up and dismantle. Moreover, the structure of lidar is relatively simple, easy to maintain and operate, and cheap.

The disadvantages of lidar are as follows.

First of all, it is greatly affected by the weather and atmosphere. Generally, the attenuation of laser is small and the propagation distance is long in clear weather. However, in heavy rain, smoke, fog and other bad weather, the attenuation increases sharply, and the propagation distance is significantly affected. The atmospheric circulation will also cause laser beam distortion and jitter, which directly impacts the measurement accuracy of lidar.

Secondly, since the beam of lidar is very narrow, searching for targets can be difficult, which directly affects the interception probability and detection efficiency of non-cooperative targets. It can only search and capture targets in a small range. Therefore, lidar is rarely used for independent search and detection.

3.2.4 Commercialization

The low price of lidar would be the foundation and core of its large-scale commercial application. At present, there is still a long way to go before lidar is widely used. The performance index and price of the commonly used vehicle borne lidar are shown in Fig. 3.5.

Velodyne, born in California, USA, started to sell its first lidar product in 2007 at the price of US $80,000. After continuous research and development, the price has now dropped to about US $8000. It is predicted that the market scale of lidar is to exceed 10 billion in 2020, and its price is to drop to about US $250. In fact, in order to reduce the cost of lidar, various manufacturers are constantly trying. The products that lower the cost by narrowing the horizontal view and limiting the detection range will be on the market. There are also some manufacturers trying to cut the cost by reducing laser ray-count or replacing sensors and processors with integrated chips.

3.3 MMW Radar

3.3.1 Technical Principles

Millimeter wave radar is a high-precision radar sensor working in millimeter wave band to measure the relative distance, relative velocity and azimuth of objects.

Model	HDL-64E	HDL-32E	VLP-16/PUCK	Ultra Puck-32A
Price (USD)	500 to 1000 thousand	100 to 300 thousand	7999	500 (mass-produced target price)
Characteristic	Good performance, high price	smaller and lighter	suitable for UAV	exclusively for vehicles
Ray-count	64	32	16	32
Dimensions (mm)	203×284	86×145	104×72	104×72
Weight (kg)	13.2	1.3	0.83/0.53	0.8 ~1.3
Laser wavelength (nm)	905	905	905	905
Horizontal view (degree)	360	360	360	360
Vertical view (degree)	26.8 (+2~-24.6)	41.34 (+10.67~-30.67)	30 (+15~-15)	28
Output frequency (ten thousand points/second)	130	70	30	70
Measuring range (m)	100~120	80~120	100	200
Range accuracy (cm)	<2	<2	<3	<2
Horizontal resolution	5 Hz: 0.08° 10 Hz: 0.17° 20 Hz: 0.35°	5 Hz: 0.08° 10 Hz: 0.17° 20 Hz: 0.35°	5 Hz: 0.1° 10 Hz: 0.2° 20 Hz: 0.4°	
Vertical resolution (degree)	0.4	1.33	2.0	
Protection Standard	IP67	IP67	IP67	IP67

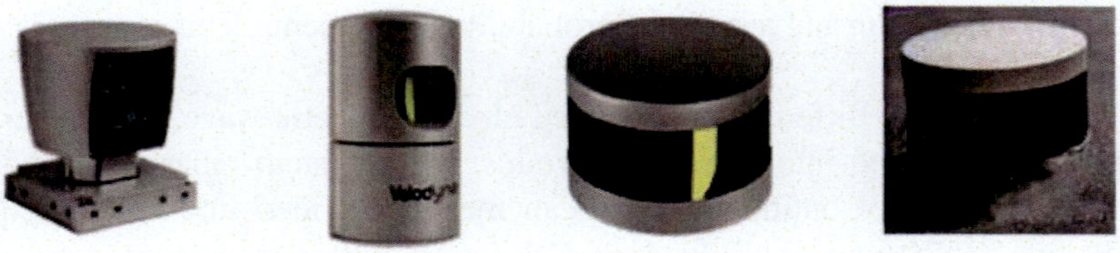

Fig. 3.5 Performance index and price of vehicle borne lidar

Millimeter wave refers to the electromagnetic wave in the frequency domain of 30–300 GHz (wavelength of 1–10 mm). Millimeter wave radar was used in the military field in the early days. With the development of radar technology, it is gradually used in many other fields, such as automotive electronics, UAV, intelligent transportation and so on. Since the late 1970s, millimeter wave radar has been used in many important military and civil systems, such as short-range high-resolution air defense system, missile guidance system, target measurement system and so on.

The working principle of the vehicle borne millimeter wave radar is as follows: the radar transmits millimeter wave outward through the antenna, receives the target reflection signal, and quickly and accurately obtains the physical environment information around the vehicle (such as the relative distance, relative speed, angle, moving direction between vehicle and objects). Then the target tracking and recognition classification are carried out according to the detected object information, and the data fusion is performed with body dynamic information. Finally, the intelligent processing is carried out through the electronic control unit. After reasonable decision-making, the driver can be informed or warned by sound, light, touch and other means, or active intervention can be made in time, so as to ensure the safety

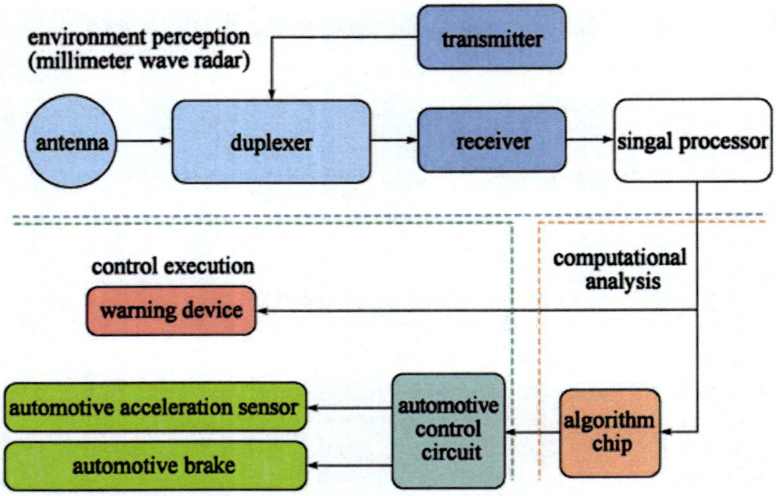

Fig. 3.6 Working process of vehicle borne millimeter wave radar

and comfort of driving and reduce the probability of accidents. This process is shown in Fig. 3.6.

According to the difference of radiated electromagnetic wave, millimeter wave radars can be divided into two major groups: pulse system radar and continuous wave system radar. Continuous wave can then be divided into FSK (frequency shift keying), PSK (phase shift keying), CW (constant frequency continuous wave), FMCW (frequency modulated continuous wave) and other modes. The features and shortcomings are shown in Fig. 3.7.

Due to its ability to measure multiple targets, high resolution, low complexity of signal processing, low cost and mature technology, FMCW radar has become the most commonly used vehicle borne millimeter wave radar. Tier 1 suppliers such as Delphi, Denso and Bosch all adopt FMCW modulation mode. FMCW radar system mainly includes transceiver antenna, RF front-end, modulation signal, signal processing module, etc. MMIC chip (monolithic microwave integrated circuit) and antenna PCB board are the hardware core. Through the correlation processing of received signal and transmitted signal, the detection of range, position and relative speed of target is conducted. The composition and working principle of FMCW radar system is shown in Fig. 3.8.

The technology of front-end monolithic microwave integrated circuit (MMIC), the key component of millimeter wave radar, is controlled by foreign semiconductor companies, while the high-frequency MMIC can only be produced by a few foreign chip manufacturers such as Infineon and Freescale. Chinese MMIC is still in its infancy. Xiamen Yixing Company and Nanjing Miller Company are developing radar MMIC, and the results remain to be verified. In addition, the only national key laboratory of millimeter wave radar at Southeast University, has been developing 77 GHz millimeter wave integrated circuits.

Group	Pulse type	Continuous wave type			
		CW	FSK	PSK	FMCW
Feature	suitable for long distance target detection; measurement process is simple and accuracy is high	velocity measurement based on Doppler shift information from target	measure the distance and speed of target	measure distance and speed by modulating video with random binary or quad phase codes	measure distance and speed of multiple targets; high resolution, low complexity of signal processing, low cost and mature technology
Shortcoming	In the short-range application of automobile anti-collision radar, it is relatively difficult to generate narrow pulse; difficult to measure multiple targets because of high peak power.	distance cannot be measured	difficult to measure multiple targets	When high resolution is required, the requirement for signal processing is very high, which is difficult to be realized by current technology.	Good LFM is not easy to obtain, which affects the range resolution.

Company	Fujitsu Ten	Pentak	Delphi	Bosch	Elesys	Denso	Hitachi
Product							
Modulation system	FMCW	FMCW	FMCW	FMCW	FMCW	FMCW	CW
EHF device	MMIC	MMIC	CUNN	MMIC	MMIC	MMIC	MMIC

Fig. 3.7 Feature and shortcoming of two types of millimeter wave radar

3.3.2 Advantage

The advantages of millimeter wave radar are as below. Millimeter wave has short length and wide frequency band and narrow beam. Therefore, it has higher resolution, and is not easily interfered. The wavelength of millimeter wave is between centimeter wave and light wave, so it has the advantages of both microwave guidance and photoelectric guidance. Compared with light wave, millimeter wave and submillimeter wave travel in the atmosphere with less attenuation, and are less affected by natural light and thermal radiation sources. Compared with centimeter wave seeker, millimeter wave seeker has the advantages of small volume, light weight and high spatial resolution. Compared with infrared, laser, television and other optical seeker, millimeter wave seeker has stronger ability to penetrate fog, smoke and dust, and has the characteristic of all-weather (except heavy rain) and all day. In addition, the anti-jamming and anti-stealth capability of millimeter wave seeker is better than that of other microwave seeker.

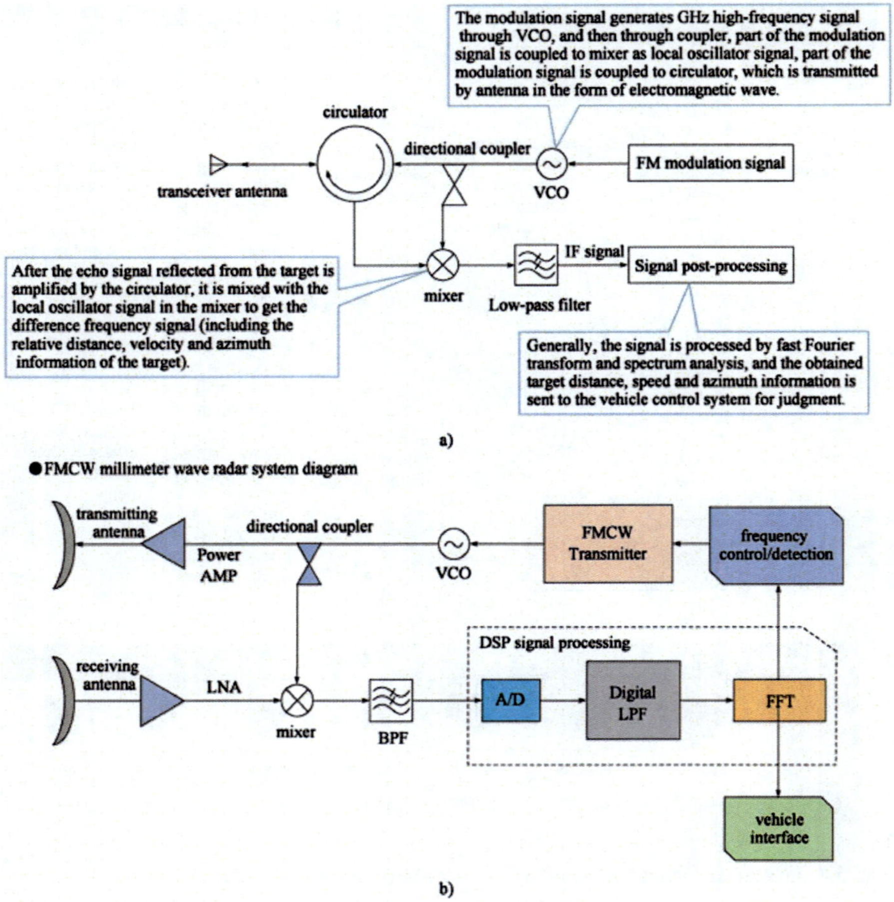

Fig. 3.8 Composition and working principle of FWCW radar system

3.3.3 *Applications on Vehicle*

At present, the frequency bands allocated by various countries to vehicle borne millimeter wave radar are shown in Fig. 3.9, which are mainly concentrated in 24 GHz

Country or region	24 GHz	60 GHz	77 GHz	79 GHz
EU	approved		approved	approved
US	approved		approved	approved
Japan		approved	pending	pending
China	approved		approved	pending

Fig. 3.9 Frequency band allocation of vehicle borne millimeter wave radar

Table 3.3 Comparison of forward and backward radar functions

Backward radar	Forward radar
Blind spot detection	Adaptive cruise control
Lane change assist	Automatic emergency braking
Rear collision warning	Forward collision warning
Reverse collision warning	–
Rear cross traffic alert	–
Door open warning	–

region (21.65–26.65 GHz) and 77 GHz region (76–81 GHz). A few countries (such as Japan) use 60 GHz band. The most common working frequencies of automotive millimeter wave radar are around 24 GHz, 77 GHz and 79 GHz. Compared with 24 GHz device, 77 GHz device has many advantages, such as smaller size of components and antenna, easier implementation of single-chip integrated structure, higher speed resolution, etc. In the future, the frequency bands of vehicle borne millimeter wave radar will converge to 77 GHz region (76–81 GHz).

When the millimeter wave radar is installed on the car, the distance, angle and relative velocity between the radar and the measured object can be measured. At present, millimeter wave radar is mainly used in high-end models, and will gradually be popularized to low-end models. According to the position of installation, vehicle borne millimeter wave radars can be divided into backward radars and forward radars. Generally, each vehicle is equipped with two backward radars, which are placed on the left and right tail of the vehicle. A forward radar is usually installed in the middle of the front bumper of the vehicle. The functions of backward radar and forward radar are shown in Table 3.3.

Millimeter wave radar can perform the functions of advanced driving assistance system (ADAS), such as adaptive cruise control, forward collision warning, blind spot detection, parking aid, and lane change assist. The 24 GHz radar system mainly performs short range reconnaissance (SRR), while the 77 GHz system mainly performs long range reconnaissance (LRR). The combined application of SRR and LRR is shown in Fig. 3.10.

In order for ADAS to fully carry out it functions, five millimeter wave radars of "1 long plus 4 mediums or shorts" are needed. At present, the new Audi A4 adopts five millimeter wave radars (1 long plus 4 shorts), and Mercedes Benz S-class adopts seven millimeter wave radars (1 long plus 6 shorts). Taking the ACC function of automatic car following as an example, three millimeter wave radars are generally required. There is a 77 GHz LRR in the middle of the car, with a detection range of 150–250 m, and an angle of about 10°. There are 24 GHz MRRs on both sides of the car, with an angle of 30°, and a detection distance of 50–70 m. AEB is the most practical function of ADAS. It will become a standard feature of medium and high-end cars in the future. It requires a 77 GHz LRR.

In the United States, anyone who wants to use wireless devices which are not FCC (Federal Communications Commission) approved, must first apply for a special temporary authority (STA) license, otherwise it is illegal. On March 17, 2017,

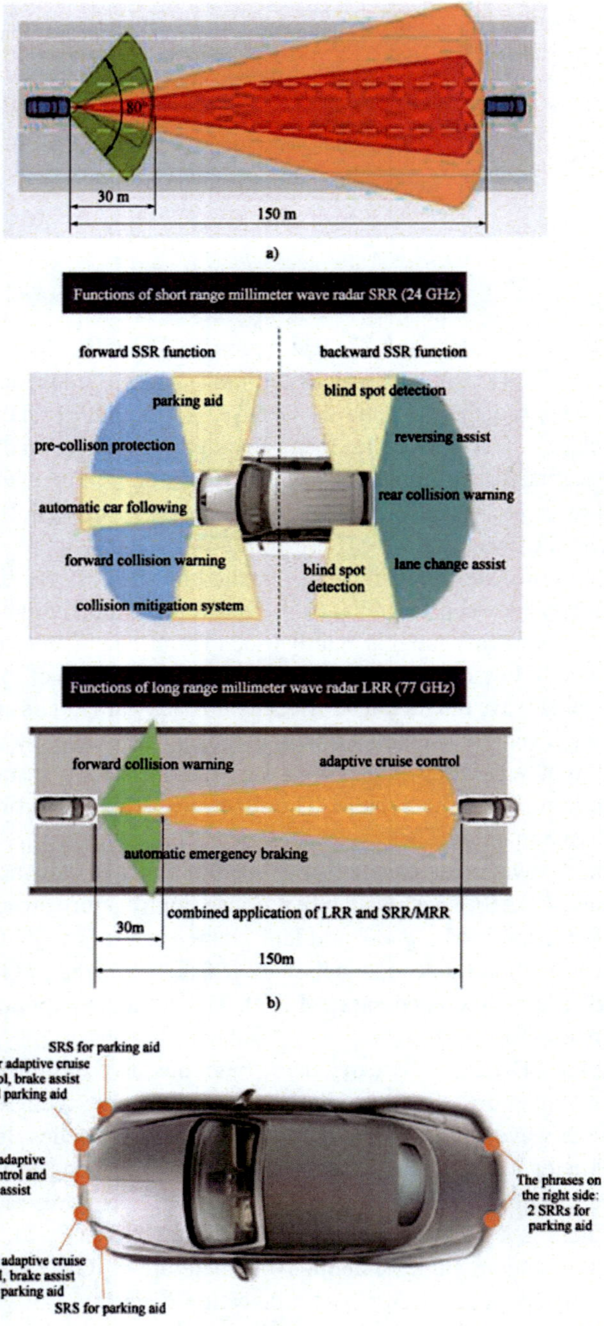

Fig. 3.10 Combined application of SRR and LRR

Japanese automotive electronics supplier Alps Electric applied for a STA license. It was testing a vehicle millimeter wave radar named Ukaza, which works at 76–81 GHz. This was the first public test of millimeter wave radar of this band. Ten days later, GM also submitted a similar STA license application to the FCC to test Ukaza radar, in preparation for its 300 autonomous vehicle test fleet.

3.3.4 Industry Structure, Mainstream Products and Development Trend

The domestic and international industry structure of vehicle borne millimeter wave radar is shown in Fig. 3.11.

At present, millimeter wave radar technology is mainly monopolized by traditional parts giants such as Continental, Bosch, Denso, Delphi and so on. In particular, 77 GHz millimeter wave radar is controlled by Bosch, Continental, Delphi, Denso, TRW, Fujitsu Ten, Hitachi and other companies. In 2015, the market shares of Bosch and Continental automotive radar were both 22%, ranking first in the world.

At present, the millimeter wave radar sensors for high-end vehicles in Chinese market are all imported, and the market is monopolized by American, Japanese and German companies. The price is high, so independent manufacture is urgently needed. Automotive millimeter wave radar is categorized as short range, medium range and long range, and their corresponding prices are also low, medium and high. The mainstream automotive millimeter wave radars on the market are shown in Fig. 3.12.

Fig. 3.11 Millimeter wave radar industry structure

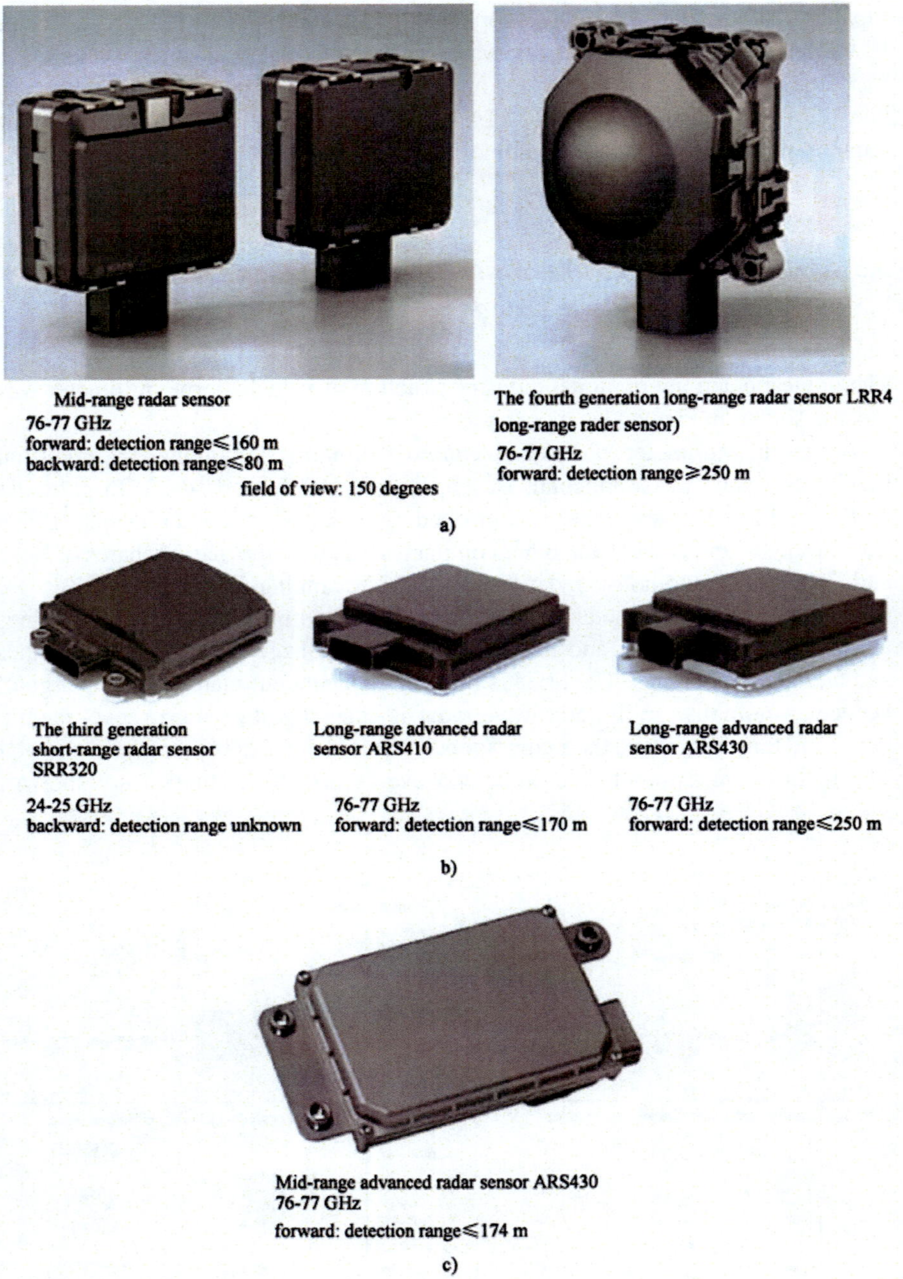

Fig. 3.12 Overview of mainstream automotive millimeter wave radars. **a** Bosch medium and long-range radar.**b** Continental short and long-range radar.**c** Delphi medium range radar

Bosch radar is small. Delphi short range radar and long range radar, ZF-TRW long range radar and Autoliv long range radar will be mass-produced in the next few years. In addition to the above companies, some companies (such as Fujitsu Ten and Nidec Elesys) also produce millimeter wave radars. Chinese vehicle borne millimeter wave radars are still in the development stage. Due to the low cost and simplicity of R&D, Chinese manufacturers mainly focus on 24 GHz radar. It is estimated that in the next five years, vehicle borne millimeter wave radar sensors of independent brands will gradually be mass-produced and installed.

In the long run around the world, the vehicle borne millimeter wave radar will eventually be unified in the 77 GHz band (76–81 GHz). This band means greater bandwidth, higher power level and longer detection range. Compared with 24 GHz, the accuracy of object resolution is improved by 2–4 times, and the accuracy of speed and distance measurement is improved by 3–5 times, which means pedestrians and bicycles can be detected. Moreover, the volume of the equipment is smaller, which makes it easier to install and deploy on the vehicle. 77 GHz frequency range is the frequency band permanently recognized globally, so it is more suitable for global vehicle platform. Within this range, 76–77 GHz is mainly used for long-range millimeter wave radar, and 77–81 GHz is mainly used for medium and short-range millimeter wave radar. In the future, 79 GHz (77–81 GHz) medium and short range millimeter wave radar will become the mainstream of medium range MRR, and is expected to replace 24 GHz short range radar. The timing of replacement depends on the industrial development, market trend and policy of each country.

3.4 MIMO Radar Virtual Array and 2D Imaging

MIMO (multiple input multiple output) is originally a concept in control system, which means a system has multiple inputs and multiple outputs. If the transmission channel of mobile communication system is regarded as a system, the transmitted signal can be seen as the input signal of mobile channel (system), and the received signal can be seen as the output signal of mobile channel. The basic meaning of MIMO radar is that the radar uses multiple transmitting antennas to transmit orthogonal signals at the same time to irradiate the target, and uses multiple receiving antennas to receive the target echo signal, and then process it comprehensively to extract the spatial position and motion state of the target.

A typical application of MIMO radar virtual array is for two-dimensional radar imaging. The range resolution of two-dimensional radar imaging mainly depends on the bandwidth of radar signal, and the azimuth resolution mainly depends on the beam width of antenna. To improve the range resolution of imaging, a relatively easy way is to increase the bandwidth of radar signal. In order to improve the azimuth resolution of radar signal, it is inevitable to increase the aperture of antenna or array, which is difficult and limited by many factors in practice. At present, the widely used solution is to use synthetic aperture technology to obtain large aperture array without increasing the physical size of the antenna. Different from the idea of synthetic aperture, MIMO radar uses the antenna structure of multiple transmitters and multiple receivers to form a virtual large aperture array to obtain high azimuth resolution. The formation of this virtual array is real-time, which can avoid the motion compensation problem in traditional ISAR imaging. Therefore, MIMO radar has its unique advantages in imaging applications.

NXP launched MR3003 and S32R274 on January 11, 2018, which is a typical cascade design. MR3003 is a millimeter wave radar transceiver with three transmitters and four receivers. The lowest cascade design of NXP is to cascade four MR3003s, with 12 transmitters and 16 receivers. With S32R274 as the processor, it can cascade up to 20 MR3003s, forming an amazing 60-transmitter and 80-receiver millimeter wave radar, which is enough for imaging. Cascading is actually similar to MIMO system.

A pair of transmitting array element and receiving array element can simulate a transmitting-receiving array element. For MIMO radar with m transmitter and n receivers, there are m*n pairs of transmitting array elements and receiving array elements. That means m*n transmitting-receiving array elements can be simulated. The number is generally much larger than n, which leads to the expansion of array aperture. Texas Instruments can work with four AWR1243 radars with three transmitters and four receivers, which are 192 virtual channels (antennas or arrays), as shown in Fig. 3.13.

Figure 3.14 shows the parameters of four AWR1243 cascaded radar of Texas Instruments. The long-range resolution is greatly improved. The azimuth resolution of 1° can be achieved at 40 m, that is, the resolution of 4.5 cm and the object resolution of about 9 cm.

With a megapixel camera of 45° FOV, the azimuth resolution is only about 20 pixels at 40 m. It cannot distinguish pedestrians from cyclists. MIMO has a very wide FOV, and the FOV of four-way cascaded radar like the one of Texas Instruments is as high as 192°(80° FOV of a camera is considered a wide angle). Therefore, there may be wide angle distortion at the edge.

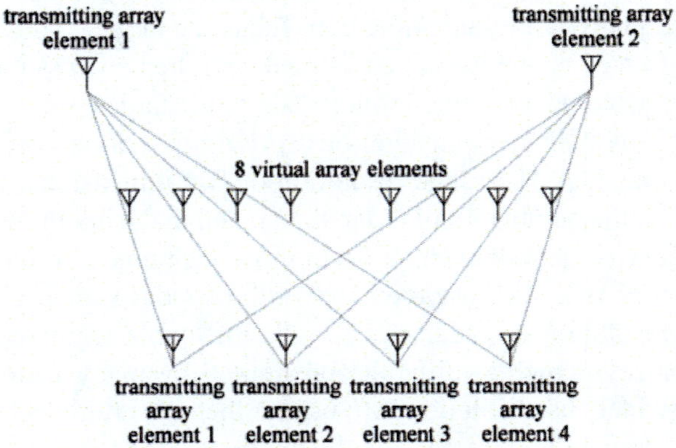

Fig. 3.13 Four TI AWR1243 radars with three transmitters and four receivers

Four AWR1243 cascaded 3D radar		
Sweep bandwidth	3333 MHz	275 MHz
Range resolution	4.5 cm←high resolution	55 cm
Max unambiguous range	40 m	250 m←long range
Azimuth angular resolution	1 °	1 °
Cross-range resolution @ max range	70 cm	4.4 m
Elevation angular resolution	14 °	14 °
Ramp slope	54 MHz/μs	9 MHz/μs
Chirp duration	61.3 μs valid(+ 13 μs inter-chirp)	30.7 μs valid(+ 7 μs inter-chirp)
Number of chirps	16	256
Max beat frequency	15 MHz	15 MHz
ADC sampling rate (I, Q)	16.7 MSPS(complex)	16.7 MSPS(complex)
Frame time	16×74.3 μs=1.2 ms	256×37.7 μs=9.7 ms
Range FFT size	1024 (complex)	512 (complex)
Radar data memory	1024×16×16 RX×4 Bytes=1 MB	512×256×16 ×4=8 MB

Fig. 3.14 Parameters of four AWR1243 cascaded radar

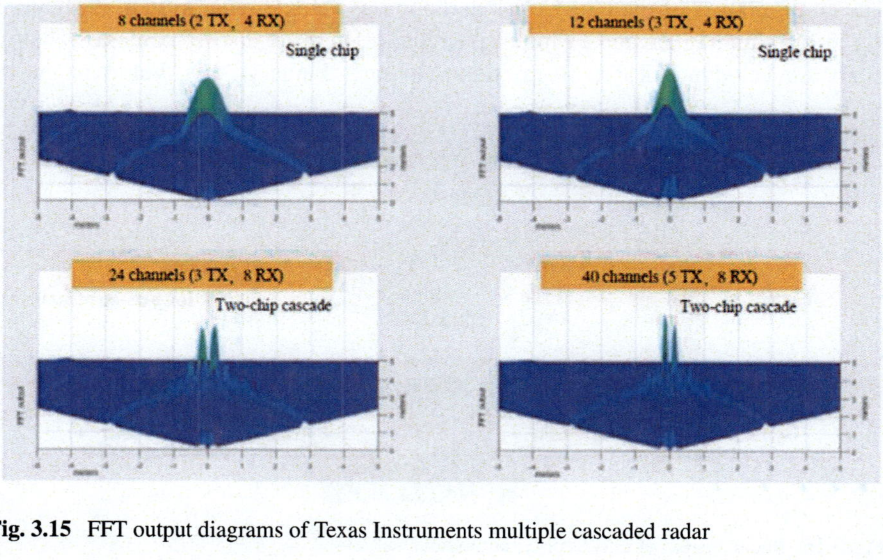

Fig. 3.15 FFT output diagrams of Texas Instruments multiple cascaded radar

Figure 3.15 shows the FFT output diagrams of a multiple cascaded radar (MIMO) tested by Texas Instruments. It can be seen that the more channels there are, the higher the precision.

3.5 Ultrasonic Radar

When the lidar is installed on the roof of the car, it forms a round area that cannot be measured at the bottom of the car, which is also the blind area of the lidar. Millimeter wave radar for adaptive cruise control is used for long-distance measurement, but the accuracy of short-range measurement is not sufficient. The accuracy for automatic parking has to reach centimeter level, which cannot be achieved by lidar and millimeter wave radar. Therefore, ultrasonic radar which is good at close range measurement will be used. The reversing radar usually is an ultrasonic radar, which is installed near the rear wheels. Google's prototype driverless car is equipped with an ultrasonic radar near the rear wheels for close range measurement, which is mainly used for automatic parking.

3.6 Examples of Autonomous Vehicle Radar System Configuration

3.6.1 GM Cruise Autonomous Vehicle

The configuration of the GM Cruise autonomous vehicle radar system is shown in Fig. 3.16, with 5 lidars and 21 mm wave radars. There are also 16 cameras. The five lidars are Velodyne VLP16 16-ray lidars. In early January 2018, Velodyne cut the

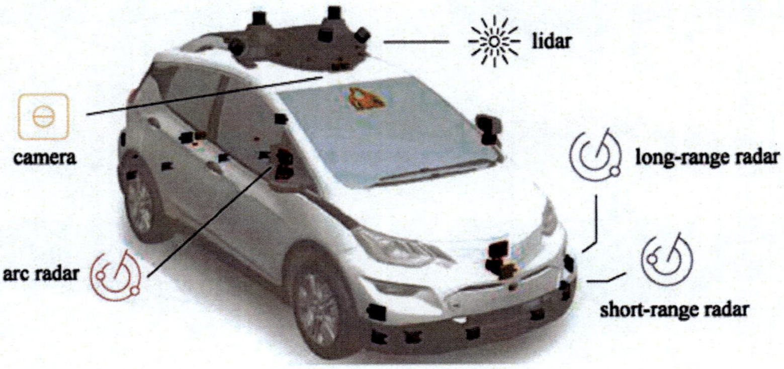

Fig. 3.16 Radar system of GM Cruise autonomous vehicle

price of VLP16 from $7999 to $3999. Among the 21 mm wave radars, twelve 79 GHz radars are provided by Alps of Japan. Two forward and two backward long-range radars are presumably provided by Germany companies, and the model is ARS-408. Five high-resolution (articulating, according to GM) millimeter wave radars are provided by Bosch. They are mainly installed on both sides and in front of the car. Their bandwidth is up to 8 times that of a traditional 77 GHz radar, so the resolution can reach 4 cm.

Twelve 79 GHz millimeter wave radar works in cascade mode, which is enough to generate a clear 360° panoramic image and track thousands of targets at the same time. Twelve 79 GHz millimeter wave radars form a redundant system. Millimeter wave radars have the strongest ability to deal with complex environment, and are most suitable for a redundant system. In the case of lidar and camera failure, they can still ensure the safe driving and parking of vehicles.

3.6.2 *Google Autonomous Vehicle*

Google Driverless Car is a fully automatic driving vehicle currently under development by Google X Laboratory. It does not require a driver to start, travel or stop. It is currently being tested and has travelled 480,000 km. Google has used seven cars for test, including six Toyota Prius and one Audi TT. They were tested on several roads in California, including Lombard street in the San Francisco Bay area. These vehicles use cameras and radars to sense the road environment and traffic conditions, and use detailed maps to carry out navigation. The actual test scene is shown in Fig. 3.17.

The configuration of Google autonomous vehicle sensor system is shown in Fig. 3.18. It is equipped with four millimeter wave radars (three in the front and one in the rear, as shown in Fig. 3.19) and one roof 3D lidar (Velodyne HDL-64E). Millimeter wave radar is used to calculate the relative speed between the other cars and the autonomous vehicle, and to predict speed changes. The HDL-64E radar continuously emits 64 laser beams with a maximum range of 120 m while rotating,

Fig. 3.17 Google autonomous vehicle

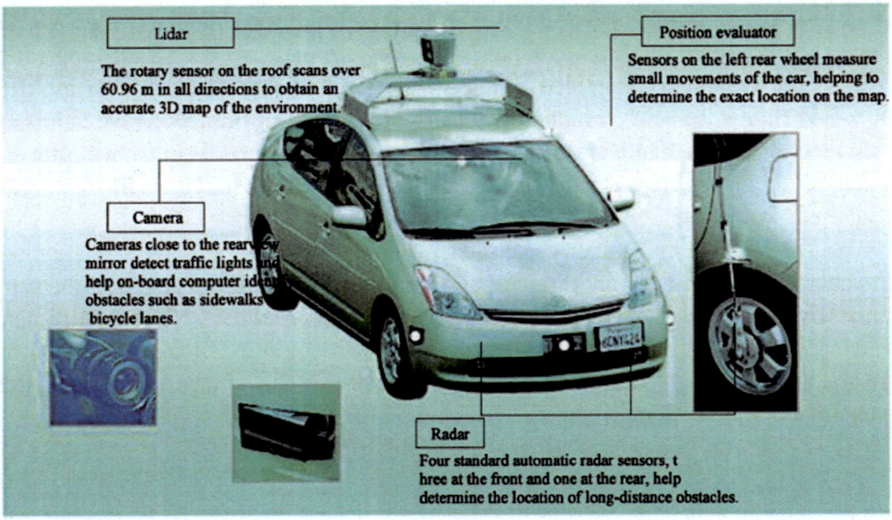

Fig. 3.18 Configuration of Google autonomous vehicle sensor system

Fig. 3.19 Vehicle front millimeter wave radar

and receives the reflected beams. According to the time difference of transmitting and receiving, it calculates the distance between the object and the car, and draws a real-time 3D topographic map around the car. After synthesizing the detection data, it can also judge the shape, size and general motion trajectory of the object. This is one of the bases for action decision-making. HDL-64E can provide 1.3 million sets of data to Google autonomous vehicle's processor per second, which means that the information provided is almost real-time. The world in the eye of HDL-64E is shown in Fig. 3.20.

In addition to radar system, Google autonomous vehicle is also equipped with lane-keeping vision system, infrared camera, stereo vision, GPS and inertial navigation system, wheel encoder.

Fig. 3.20 World in eye of Velodyne HDL-64E lidar

3.6.3 *Waymo Autonomous Vehicle*

Since the middle of October, 2017, Google's Waymo autonomous vehicle has started open road test in Arizona. The test scene and sensor configuration are shown in Fig. 3.21.

The sensor configuration of Waymo autonomous vehicle is as follows:

Lidar system: it includes three kinds of lidars: short-range lidar which can provide 360° coherent field of vision, high-definition mid-range lidar, and a new generation of long-range lidar which can see through almost three football fields. The system works fine day and night.

Millimeter wave radar system: it uses multiple wavelengths to detect objects and motion. These radar waves can bypass objects such as raindrops, so they can also work in rainy, snowy and foggy days. Nighttime does not affect it. Waymo's radar system has a 360° continuous field of view, so it can measure the speed of road participants in front of, behind and on both sides of the vehicle.

Vision (camera) system: the vision system includes cameras designed to see the world around like human beings, but it has a 360° synchronous field of vision, while human drivers can only see 120° of vision. Because the HD vision system can detect colors, it can help the system identify the strobe lights of traffic lights, construction areas, school buses and ambulances. Waymo's vision system consists of a number of high-definition cameras designed to see distant objects in the daytime and low light conditions.

Additional sensors: Waymo has many additional sensors, including an audio detection system that can hear sirens from police cars and ambulances.

Relying on the cooperation of sensors and software, Waymo can sense the surrounding world. It can recognize vehicles, pedestrians, bicycles, obstacles, and also distinguish the color of traffic lights, temporary parking signs, etc. The world seen by Waymo autonomous vehicle is shown in Fig. 3.22.

a)

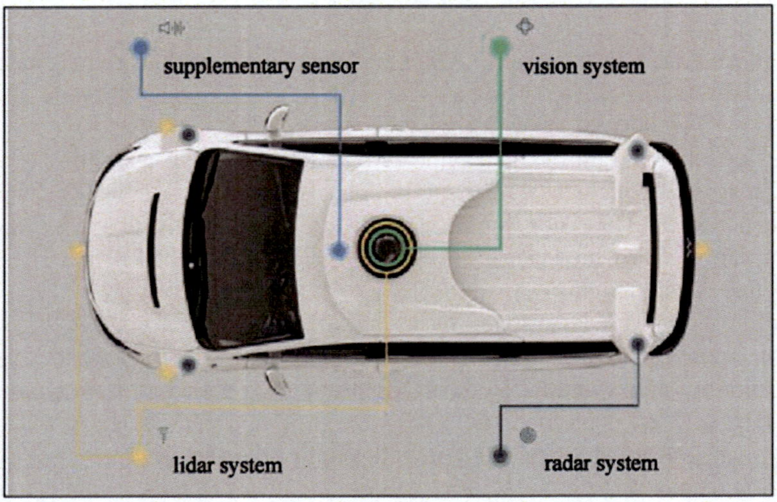

b)

Fig. 3.21 Test scene and sensor configuration of Waymo autonomous vehicle. **a** road test; **b** sensor configuration

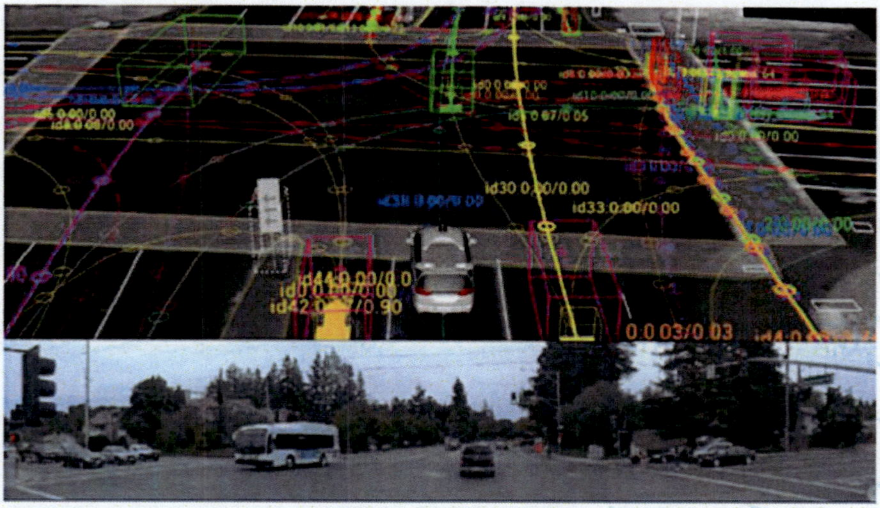

Fig. 3.22 World seen by Waymo autonomous vehicle

Chapter 4
Lane Visual Detection and Recognition Technology

4.1 Fundamentals of Visual Navigation

The purpose of lane image understanding is to perform visual navigation. Most of the early outdoor mobile robot visual navigation methods are no map methods, and the research background is mainly for structured environment, including highway vehicle automatic cruise and factory automatic guided vehicle (AGV). According to the structure of road environment, outdoor visual navigation can be divided into structured environment navigation and unstructured environment navigation.

Outdoor visual navigation in structured environment mainly refers to road detection and tracking. The most representative research achievement in road detection and tracking is the Navlab Project of CMU (Carnegie Mellon University). Since 1984, a series of auto robots have been developed successfully. The latest Navlab family product is Navlab 11, which is refitted from a SUV and can complete near and medium range obstacle detection. The single image algorithm in Navlab Project includes three stages: first, Gaussian distribution is used for color and texture classification of road and non-road pixels. Second, Hough transform is performed on road pixels to obtain road vanishing point and orientation parameters. The third is to reclassify the pixels according to the determined road edge. The scenes of Navlab driving on the road with lane line are shown in Fig. 4.1, and its lane keeping system is shown in Fig. 4.2. Many researches on structured outdoor road tracking actually follow the idea of Navlab project.

The general algorithm framework of visual navigation system usually includes feature extraction, environment parameter estimation and system state tracking, as shown in Fig. 4.3.

According to the distance between the road and the camera, the road images can be divided into near field images and far field images. In the near field where the road boundary curvature is small and the vehicle field of vision is small, the road boundary and lane line can be described by a linear model. When the curvature of

M. Du, *Autonomous Vehicle Technology*, Unmanned System Technologies,
https://doi.org/10.1007/978-981-19-4143-6_4

Fig. 4.1 Navlab on the road **a** normal road No. 1 with lane lines **b** normal road No. 2 with lane lines 2 **c** slippery road with lane lines

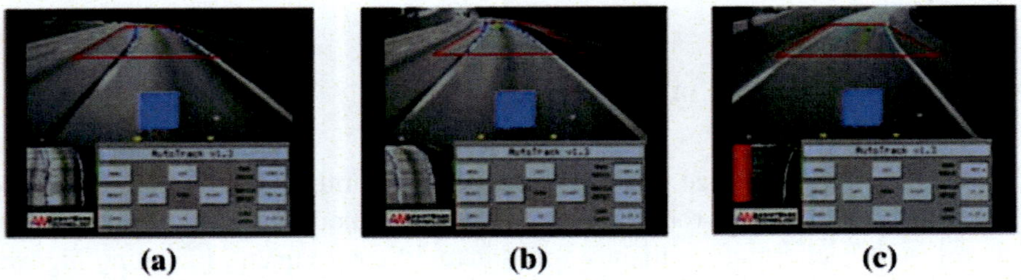

Fig. 4.2 Navlab's lane keeping system **a** segment 1; **b** segment 2; **c** segment 3

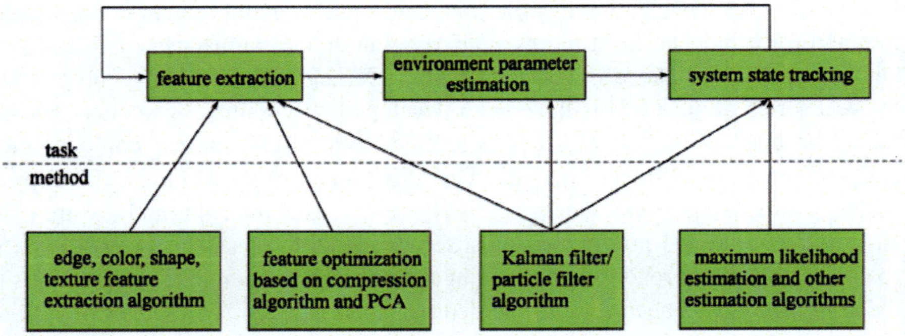

Fig. 4.3 General visual navigation algorithm framework

road boundary is large, it can be described by a conic model. If the linear model is used, the computation is small and the real-time performance is high. The typical situations of the actual road in the vehicle field of vision are shown in Fig. 4.4.

Due to the perspective projection transformation, the parallel road boundaries and lane lines on the in the two-dimensional image coordinate system are distorted into the intersection lines in the three-dimensional world coordinate system. Taking the top left vertex of the image as the coordinate origin, in the image coordinate system, the road geometric model is defined as:

Fig. 4.4 Situations of the actual road in the vehicle field of vision **a** straight line; **b** near field straight line, far field curve; **c** curve

1. Linear model

The lane lines and road boundaries are straight lines in the image, and the mathematical expression is:

$$y = kx + b \tag{4.1}$$

Specifically, the left and right lane lines L_L and L_R can be described as follows:

$$y_L = k_{PL} \cdot x_L + b_{PL} \tag{4.2}$$

$$y_R = k_{PR} \cdot x_R + b_{PR} \tag{4.3}$$

$$k_{PL} \neq k_{PR} \tag{4.4}$$

Let L be the virtual vehicle navigation reference line, then L can be expressed as:

$$y = \tan\left(\frac{\arctan k_{PL}}{2} + \frac{\arctan k_{PR}}{2}\right) \cdot x + b \tag{4.5}$$

2. Parabolic model

The lane lines and road boundaries are parabolas in the image, with expression as below:

$$y = \alpha x^2 + \beta x + \gamma \tag{4.6}$$

In the above models, x and y are pixel coordinates, (k, b) and (α, β, γ) are road parameters and all real numbers. After these parameters are determined, the position of lane line and road boundary can be determined to achieve road tracking. In practice, the detected road feature data have to go through numerical fitting or Hough transform to obtain the candidate line or curve set with the highest matching degree with the road feature points.

The ideal situation is that the autonomous vehicle can keep in the middle of the two lane lines or the road boundaries when driving on the road, that is, the ideal path is the centerline. A point model is used to represent the car body (point B), and an arrow is extended from the point to represent the actual movement direction of the car. The ideal position of the car is assumed to be point A. The deviation of vehicle from the road is described by a two-dimensional vector (θ, d), where θ represents the error of direction angle, i.e., the angle between the actual motion direction of vehicle and the tangent line of centerline; and d indicates the distance between the car and the centerline, i.e., the lateral position error. Then the road and tracking error model as shown in Fig. 4.5 can be established (the dotted area is the region of interest).

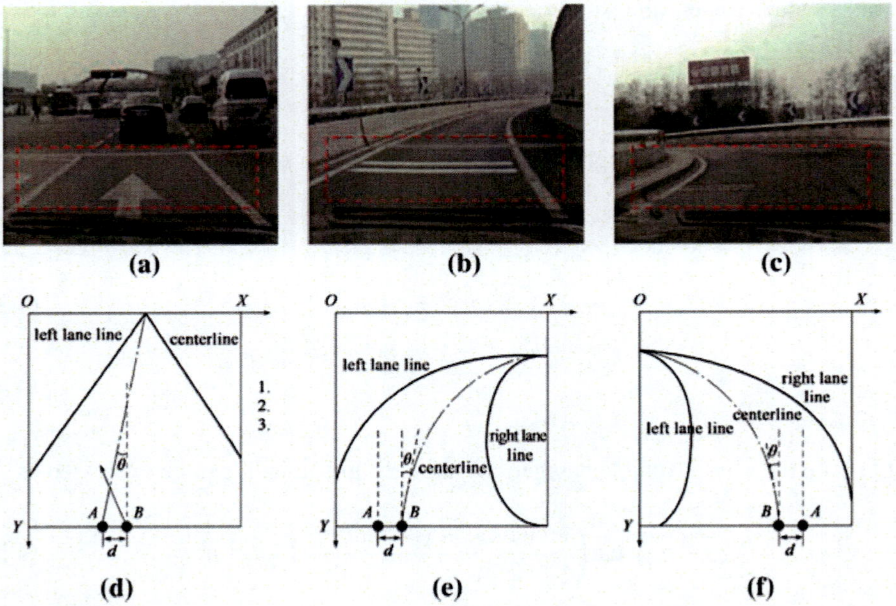

Fig. 4.5 Road and tracking error model (actual photos of roads in Beijing) **a** video frame of a straight road **b** video frame of a right curve road **c** video frame of a left curve road **d** straight road and tracking error model **e** right curve road and tracking error model **f** left curve road and tracking error model

4.2 Lane Detection and Tracking

4.2.1 *Hough Transform and Relevant Improved Algorithms*

The classical Hough transform could detect straight lines in the image at first, and later it could detect objects of any shape. In order to improve the detection accuracy and speed, some improved Hough transform algorithms are proposed, such as probabilistic Hough transform, quad tree Hough transform, hierarchical Hough transform and so on.

In practice, Hough transform algorithm needs to be combined with other image processing steps to work better, and it needs to be optimized according to the characteristics of the system. Transforming front view image into IPM image, and only detecting lane lines in the region of interest, can effectively improve the performance.

According to the application features of lane image detection, this book proposes a fast and robust lane line detection method based on IPM lane image. Firstly, the wavelet transform is used to preprocess the lane image, reducing the polar coordinate parameter ρ in Hough transform. Then the fast Hough transform is used for lane line detection. For an image with m × n pixels, the usual range of ρ is $[-\sqrt{2}m/2, \sqrt{2}n/2]$, and the value range of θ is $[-90°, 90°]$. After a wavelet transform, the pixel of image is (m/2) × (n/2). In the reconstruction of low resolution image, Hough transform is used, and θ keeps unchanged while ρ is halved, which effectively improves the real-time performance. In order to reduce the influence of interference pixels after preprocessing, the Hough transform based on polar constraint region is used to detect the candidate lane lines. Then the Hough transform based on dynamic ROI is used to track the lane according to the previous image information. The detection steps are as follows:

1. Determining the rough direction of the lane lines

In IPM, the direction of the candidate lane lines after preprocessing is approximately $\theta \approx 0°$.

2. Using Hough transform to detect straight line in rough direction 6 neighborhood

A point on a straight line in image space corresponds to a parameter space (ρ, θ) accumulator $H(\rho, \theta)$. In fact, $H(\rho, \theta)$ is the number of target points in a certain direction. The traditional Hough transform method does coordinate space transformation from 0° to 179° for each pixel, so the calculation is huge. In this book, we only carry out cumulative voting in the $[\theta - 30°, \theta + 30°]$ interval. At the same time, the idea of layering is adopted. First, the direction corresponding to the maximum $H(\rho, \theta)$ value is selected by calculating the $H(\rho, \theta)$ value at a 10° interval. Then the line is accurately located by calculating $H(\rho, \theta)$ value at the interval of 1° in the 10° region of that direction. This greatly reduces the computation compared to direct detection.

4.2.2 Self-tuning Closed-Loop Vision Detector

The idea of closed-loop is fundamental in the field of control, and its main purpose is to weaken and eliminate the error in real time according to the feedback value. However, in the field of detection, a perfect theoretical system of closed-loop detection has not been established, and is seldom mentioned. This book believes that the idea of closed-loop detection is very suitable for the complex detection system with dynamic vision. The previous lane line feature detection is actually an open-loop detection, and there is no feedback in the system. In view of this, this book proposes a self-tuning closed-loop lane vision detector architecture based on the self-tuning adaptive control system theory and the actual situation of autonomous vehicle road perception, as shown in Fig. 4.6.

The system parameters include the geometric parameters of road model and the deviation of road tracking are:

$$P_{road} = \{k_{PL}, k_{PR}, b_{PL}, b_{PR}; \theta, d\} \tag{4.7}$$

Different from the adaptive control system, the quantity imposed by the feature detector on the detected object, that is, the actual road, is the detection algorithm under certain parameters, not the control quantity. The control quantity has an impact on the state of the controlled object, and the detection algorithm also has an impact on the future state of the detected object. In other words, the detectable field of vision presented by the actual road to the detector is affected. From this point of view, the controller and the detector are the same. Therefore, different from the flow diagram of traditional system architecture (where the arrow of source information flow is from the detection object to the detection organization), this book uses the opposite arrow direction between the detector and the detected object.

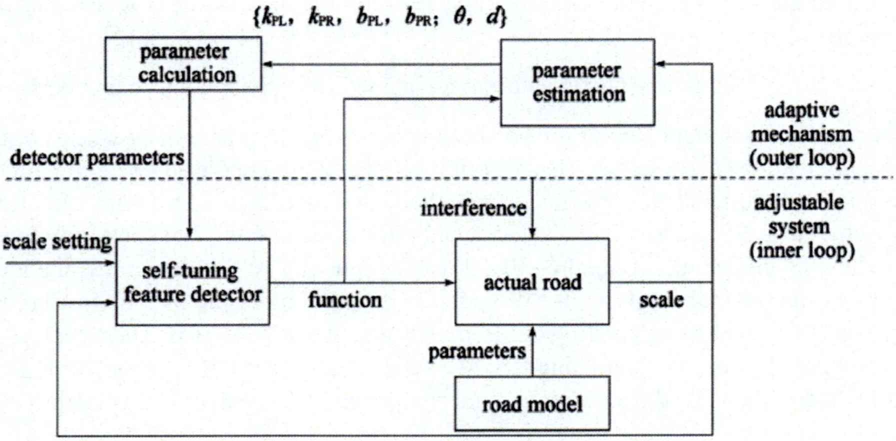

Fig. 4.6 Architecture of self-tuning closed-loop lane vision detector

Interference refers to illumination change, shadow, scale change, rotation, angle change, weather change, dust pollution, pseudo obstacles and other abnormal detection conditions. These conditions are the basis for setting the resolution of the detector. For example, in good detection conditions, in order to improve the real-time performance of the system, we can choose a lower resolution, as long as the feedback confidence indicates that it can track the navigation mark stably.

The algorithm of self-tuning vision detector in the system is as follows: after de-noising by wavelet packet, multi-resolution edge detection of lane line is carried out by using wavelet modulus maxima, the confidence of edge detection is output at the key frame, which is converted into corresponding scale and fed back to the input of detector as the basis of scale space setting. Based on the self-tuning closed-loop lane vision detector architecture, this book proposed a detector algorithm based on optimal wavelet packet. Based on the principle of sparse description of image by wavelet packet, a real-time compression sensing algorithm for road environment image is proposed, which is composed of compression denoising and edge feature extraction.

From the perspective of signal, the edge of lane line is a type of obvious signal mutation point. In the autonomous vehicle dynamic image system, the signal mutation is very fast, which is reflected in the change in time and space of the local extreme points of the signal. With the help of the good local analysis function of wavelet analysis, the main feature points can be quickly captured.

4.2.3 Lane Detection Based on Inverse Perspective Mapping

Inverse perspective mapping (IPM) is the inverse process of perspective mapping. It can use the camera angle, height and other location information to establish a three-dimensional coordinate system, eliminate the perspective effect, and get the top view of the scene. After reverse perspective mapping, the lane lines with intersection tendency are transformed into parallel lines in the top view, which are more convenient for detection.

The typical IPM formula in the field of inverse perspective mapping technology was derived by M. Bertozzi and others, and has been successfully applied to GOLD autonomous vehicle. The IPM formula is as follows:

$$\begin{aligned} x(u, v) &= h \times \cot\left[(\theta - \alpha) + u \times \frac{2\alpha}{m-1}\right] \times \cos\left[(\gamma - \beta) + v \times \frac{2\beta}{n-1}\right] + l \\ y(u, v) &= h \times \cot\left[(\theta - \alpha) + u \times \frac{2\alpha}{m-1}\right] \times \sin\left[(\gamma - \beta) + v \times \frac{2\beta}{n-1}\right] + d \end{aligned} \tag{4.8}$$

From Eq. 4.8, the relationship between road surface coordinates $(x, y, 0)$ and image coordinates (u, v) in world coordinate system can be deduced as follows:

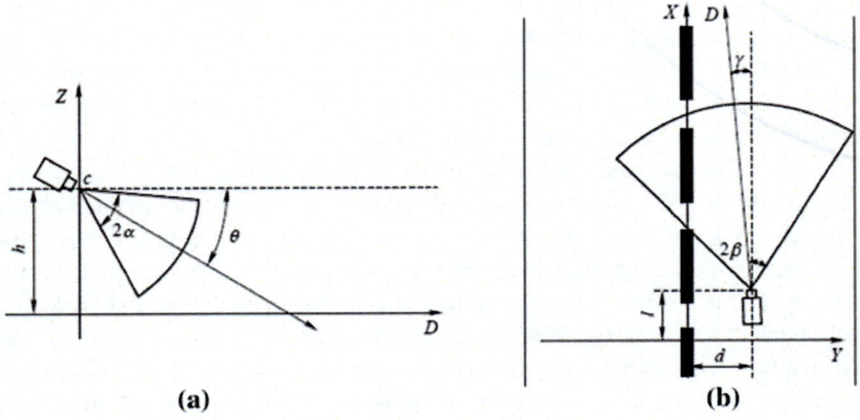

Fig. 4.7 Camera position parameters **a** Side view; **b** Top view

$$u(x, y, 0) = \frac{\cot\left\{\frac{h\times\sin\left[\cot\left(\frac{y-d}{x-l}\right)\right]}{y-d}\right\} - (\theta - \alpha)}{2\alpha/m - 1} \tag{4.9}$$

$$v(x, y, 0) = \frac{\cot\left(\frac{y-d}{x-l}\right) - (\gamma - \beta)}{2\beta/n - 1} \tag{4.10}$$

The engineering background of formula derivation is camera position parameters, as shown in Fig. 4.7. The position coordinate of the camera in the world coordinate system is (l, d, h) meters. The resolution of the camera is $m \times n$ pixels. The field of view angle is $2\alpha \times 2\beta$ radian. The yaw angle is γ radians, and the pitch angle is θ radians.

The relationship between the coordinates of a point $Q(x, y, z)$ and its image point $q(u, v)$ in the world coordinate system is as follows:

$$Z\begin{bmatrix} u \\ v \\ 1 \end{bmatrix} = \begin{bmatrix} \frac{1}{d_x} & 0 & u_0 \\ 0 & \frac{1}{d_y} & v_0 \\ 0 & 0 & 1 \end{bmatrix}\begin{bmatrix} f & 0 & 0 & 0 \\ 0 & f & 0 & 0 \\ 0 & 0 & 1 & 0 \end{bmatrix}\begin{bmatrix} \mathbf{R} & \mathbf{t} \\ 0^T & 1 \end{bmatrix}\begin{bmatrix} x \\ y \\ z \\ 1 \end{bmatrix}$$

$$= \begin{bmatrix} a_x & 0 & u_0 & 0 \\ 0 & a_y & v_0 & 0 \\ 0 & 0 & 1 & 0 \end{bmatrix}\begin{bmatrix} \mathbf{R} & \mathbf{t} \\ 0^T & 1 \end{bmatrix}\begin{bmatrix} x \\ y \\ z \\ 1 \end{bmatrix} = \mathbf{M}_t\mathbf{M}_0\begin{bmatrix} x \\ y \\ z \\ 1 \end{bmatrix} = \mathbf{M}\begin{bmatrix} x \\ y \\ z \\ 1 \end{bmatrix} \tag{4.11}$$

where,

Z is the normalization coefficient;

u_0, v_0 is the pixel coordinates of the image center point;

d_x, d_y is the physical dimensions of the pixels in the horizontal and vertical directions of the sensor;

f is lens focal length, $a_x = f/d_x, a_y = f/d_y, 0^T = (0, 0, 0)$;

Vector **t** is the relative position relationship between the camera coordinate system and the origin of the world coordinate system;

Matrix **R** is the attitude relationship between camera coordinate system and world coordinate system, called camera rotation matrix;

$\mathbf{M}_t$ is the internal parameter matrix (determined as soon as the camera is given);

$\mathbf{M}_0$ is the camera external parameter matrix;

M is the projection matrix.

If the camera coordinate system coincides with the world coordinate system, the external parameter matrix $\mathbf{M}_0$ only needs to consider the influence of rotation matrix **R**. In the pinhole perspective projection model with the assumption of a flat road surface, in the plane of $z = 0$, the transformation relationship between the world coordinate system coordinates of the scene point (x, y, z) and the image coordinate system coordinates of the projection point (u, v) is as follows:

$$x = \frac{\beta h(u - u_0)}{\alpha[(v - v_0)sin\theta + \beta cos\theta]} \tag{4.12}$$

$$y = -\frac{h(v - v_0)}{(v - v_0)sin\theta \cos\theta + \beta cos\theta cos\theta} \tag{4.13}$$

The lane in the image coordinate system is back projected to the road world coordinate system to get the actual road lane. After back projection, the lane equations of near view region and far view region are obtained as follows:

$$x = \frac{((kv_0 - u_0 + b)sin\theta - k\beta cos\theta)}{\alpha} y_w + \frac{h(kv_0 - u_0 + b)}{\alpha \cos\theta} \tag{4.14}$$

$$x = k_1 \frac{(-y\beta cos\theta)^3}{\alpha(ysin\theta + \frac{h}{\cos\theta})^2} + k_2 \frac{(-y\beta cos\theta)^2}{\alpha(ysin\theta + \frac{h}{\cos\theta})} - k_3 \frac{y\beta cos\theta}{\alpha} + k_4 \frac{y \sin\theta + \frac{h}{\cos\theta}}{\alpha} \tag{4.15}$$

In the actual dynamic image sequence obtained during driving, the inverse perspective transformation of the detected lane ROI does not provide a clean IPM lane image with only lane line and road surface. Especially under the imperfect conditions such as vehicle turning, the acquisition of straight lane line model is very unstable, and there is always interference in the lane image. Therefore, the lines detected by Hough transform will inevitably come with other components. The removal of these components will lead to a lot of post-processing calculation. Therefore, it is necessary to improve the image acquisition method and optimize the preprocessing of IPM image.

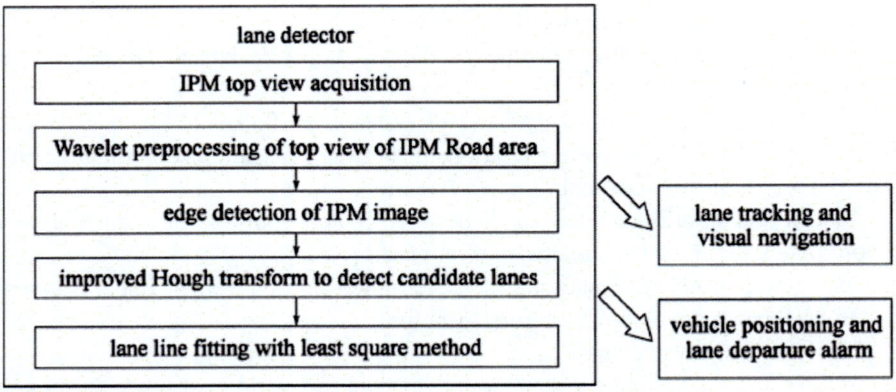

Fig. 4.8 Lane detection algorithm process and its subsequent application

The process of lane detection algorithm based on inverse perspective transformation used in this book and its subsequent application is shown in Fig. 4.8.

Caltech Lanes Dataset is currently a public data platform for research and verification of lane detection around the world. This book chooses this dataset as the experimental platform for algorithm testing.

If the edge of the original image (front view) is extracted directly, a large number of non-target edges outside the lane will also be extracted, such as trees, sky, traffic signs, vehicles, pedestrians and so on, which interfere with lane line detection severely. There is almost no such problem in the edge extraction of the selected road area on the IPM image. This is because the shape of the road area in the top view is regular, easy to be segmented, and the lane lines are parallel and clear. In the original image (640×480 pixels), the rectangular area (226×122 pixels in two-lane mode and 560×130 pixels in four-lane mode) in front of the vehicle is initially selected as the region of interest of driving.

It can be seen from Fig. 4.9 that the perspective mapping process of 3D road scene in the 2D image plane brings disadvantages to the subsequent edge extraction and lane line modeling.

After obtaining the projection matrix **M**, the IPM transformation is performed to obtain the top view of the road. The unshaded four-lane original image ROI and IPM top view from Caltech Lanes Dataset are shown in Fig. 4.10. Compared with the original image ROI, it is obvious that the lane lines in the IPM top view are more distinct (basically parallel), and the road panorama is clear. The useful road area can be selected manually, so it is more convenient for detection. The shaded lane IPM is shown in Fig. 4.11.

The installation height of the camera has a direct impact on the furthest distance that camera can see. The experimental IPM results with different installation heights are shown in Fig. 4.12.

The experimental results of reconstructing the original image with low frequency subgraph are shown in Fig. 4.13.

Fig. 4.9 Lane ROIs **a** two-lane mode; **b** four-lane mode

Fig. 4.10 Unshaded four-lane original image ROI and IPM top view **a** original image ROI; **b** IPM top view

Fig. 4.11 Shaded lane IPM

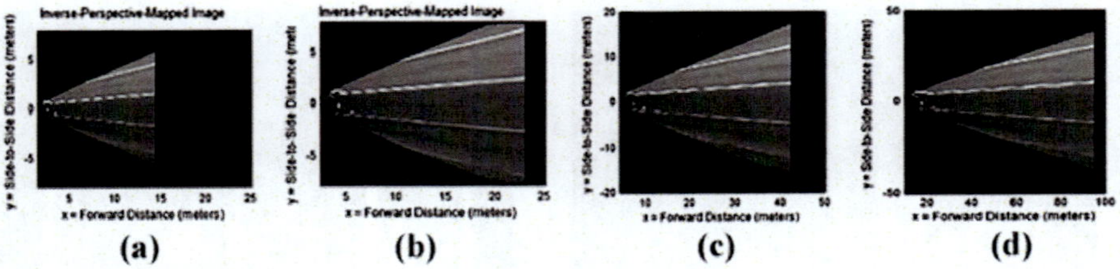

Fig. 4.12 IPM results with different installation heights **a** $h = 0.5$ m; **b** $h = 0.8$ m; **c** $h = 1.5$ m; **d** $h = 3.3$ m

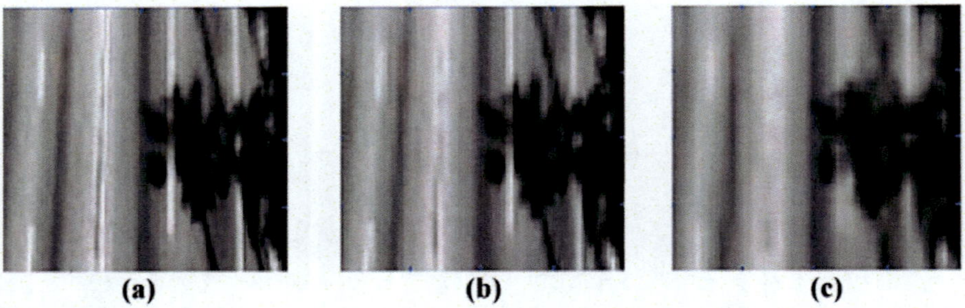

Fig. 4.13 Results of reconstructing the original image with low frequency subgraph **a** Layer 1 reconstruction; **b** Layer 2 reconstruction; **c** Layer 3 reconstruction

The advantage of using wavelet method for image compression is that it can be carried out in layers. The maximum number of layers can be selected adaptively according to the characteristics of the specific image, and any compression ratio can be achieved on the premise of meeting application requirements.

For lane feature extraction and candidate lane location, a method proposed in this book is as follows: firstly, wavelet multi-scale edge extraction method is used to extract the edge features of the lane image, and then probability Hough method is used to accurately locate lane lines.

The actual road detection experiments show that the natural road image is greatly affected by the light intensity and the clarity of lane lines. The timeliness and accuracy of the detection are mutually restricted and have to be comprehensively considered. Therefore, the ideal edge detection algorithm should be able to adjust the accuracy in real time according to road conditions, and the multi-scale edge detection algorithm based on the wavelet modulus maxima meets this requirement. In this book, a fast adaptive multi-scale lane edge detection algorithm based on binary spline wavelet is proposed, and the IMP road image is taken as an example for research and experiment.

Two dimensional spline filter $h(n, m)$ is used for digital filtering of the original image, and the modulus of wavelet transform is calculated level by level. In this experiment, a total of 5 levels are calculated, and the results of Caltech lane line multi-scale edge detection results are shown in Fig. 4.14.

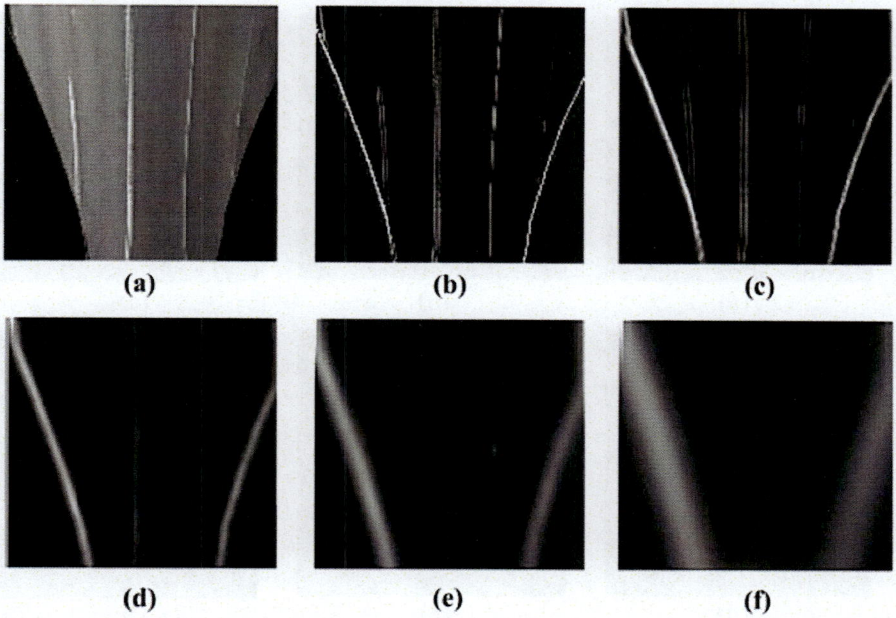

Fig. 4.14 Caltech lane line multi-scale edge detection results

Figure 4.14a–c are the original image, level 1 edge and level 2 edge. Figure 4.14d–f are level 3 edge, level 4 edge and level 5 edge. It can be seen that the level 1 edge extraction is the best, which is suitable for all kinds of road conditions (sunny, rainy, with or without shades). The level 2 edge can also be used when road conditions are good (normal light, clear original image, less dynamic obstacle interference, etc.). The edges of level 3 and above are not clear and should not be used for high navigation accuracy. The detection results of level 1 edge and level 2 edge show that, firstly, there is no missing lane line in this method. Secondly, the interference points outside the lane are very few, and clean lane lines can be obtained by further filtering. Thirdly, the results of level 1 edge detection and level 2 edge detection are usable, and the level can be adaptively selected according to road conditions. When level 2 is selected, it is a compressed sensing method, which will greatly improve the processing speed and real-time performance of the system.

In order to compare the detection results, Sobel operator with adaptive threshold is used to extract the edge features of the same image, and the results are shown in Fig. 4.15.

Sobel operator edge detection results show that, when the threshold value is 0.02, all the lane lines can be detected, but there are more interference points detected at the same time, which is much worse than the wavelet multi-scale method. Under the adaptive threshold, not all lane lines can be detected, which does not happen for the wavelet method. Additionally, the detection speed of Sobel operator is very fast. For

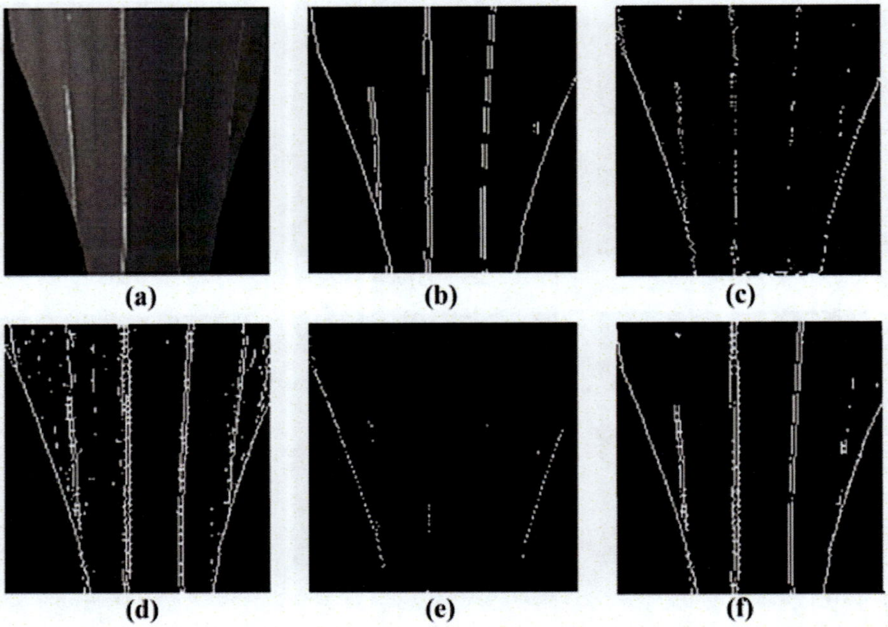

Fig. 4.15 Sobel operator edge detection results of unshaded IPM **a** original image; **b** adaptive threshold of 0.0831; **c** threshold of 0.02, horizontal **d** threshold of 0.02, vertical; **e** threshold of 0.05, horizontal; **f** threshold of 0.05, vertical

the image sequence of real road, it is difficult to determine the adaptive threshold without undetected lane lines, so Sobel operator method has limitations.

Based on the wavelet multi-scale edge detection, the probability Hough transform is used to detect the lane lines in level 1 and level 2 IPM images. The results are shown in Fig. 4.16.

Another solution can also be used: Canny edge extraction plus probability Hough method. The specific implementation method is as follows.

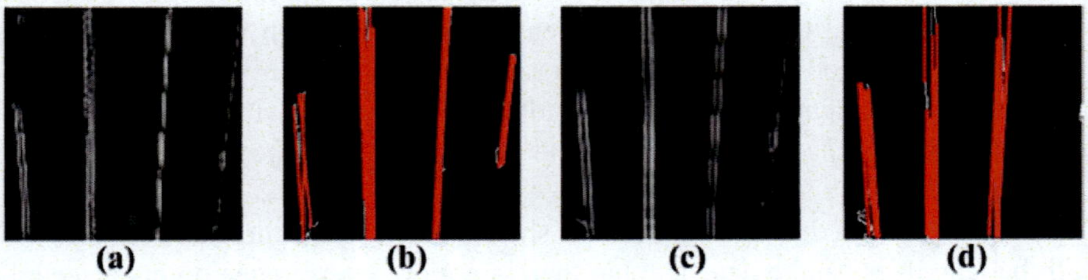

Fig. 4.16 Hough transform lane detection results of IPM road images **a** Level 1 edge image; **b** Level 1 edge image lane line **c** Level 2 edge image; **d** Level 2 edge image lane line

Canny operator in OpenCV library function cvCanny is used to detect the edge of unshaded IPM road image, and then the probability Hough transform is used to detect the lane line. The results are shown in Fig. 4.17.

Canny edge detection algorithm uses double thresholds. According to the high threshold value, it gets an edge image with few false edges, but since the threshold is high, the detected edge may be discontinuous. Therefore, the low threshold value is set.

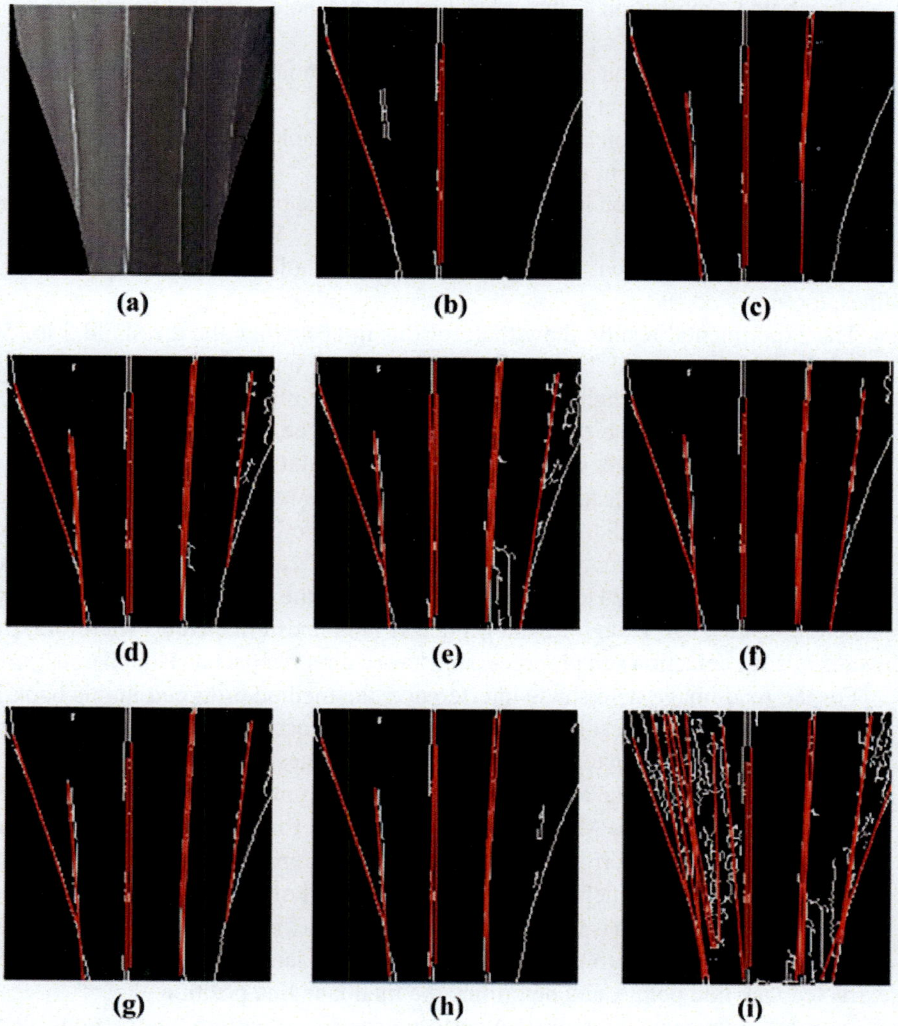

Fig. 4.17 Results of Canny edge detection and Hough transform lane detection **a** original image; **b** Threshold Combination 1; **c** Threshold Combination 2 **d** Threshold Combination 3; **e** Threshold Combination 4; **f** Threshold Combination 5 **g** Threshold Combination 6; **h** Threshold Combination 7; **i** Threshold Combination 8

For the IPM road image, the threshold combinations and detection results in Figure b to i are as below.

Threshold Combination 1: threshold 1 = 30, threshold 2 = 450, many missing lane lines;

Threshold Combination 2: threshold 1 = 30, threshold 2 = 350, some missing lane lines;

Threshold Combination 3: threshold 1 = 20, threshold 2 = 150, no missing lane line;

Threshold Combination 4: threshold 1 = 10, threshold 2 = 150, no missing lane line, more fine edges;

Threshold Combination 5: threshold 1 = 50, threshold 2 = 150, no missing lane line, less fine edges, decent result;

Threshold Combination 6: threshold 1 = 50, threshold 2 = 100, no missing lane line, less fine edges, decent result;

Threshold Combination 7: threshold 1 = 75, threshold 2 = 150, some missing lane lines, few fine edges;

Threshold Combination 8: threshold 1 = 5, threshold 2 = 50, many false lane lines, too many edges.

The experimental results show that, for this image, when the threshold 2 is 150 and the ratio of threshold 2/threshold 1 is between 2 and 3, the detection result is the best. Moreover, the higher the threshold 1 is, the more the fine edges are, and vice versa. The higher the threshold 2 is, the fewer the edges are, and vice versa. Finally, the more the edges are, the larger the computation of Hough transform is, and too many edges will lead to false detection. The fewer the edges are, the smaller the computation of Hough transform will be, and too few edges will lead to missing lane line. In a word, too many or too few detected edges are neither beneficial for the detection of lane lines by Hough transform. Only in the appropriate image of edge detection results, can Hough transform detect lane lines effectively. Therefore, the results of edge detection and preprocessing have a direct impact on Hough transform.

For the road image with shade interference, the method proposed in this book is also applicable, and the result is good. Different preprocessing methods are used for the IPM image, and the detection results of shaded lanes are shown in Fig. 4.18.

The above experimental results show that the result of lane detection is better on the binary preprocessed IPM image (Fig. 4.18k and l) for instance). In addition, the comparison between the original IPM image and the preprocessed IPM image shows that the preprocessing benefits the accuracy and speed of lane detection.

Because the extraction is for all the lane line points existing in a certain parameter subspace, it is necessary to get the parameters of the lane line model by fitting the extracted lane line points, and determine the final lane line position.

In order to improve the real-time performance of the system, on the premise of meeting the navigation accuracy, only the main control points (6 points per lane) on each lane line are selected for fitting. Each lane line is framed with a rectangle, and the lane lines are fitted in four rectangles in turn. The selection of main control points is shown in Fig. 4.19.

The fitting results of different functions are shown in Fig. 4.20.

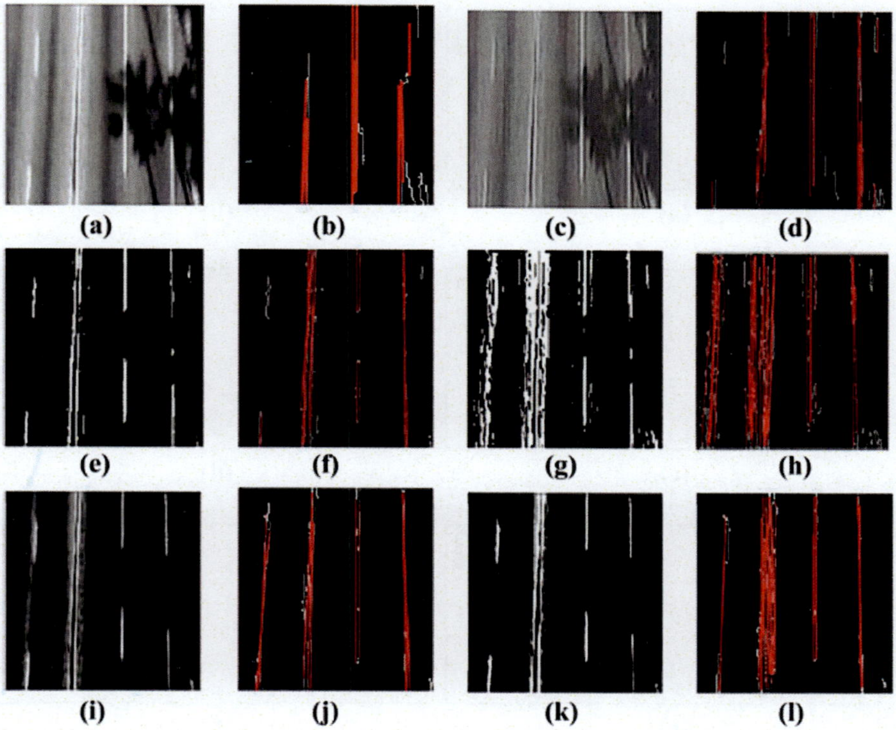

Fig. 4.18 Hough transform lane detection results of shaded IPM road image **a** original image **b** Hough transform detection results of original image, some missing lane lines **c** wavelet coefficients enhanced image **d** Hough transform detection results after wavelet coefficients enhancement **e** higher gray threshold segmentation image **f** Hough transform detection results after segmentation with higher gray threshold **g** lower gray threshold segmentation image **h** Hough transform detection results after segmentation with lower gray threshold **i** brightness = −99, contrast = 68, preprocessed image **j** Hough transform detection results, brightness = −99, contrast = 68, preprocessed image **k** brightness = −98, contrast = 90, preprocessed image **l** Hough transform detection results, brightness = −98, contrast = 90, preprocessed image

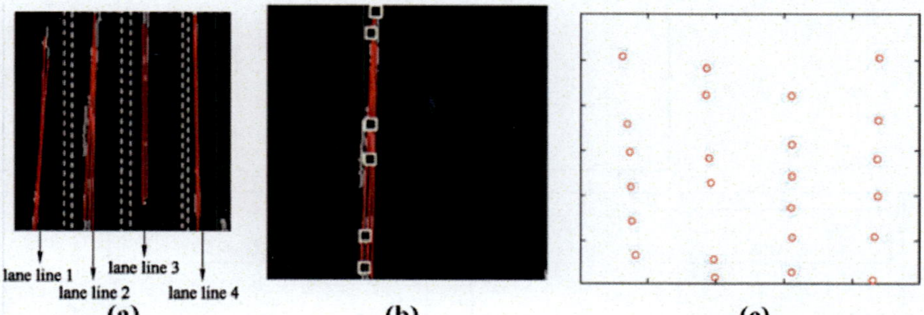

Fig. 4.19 Selection of main control points of lane **a** rectangular division **b** main control points of a lane line **c** main control points of four lanes

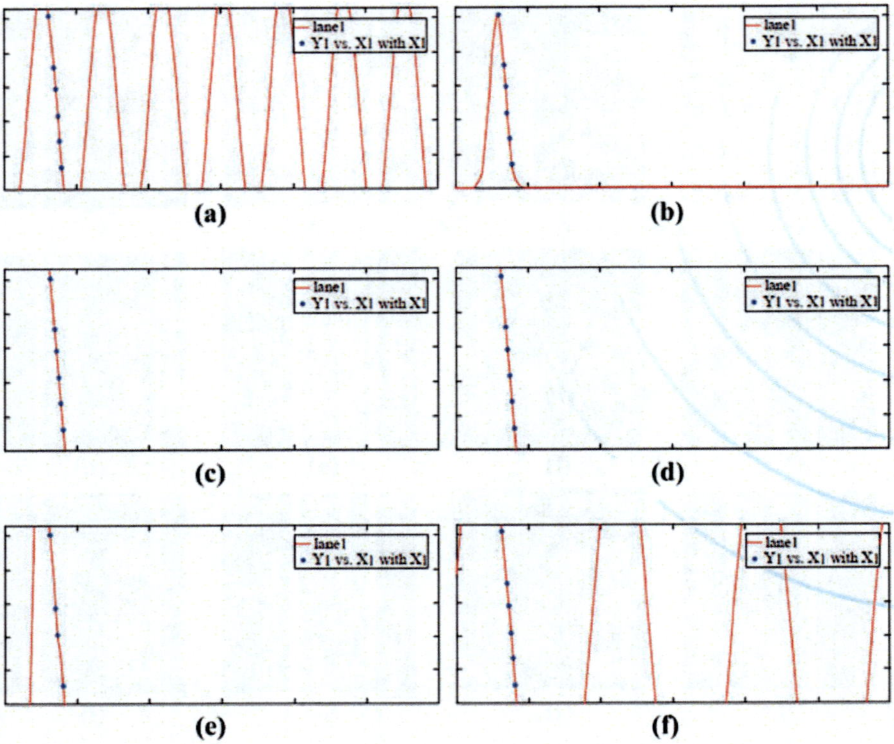

Fig. 4.20 Fitting results of different functions **a** Fourier function; **b** Gaussian function **c** Interpolation function; **d** Polynomial function **e** Smooth spline fitting function; **f** Sine function

After comparison, the Polynomial function with the best result is selected as the lane line fitting function. The fitting results of lane line 2, 3 and 4 are shown in Fig. 4.21.

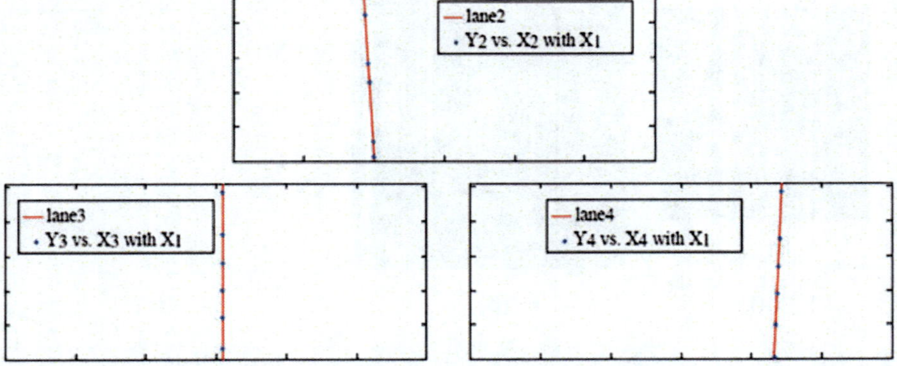

Fig. 4.21 Fitting results of lane line 2, 3 and 4

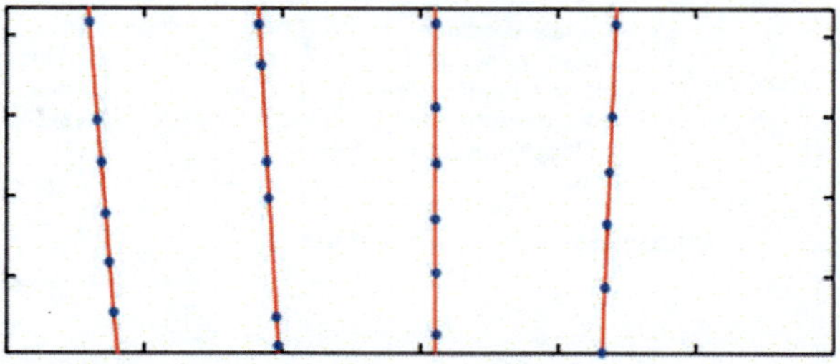

Fig. 4.22 Fitting results of four lanes

The fitting results of four lanes are shown in Fig. 4.22.

By fitting multiple candidate lane lines, the accurate position of the lane line is obtained, and the lane line detection is completed.

4.2.4 Lane Tracking and Lane Departure Warning System

4.2.4.1 General Idea

In order to verify the lane detection results, the most direct way is to input the detection results into the subsequent application. This book combines the lane detection module with the lane tracking module. On the one hand, it uses the lane tracking and departure correction experimental results to verify the lane detection method. On the other hand, the lane detection algorithm is practical and can be embedded in the lane departure warning system.

The core technology of lane departure warning system is lane detection and tracking. Lane tracking of real road is a nonlinear target tracking task, which mainly involves three core issues, namely model, method and evaluation. The models include the motion model and the observation model of the target. The motion model depends on the type and motion state of the target, and the observation model is determined by the type of sensor and the observed quantity. In this book, a lane tracking and automatic departure correction method based on wavelet domain multi-scale feature detection and adaptive scale Kalman filter is proposed.

4.2.4.2 Lane Correction Mechanism

The working steps of lane departure warning system based on machine vision are as follows:

1. The lane lines are detected in the current video frame.
2. According to the environmental conditions and detection quality, the applicability of the current scale is evaluated, and the best scale value is given.
3. Kalman filtering is performed on the optimal scale to match the lane lines in the past video frame.

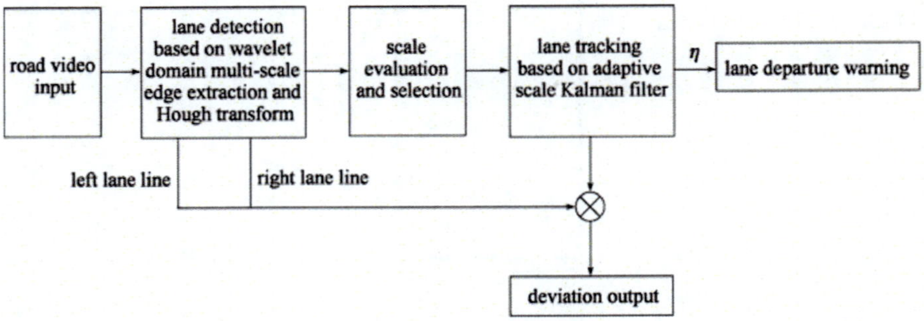

Fig. 4.23 Algorithm flow of lane departure warning system

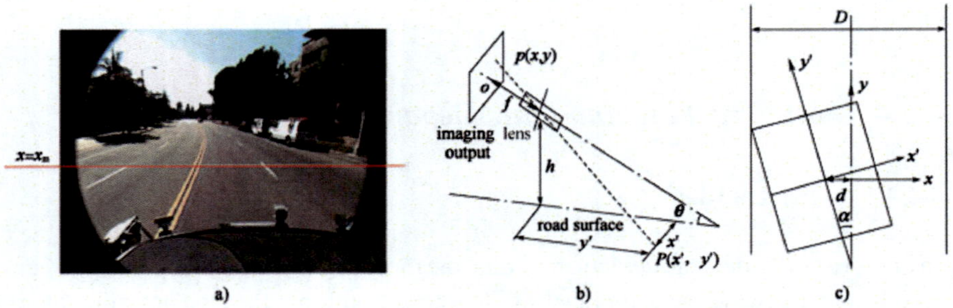

Fig. 4.24 Lane model and deviation correction mechanism **a** lanes in machine vision; **b** CCD imaging model; **c** mechanism of lane correction

4. The left and right lane lines are located by combining the actual detection value with the estimated value.
5. If the vehicle deviates from the lane line, an alarm will be given.

The overall algorithm flow is shown in Fig. 4.23.

The lane model and deviation correction mechanism are shown in Fig. 4.24.

Figure 4.24a shows the lane image in the view of the machine. With x_m as the boundary, each lane above $x = x_m$ can be simulated by quadratic function $y = ax^2 + bx + c$, and the part below $x = x_m$, or the near field part, is approximately straight lines, which can be represented by $y = kx + d$. In fact, whether the vehicle deviates from the lane can be decided according to the lane line in the near view.

Figure 4.24b shows the vehicle CCD imaging model. The installation height of the camera is h, the angle between the camera and the horizontal plane is θ, and the focal length of the camera is f. The radial and transverse distances between a point $P(x', y')$ on the road and the center of the camera lens are x' and y', respectively. In the image coordinate system, the coordinates of the corresponding point $P(x, y)$ are:

$$x = \frac{x'f}{\sqrt{h^2 + y'2 - (hcos\theta - y'sin\theta)^2}} \tag{4.16}$$

$$y = \frac{(hcos\theta - y'sin\theta)f}{\sqrt{h^2 + y'2 - (hcos\theta - y'sin\theta)^2}} \tag{4.17}$$

Figure 4.24c shows the mechanism of lane correction. In the road coordinate system, the distance between two lane lines is D, and the functions of left and right lane lines are $x = -\frac{D}{2}$ and $x = \frac{D}{2}$ respectively. According to the slope of lane line in road image, the distance between the vehicle and two lane lines can be calculated. When the vehicle deviates from lane, the slope value of the lane line in the CCD image will also change. Through the slope parameter of lane line, it is easy to judge whether the vehicle deviates. The degree of deviation η can be obtained from the following formula:

$$\eta = \arctan \frac{-h \cos \alpha}{D/2 - d} + \arctan \frac{h \cos \alpha}{D/2 + d} \tag{4.18}$$

The value η is usually near π when the vehicle is driving normally. If it is too large, there is a tendency of deviation to the right. If it is too small, there is a tendency of deviation to the left.

4.2.4.3 Adaptive Scale Kalman Predictor

The multi-scale estimation theory should be used when the sampling rates are different, the observation is not synchronous and the sampling is not uniform. Taking the characteristics of autonomous vehicle visual navigation system into consideration, on the basis of stochastic linear discrete system Kalman predictor and multi-scale estimation theory, this book proposes an adaptive scale Kalman filtering method based on wavelet domain feature detection. The adaptive scale Kalman filtering algorithm is more concise and practical than the multi-scale Kalman filtering algorithm.

The dynamic system model at each scale $i - 1 (L \leq i \leq N)$ is:

$$X_{i-1,k+1} = A_{i-1} X_{i-1,k} + \Gamma_{i-1} W_{i-1,k} \tag{4.19}$$

$$Z_{i-1,k} = H_{i-1} X_{i-1,k} + V_{i-1,k} \tag{4.20}$$

The dynamic system on the coarsest scale L (the highest resolution level) is

$$X_{L,k+1} = A_L X_{L,k} + W_{L,k} \tag{4.21}$$

where

$$A_L = A_{L+1}^2 \tag{4.22}$$

The observation equation on this scale is

$$Z_{L,k} = H_L X_{L,k} + V_{L,k} \tag{4.23}$$

where $V_{L,k}$ is the Gaussian white noise uncorrelated to $W_{L,k}$ with zero mean, and its variance is R_L. After the system reviews the hypotheses, the wavelet domain feature detection results with the best real-time performance and accuracy are located on the scale $L_1 (L \leq L_1 \leq N)$, and Kalman filtering is carried out:

$$\hat{X}_{L_1,k+1} = \hat{X}_{L_1,(k+1,k)} + K_{L_1,k+1}[Z_{L_1,k+1} - H_{L_1}\hat{X}_{L_1,(k+1,k)}] \tag{4.24}$$

$$\hat{X}_{L_1,(k+1,k)} = A_{L_1}\hat{X}_{L_1,k} \tag{4.25}$$

$$P_{L_1,(k+1,k)} = A_{L_1} P_{L_1,k} A_{L_1}^T + \Gamma_{L_1} Q_{L_1} \Gamma_{L_1}^T \tag{4.26}$$

$$K_{L_1,k+1\_} = P_{L_1,(k+1,k)} H_{L_1}^T [H_{L_1} P_{L_1,(k+1,k)} H_{L_1}^T + R_{L_1}]^{-1} \tag{4.27}$$

$$P_{L_1,k+1} = [I - K_{L_1,k+1} H_L] P_{L_1,(k+1,k)} \tag{4.28}$$

Starting from $\hat{X}_{L_1,k}$ and $P_{L_1,k}$, using the established multi-scale model, with the observation information on the most suitable scale obtained from evaluation, that is, the output results of lane detection in wavelet domain on a certain scale, the Kalman filtering algorithm on the corresponding scale is selected to recursively estimate the target state.

4.2.4.4 Experiment and Analysis

Firstly, the lane tracking method stores the detected lane lines as a knowledge base, and counts the number of detected lane lines. The tracking system matches the lane lines detected in the current video frame with the lane lines in the knowledge base. If the current lane lines are sufficiently similar to ones in the knowledge base, the newly detected lane lines will replace the ones in the knowledge base. The adaptive scale Kalman filter is used to predict the position of lane lines in the knowledge base, so as to improve the accuracy of lane tracking. In the lane departure warning system, Hough transform lane line detection module is used to convert the polar coordinates into Cartesian coordinates, and then the distance between the two lane lines and the position of the centerline are calculated. If the drift distance is greater than the set threshold, the system will give an alarm.

The basic experimental parameters are set as follows: the minimum number of frames detected and allowed to be lane lines is 5, and the maximum number of frames missed because lane lines cannot be matched correctly is 20. The lane departure alarm results are shown in Fig. 4.25.

The safety limit is expressed by the distance between the vehicle and the nearest lane line. The curves in Fig. 4.26 represent the safety limit. When the curve is below the zero baseline, the vehicle is in the state of departure, and the degree is determined by the departure of the curve from the zero baseline. Otherwise, the vehicle is in normal driving state.

Taking the average value of 160 experimental results of 1000 road sequence images, the processing results on different scales are shown in Table 4.1.

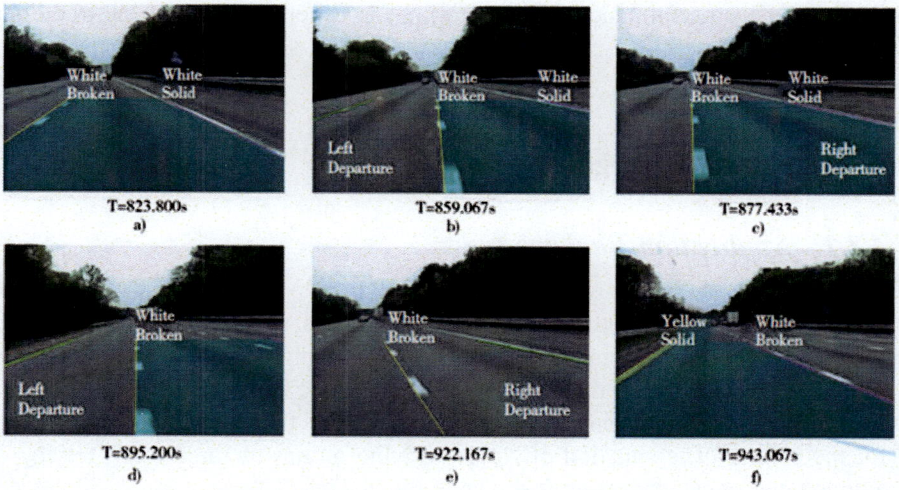

Fig. 4.25 Experimental results of lane departure alarm

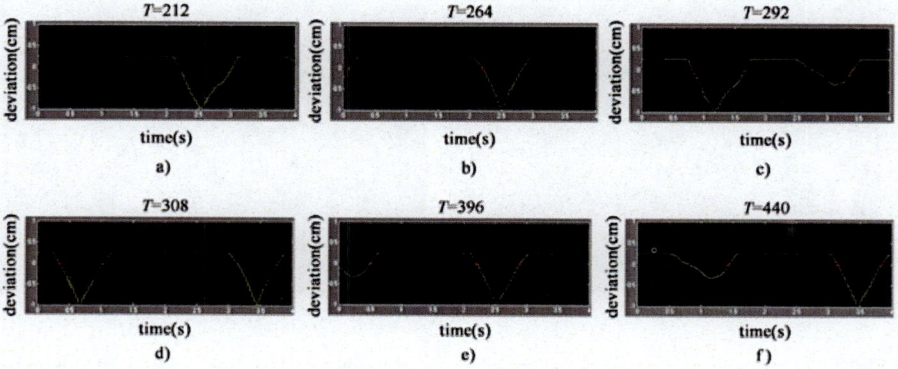

Fig. 4.26 Lane departure measurement results

Table 4.1 Experimental results on different scales

Scale	Image preprocessing speed (MS/frame)	Missing alarm rate (%)	False alarm rate (%)
1	22	1	2
2	18	1.2	2.5
3	11	1.5	3
4	6	2	3.5
Adaptive scale	6	1.6	1.8

Experimental results show that when the false alarm rate is the same, the time consumption of the method proposed in this book is significantly lower, and the comprehensive performance of the algorithm is better than that of the traditional method.

4.3 Lane Image Segmentation

4.3.1 Application Background

In suburban, rural and other areas, the autonomous vehicle will inevitably face unstructured road environment without lane line navigation. Under these circumstances, the road model based on straight line or conic will no longer be applicable. It is necessary to understand the image structure, and then locate the passable area for the vehicle. The situation of Navlab driving on unstructured roads relying on autonomous visual navigation is shown in Fig. 4.27.

Fig. 4.27 Navlab's visual navigation on unstructured roads **a** no lane line, forest road 1; **b** no lane line, forest road 2

Visual saliency feature and terrain learning are commonly used in the field of unstructured road understanding. Unstructured road understanding involves two key tasks. One is feature selection and construction, and the other is pixel classification algorithm.

There are two types of unstructured roads: bounded roads and unbounded generalized roads. This section only studies the understanding of bounded roads. The basic idea is that on the unstructured road with boundary where there is no lane line or the lane line is not enough to be used as the basis for navigation, the road navigation sign degrades into the virtual road centerline. Therefore, the algorithm in this chapter aims to detect the road boundary through fast and robust image segmentation and then calculate the virtual road centerline from the boundary position. In this chapter, compressed sensing theory is introduced into the field of autonomous vehicle road image understanding. The adaptive genetic algorithm is used to optimize the image segmentation process, and an unstructured road understanding method based on single-layer wavelet packet approximation compressed sensing is proposed.

4.3.2 Image Compressed Sensing Theory

Signal compression is a necessary means to improve the efficiency of system storage, transmission and application. However, in the process of compression, it is inevitable that a large number of useful sampling data are discarded. In other words, the redundant information is abandoned along with some useful information. This shows that bandwidth cannot essentially express the information of signal, and the Nyquist sampling theorem based on signal bandwidth is redundant.

Compared with signal bandwidth, sparsity can express the information of signal intuitively and essentially. Sparsity plays an important role in modern signal processing, such as approximation, estimation, compression and dimensionality reduction based on sparsity, etc.

Candès and his team have proved the following principle: by randomly selecting $M \geq 2K$ Fourier coefficients of the signal, the original image can be reconstructed uniquely and accurately. It is this discovery that triggered the thought of compressed sensing theory. Candès and Donoho proposed the theory of compressed sensing (or compressed sampling) in 2004. The theory can be understood as compressing analog signal into digital signal, finding the least coefficient to represent the signal in the sampling process, and recovering the original signal from the compressed data with an appropriate reconstruction algorithm.

The three core issues of compressed sensing theory research are as follows:

1. Over-complete dictionary design with sparse representation ability;
2. The measurement matrix design which meets the incoherence or equidistant constraint criteria;
3. Signal reconstruction algorithm design.

After Rice University successfully designed a new single pixel camera based on compressed sensing, the theory has been applied to many research fields, including synthetic aperture radar imaging, remote sensing imaging, nuclear magnetic resonance imaging, deep space exploration imaging, wireless sensor network, source coding, image recognition, speech recognition, ground penetrating radar imaging, etc.

When the signal is sparse or compressible in a certain transform domain, the transform coefficients can be linearly projected into the low dimensional observation vector by using the measurement matrix which is not coherent with the transform matrix. Meanwhile, this projection keeps the information needed to reconstruct the signal. By further solving the sparse optimization problem, the original high-dimensional signal can be reconstructed from the low dimensional observation vector with accuracy or high probability accuracy. In this framework, the sampling rate no longer depends on the bandwidth of the signal, but largely depends on two basic criteria: sparsity and incoherence, or sparsity and equidistant constraint.

Many natural signals have a compact representation on a specific basis Ψ, that is, they are considered to be sparse or compressible. Natural images are sparse in wavelet domain, so the compressed sensing theory based on linear projection can be applied to natural images. At present, almost all of the researches on image compressed sensing fail to consider the inherent characteristics of natural images, such as geometric structure, high-order statistics, noise, etc., let alone the human eye mechanism of robot visual cognitive process. Therefore, it is worthwhile to introduce compressed sensing theory into the natural image processing of robot visual system.

4.3.3 Adaptive Genetic Algorithm for Image Segmentation

Genetic algorithm (GA) is an adaptive global optimization probability search method, which is formed by simulating the genetic and evolutionary process of organisms in the natural environment. Its function prototype is:

$$GA(Fitfun, Fit\_threshold, n, c, m) \tag{4.29}$$

where, $Fitfun$ is the fitness function; $Fit\_threshold$ refers to the threshold of fitness when the algorithm is terminated; n is the population size; c is the crossover probability; m is the mutation probability.

The traditional genetic algorithm in solving the problem of image segmentation has the following defects. Firstly, the model is complex with many parameters and low optimization speed. It cannot reach the optimal solution, and settles for the suboptimal solution instead. Secondly the constraints are complex, so it is difficult to meet the constraints when the algorithm converges. Therefore, it is necessary to improve the genetic algorithm in the processing of robot road image understanding to achieve fast, intelligent and adaptive results. In this case, an adaptive genetic algorithm for image segmentation is proposed.

The idea of image segmentation using genetic algorithm is as follows. In order to reduce the amount of calculation, the gray level of a single image is quantized from 256 to M (M < 256). According to the gray level value, the pixels in the image are divided into target O and background B by threshold μ. O is composed of pixels with gray level value between 0 and μ, and B is composed of pixels with gray level value between $(\mu + 1)$ and (M−1). The research of video segmentation using genetic algorithm has been reported, but adaptive genetic algorithm is rare. In this book, an adaptive genetic algorithm is used. The idea is as follows. When the fitness of each individual of the population tends to be uniform or to be locally optimal, c and m are increased. When the population fitness is dispersed, c and m are decrease. Individuals with higher fitness correspond to lower c and m, while individuals with lower fitness correspond to higher c and m. The corresponding formulas are:

$$c = \begin{cases} \frac{k_1(f_{\max} - f')}{f_{\max} - f_{avg}},\ f \geq f_{avg} \\ k_2,\ f < f_{avg} \end{cases} \tag{4.30}$$

$$m = \begin{cases} \frac{k_3(f_{\max} - f)}{f_{\max} - f_{avg}},\ f \geq f_{avg} \\ k_4,\ f < f_{avg} \end{cases} \tag{4.31}$$

where, $k_1 k_2 k_3 k_4 \in (0, 1)$; $f_{\max}$ is the largest fitness value in the population; f_{avg} is the average fitness value of each generation; f' is the larger fitness value of the two individuals to be crossed; f is the fitness value of individual to be crossed or mutated. The algorithm flow is shown in Fig. 4.28.

The selection uses fitness calculation and truncation selection algorithm based on sorting, in order to overcome the scale problem of proportional fitness calculation. The crossing process uses even crossing, with a certain threshold. When the distance between individuals is lower than the threshold, the crossing process is not carried out. While the evolution converges, the threshold is gradually reduced. When the population evolves to a certain convergence period, some individuals are selected from the optimal individuals for initialization.

For autonomous vehicle road perception system, it is the key task to obtain the adaptive optimal threshold to achieve the robust segmentation of road image. According to the application characteristics, the fitness function is constructed. The optimal threshold should ensure that each class have the minimum intra-class variance and the maximum inter-class variance, which is actually a two-objective optimization problem. In this book, the two objectives are combined into one target, which is achieved by taking the intra-class variance of gray value as negative in the fitness function.

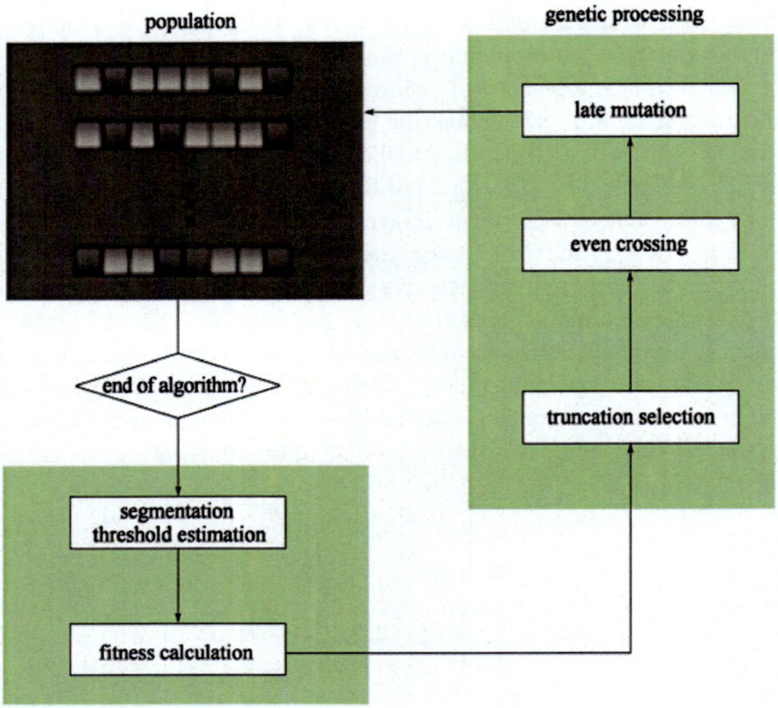

Fig. 4.28 Adaptive genetic algorithm flow for image segmentation

4.3.4 Lane Image Segmentation

4.3.4.1 Approximate Wavelet Compressed Sensing Algorithm for Road Image

1. Wavelet Coefficient Vector Field

The single road image $\boldsymbol{I}$ detected by autonomous vehicle is decomposed by layer wavelet packet at S layers, and each wavelet scale space is represented by layer number s ($s \in [0, ..., S-1]$). Firstly, the road image of original resolution is decomposed into low frequency approximation sub image (LL), and detail sub images including horizontal (HL), vertical (LH), diagonal (HH). Then the sub image of the upper layer is decomposed into four sub images layer by layer, so as to form an image pyramid with quad tree structure, as shown in Fig. 4.29.

Each node of the quad tree corresponds to a two-dimensional function space $U_s^{p,q}$ (p and q are wavelet packet parameters), and can be decomposed into the direct sum of four orthogonal subspaces:

$$U_s^{p,q} = U_{s+1}^{2p,2q} \oplus U_{s+1}^{2p,2q+1} \oplus U_{s+1}^{2p+1,2q} \oplus U_{s+1}^{2p+1,2q+1} \tag{4.32}$$

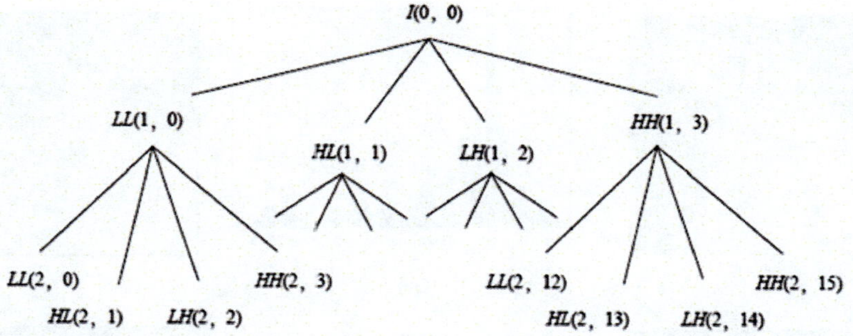

Fig. 4.29 Quad tree image pyramid

The set of road image pyramid layers is represented as:

$$\vec{I} = \{L_0, L_1, ..., L_{S-1}\} \tag{4.33}$$

The wavelet coefficients in different frequency bands of each resolution in multi-scale space are combined into coefficient eigenvectors to form a wavelet coefficient vector image at that resolution. In this way, sub image sequences of S resolutions can be obtained. The sub image sequence sorted by importance is represented as a set:

$$\begin{aligned}\vec{I} = \{&LL(1,0), HL(1,1), LH(1,2), HH(1,3); LL(2,0), HL(2,1), LH(2,3), HH(2,4),\\ &..., LL(2,12), HL(2,13), LH(2,14), HH(2,15); ..., LL(S-1,0), LH(S-1,1),\\ &HL(S-1,2), HH(S-1,3), ..., LL(S-1,4^S-4), LH(S-1,4^S-3),\\ &HL(S-1,4^S-2), HH(S-1,4^S-1)\}\end{aligned} \tag{4.34}$$

The coefficient eigenvector corresponding to the scale $s\,(0 < s \leq S-1)$ is expressed as:

$$\omega_{ij}^{s} = [\omega_{ij}^{s,0}, \omega_{ij}^{s,1}, \omega_{ij}^{s,2}, \omega_{ij}^{s,3}; ...; \omega_{ij}^{s,0}, \omega_{ij}^{s,1}, \omega_{ij}^{s,2}, \omega_{ij}^{s,3}, ..., \omega_{ij}^{s,4^s-4}, \omega_{ij}^{s,4^s-3}, \omega_{ij}^{s,4^s-2}, \omega_{ij}^{s,4^s-1}]^T \tag{4.35}$$

where i and j represent the row value and column value of the pixel in the image. The dimension of coefficient eigenvector is $d = 4^s$

2. Characteristic Analysis of Wavelet Coefficients

The road image is decomposed on four levels. Taking the fourth level as an example, the experimental results and reconstruction images of wavelet coefficient distribution in low frequency, high frequency horizontal direction, high frequency vertical direction and high frequency diagonal direction are shown in Fig. 4.30.

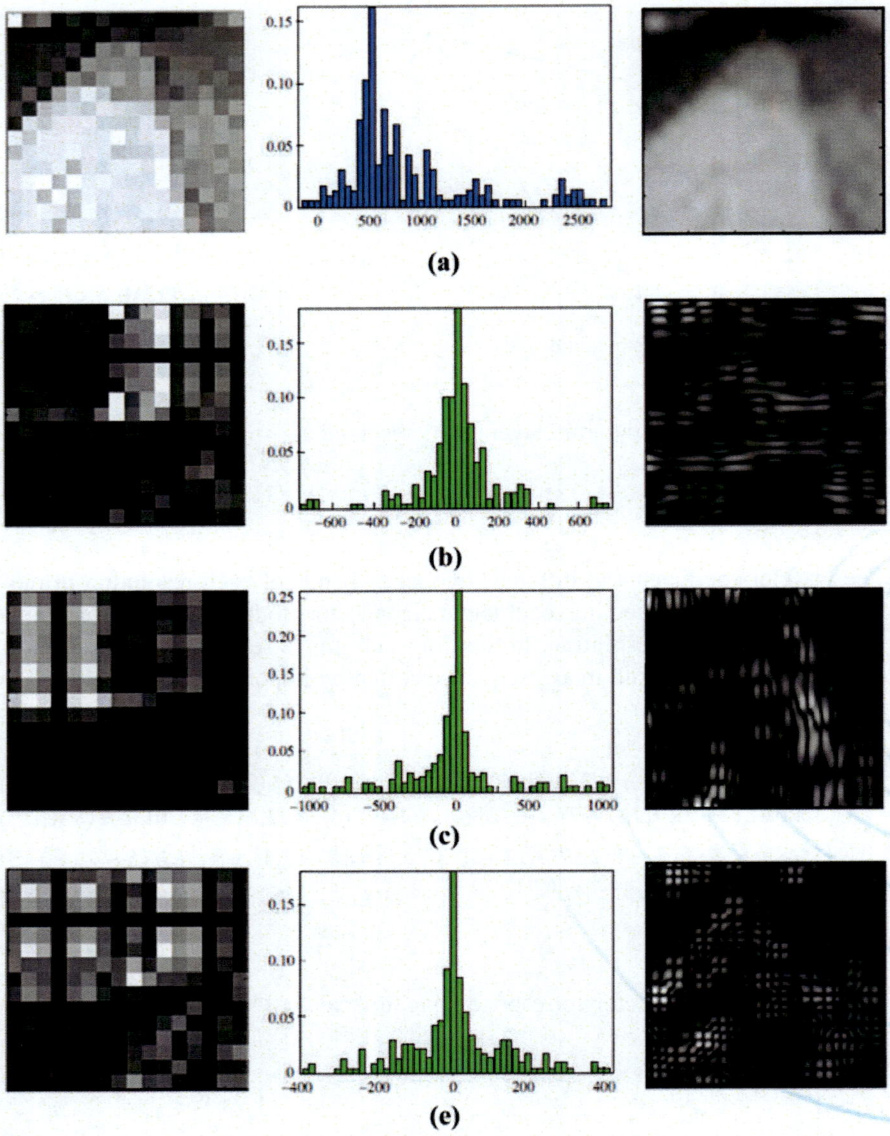

Fig. 4.30 Experimental results and reconstruction images of wavelet coefficient distribution **a** low frequency coefficient, low frequency coefficient distribution, low frequency reconstruction image (LL4) **b** H-direction detail coefficient, H-direction coefficient distribution, H-direction reconstruction detail image (HL4) **c** V-direction detail coefficient, V-direction coefficient distribution, V-direction reconstruction detail image (LH4) **d** D-direction detail coefficient, D-direction coefficient distribution, D-direction reconstruction detail image (HH4)

Figure 4.30 shows that the amplitudes of most wavelet coefficients are small after the wavelet decomposition of road image, which reflects the signal smoothness in local area of image. A small number of wavelet coefficients have large amplitudes, which reflects the signal variation in local structure of the image, corresponding to texture, edge and other important structural features. The wavelet coefficients of road image have enough sparsity, which can meet the requirements of compressed sensing theory to recover the signal. It can help solve the problem of navigation image big data processing in practice.

3. Single-Layer Wavelet Approximate Compressed Sensing Algorithm

In the past, most of the image compressed sensing algorithms in wavelet domain were aimed at still images. It is actually a multi-resolution data analysis and processing method. In this book, taking the characteristics of high real-time requirements of mobile robot environment sensing system into consideration, an approximate compressed sensing method is proposed based on the traditional compressed sensing theory. It decomposes and reconstructs the road navigation image only in the optimal scale space of specific low resolution level, and sets a wavelet coefficient threshold on the resolution. The high frequency wavelet coefficients less than the threshold are forced to be 0 to reduce the dimension of the coefficient vector. The scale of the optimal scale space is determined by the approximate image rough measurement test. The algorithm framework of single-layer wavelet road image compressed sensing is shown in Fig. 4.31.

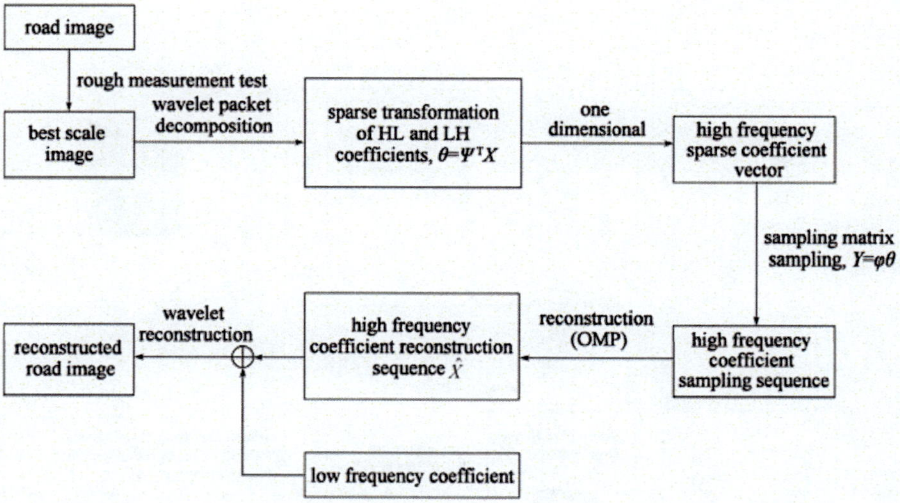

Fig. 4.31 Algorithm framework of single-layer wavelet road image compressed sensing

4.3.4.2 Road Image Segmentation Experiment

The vision sensor is connected with the autonomous vehicle body through the interface, and fixed on the body. After calibration, the experimental research described in this book can be carried out. It uses image segmentation technology to get the road boundary coordinates, and then converts the road 3D world coordinate system to the autonomous vehicle image coordinate system. The visual servo motion control of the autonomous vehicle can be performed this way. The whole control process is driven by the results of visual perception and completed by the autonomous vehicle, so it is called autonomous visual servo control. The system architecture is shown in Fig. 4.32.

Where (u, v) is the coordinates of the pixel in the image coordinate system; (X_w, Y_w, Z_w) is the three-dimensional coordinates of any point in the environment space; $\vec{K}$ is the internal parameter matrix of the camera, which is only related to the internal structure of the camera; $[\vec{R}\ \vec{t}]$ is the camera external parameter matrix, which is composed of a rotation matrix and a translation vector.

For a monocular vision system, the mapping relationship between image coordinate and road coordinate system is as follows:

$$\begin{cases} v = h \cdot \left(\dfrac{H \cdot \tan\gamma + 2v \tan\alpha}{H - \tan\gamma + \tan\alpha + 2v} \right) \\ u = \sqrt{h^2 + v^2} \cdot \dfrac{2 \cdot u \tan\beta}{W} \end{cases} \tag{4.36}$$

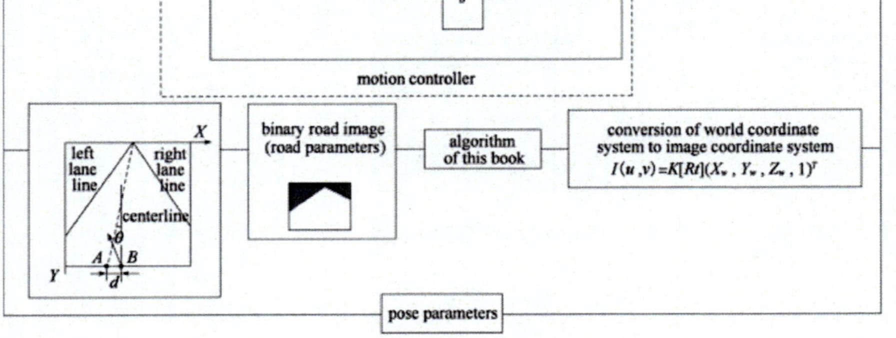

Fig. 4.32 System architecture of autonomous visual servo control

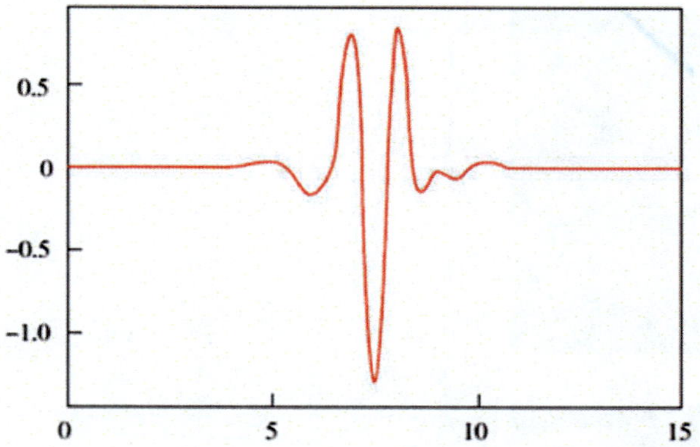

Fig. 4.33 Sym8 wavelet

where, h is the installation height of the camera lens center; γ is the angle between the connecting line between the nearest visible point on the ground plane and the camera lens center, and the perpendicular to ground plane; α is half of the vertical field angle of the camera lens; β is half of the vertical field angle of the camera lens; H and W are the height and width of the image.

The platform of the algorithm program is a laptop with 32-bit Win7 operating system. The CPU frequency is 2.53 GHz and the memory is 1 GB.

The higher the vanishing moment of wavelet, the higher the proportion of zero or very small value of wavelet transform amplitude. In order to achieve the goal of approximating the original image with as few wavelet coefficients as possible, Sym8 wavelet in Symlets function system is selected for road image wavelet packet decomposition and reconstruction experiment. Sym8 is a compactly supported orthogonal wavelet with good tradeoff between vanishing moment order and support length, which is suitable for image analysis. Sym8 wavelet is shown in Fig. 4.33.

The road image approximate compressed sensing experiment is carried out on scale 2, and the high-frequency coefficients are quantized by global threshold. The road images reconstructed by OMP algorithm under various thresholds, the energy retention rates and zero coefficient rates are shown in Fig. 4.34.

The experimental results show that after the original image is reconstructed, the energy retention rate is high enough, the image quality is almost not affected, and the road image noise is effectively suppressed, so the impact on the segmentation accuracy is very small.

The experimental result of the actual campus road images is shown in Fig. 4.35.

The algorithm of this book is validated on real road image sequential dataset of outdoor mobile robot provided by CMU (256 × 240). The 9th, 11th, 13th, 15th, 17th and 19th frames in the image sequence are randomly selected. These frames have different light features, as well as shadows, cracks and potholes.

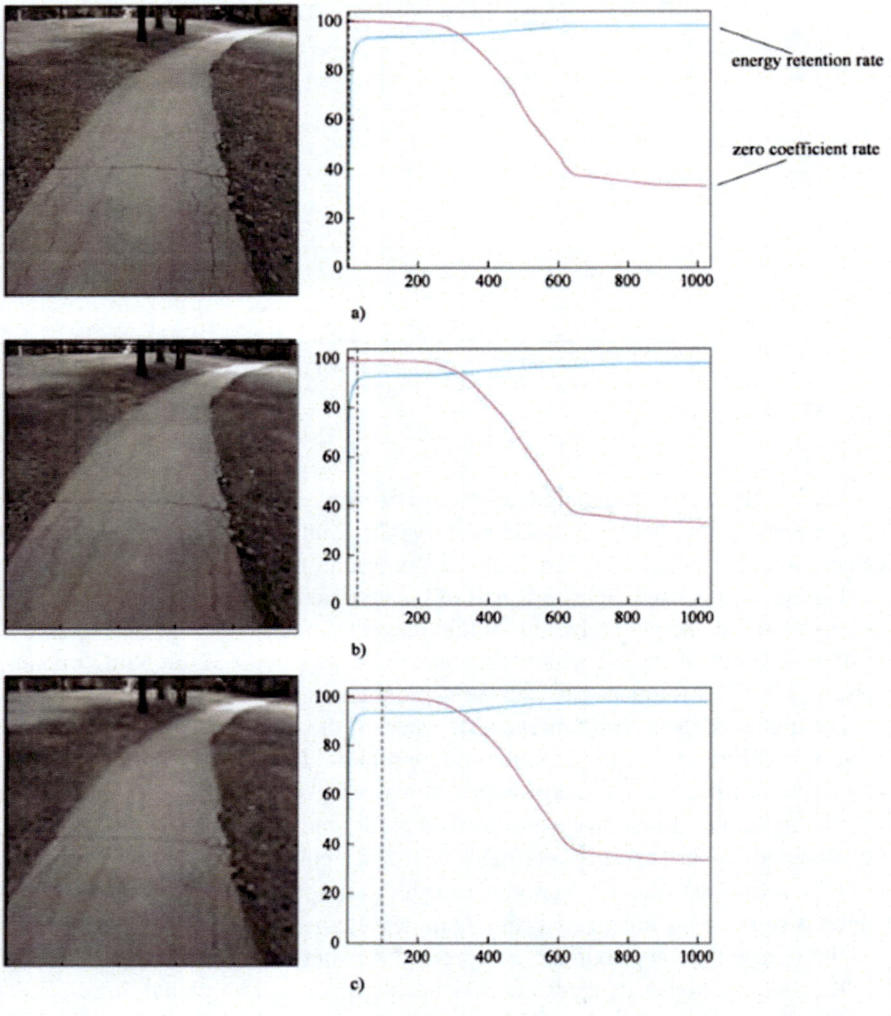

Fig. 4.34 Reconstructed road images, the energy retention rates and zero coefficient rates **a** $\varepsilon = 2.5$: energy retention rate = 99.99%, zero coefficient rate = 49.48% **b** $\varepsilon = 30$: energy retention rate = 99.72%, zero coefficient rate = 91.85% **c** $\varepsilon = 100$: energy retention rate = 99.46%, zero coefficient rate = 93.37%

The parameters of genetic algorithm are: population size 20, crossover probability 0.7, mutation probability 0.14, maximum algebra 60. Bayesian segmentation model which models pixels as a MRF (Markov random field) is a popular image segmentation method in recent years, and it has strong robustness. The experimental results of the algorithm of this book and ICM algorithm optimized MRF model segmentation method are compared as shown in Fig. 4.36.

Fig. 4.35 Result of the campus road images **a** compressed sensing images of several campus road scenes **b** road understanding results

The evolution results of the optimal threshold for each generation of the corresponding frame are shown in Fig. 4.37.

Almost all frames evolve to a relatively stable threshold during the 10th to 20th generations. Some frames evolve faster, and only a few frames evolve slower. Overall, the fitness function converges faster. In each frame, the adaptive genetic algorithm makes the threshold evolve continuously through genetic operation, and finally reaches the optimal value. The optimal segmentation threshold of each frame in the image sequence is adaptive. After this calculation, each frame in the image sequence is segmented with the optimal threshold, so as to achieve the global optimization of spatiotemporal segmentation.

The experimental results show that the algorithm has strong robustness to the unstructured road with severe shadow interference, cracks and random illumination changes, and can segment the road area accurately and quickly. In addition, the potholes in the road can also be effectively detected. MRF model segmentation method performs well with shadow interference, but it is not good at suppressing the interference of cracks, and may confuse road with non-road. Although this method has been successfully applied in remote sensing image and laser image segmentation, it is not suitable for dynamic navigation image segmentation because of the essential opposition of the two objective functions and the huge time consumption of iterative calculation.

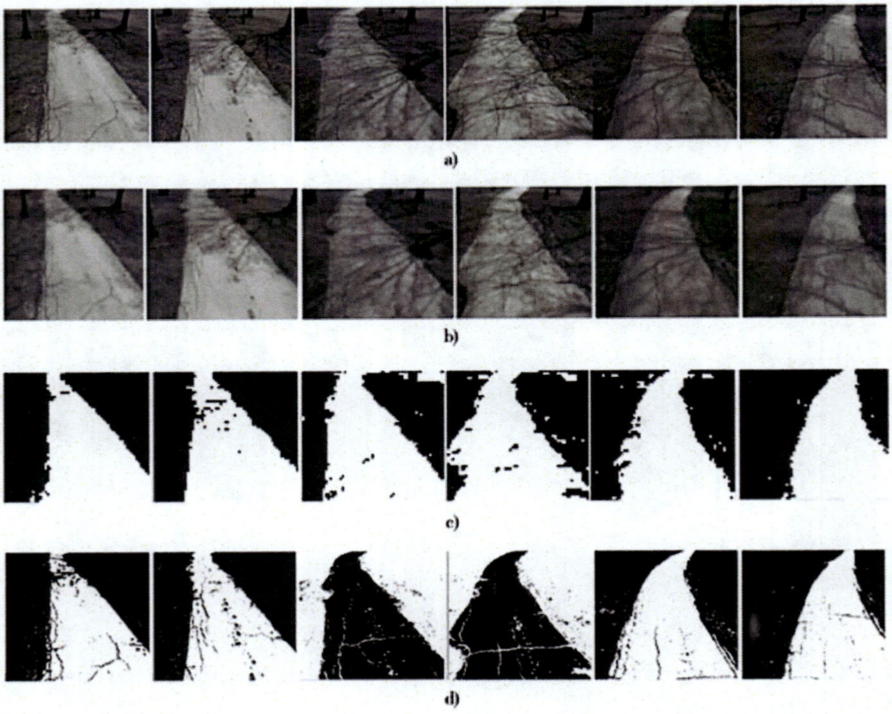

Fig. 4.36 Experimental results of the algorithm of this book and ICM algorithm optimized MRF model segmentation method **a** original image sequence (frame 9, 11, 13, 15, 17, 19) **b** wavelet reconstruction image sequence (frame 9,11,13,15,17,19) **c** segmented image sequence based on adaptive genetic algorithm (frame 9,11,13,15,17,19) **d** segmented image sequence based on ICM algorithm optimized MRF model (frame 9, 11, 13, 15, 17, 19)

3 road image frames (other frames are similar in performance) are selected for road understanding algorithm performance comparison, as shown in Table 4.2.

It can be seen that the amount of data and storage involved in the method of image understanding without wavelet domain is much larger than that of the method of this book. In the premise of not reducing the accuracy, the algorithm of this book has an obvious advantage in the total time consumption of single frame understanding. Even the method of image understanding in the wavelet domain does not fully consider the problem of maximum reduction of data. Based on experimental verification, this book proposes a wavelet compressed sensing algorithm from the practical point of view, which only retains a single layer, and greatly reduces the amount of calculation and storage.

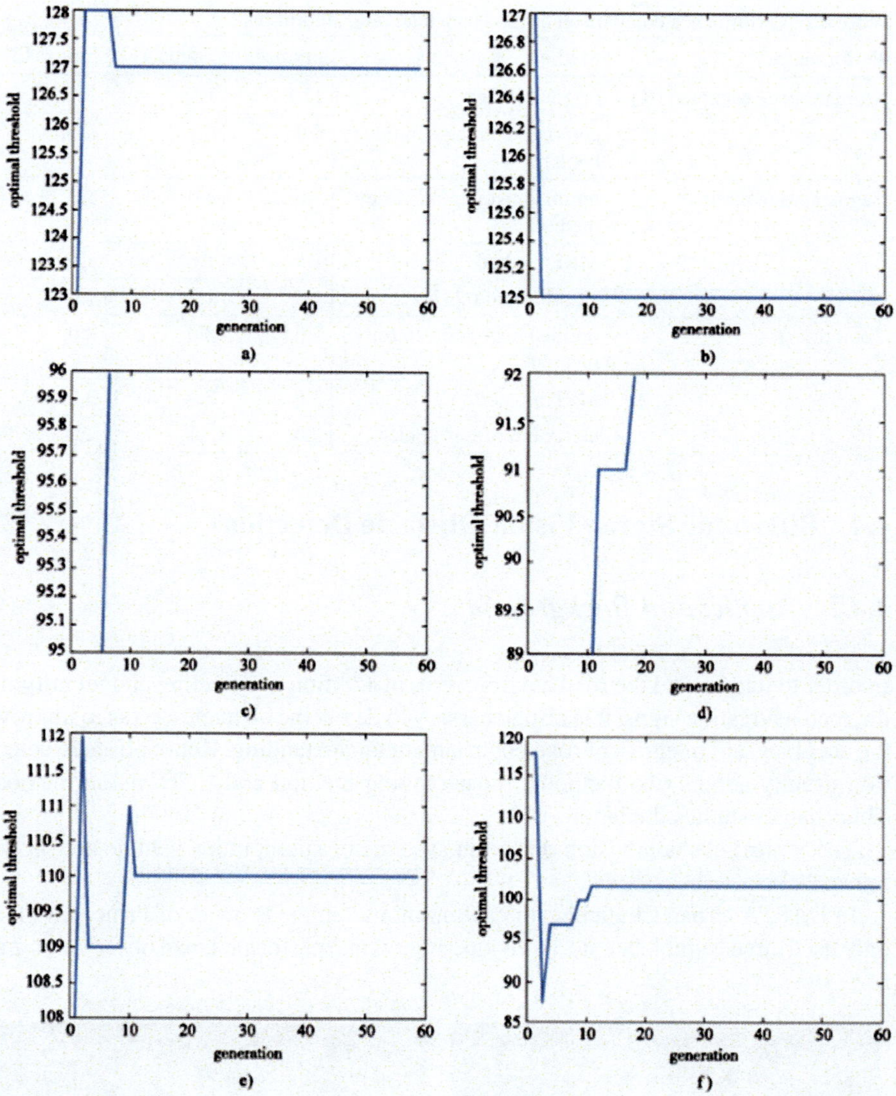

Fig. 4.37 Evolution results of the optimal threshold for each generation **a** Frame 9; **b** Frame 11; **c** Frame 13 **d** Frame 15; **e** Frame 17; **f** Frame 19

Table 4.2 Performance comparison of road understanding algorithms

Performance	Method	Frame 9	Frame 11	Frame 13
Storage space occupied (K)	original image	110	107	108
	compressed image in this book	6.03	6	6.01
Segmentation time (s)	optimal histogram entropy method	7.6	8.0	8.0
	MRF method	1.8	1.6	1.5
	algorithm of this book	1.5	1.5	1.5
Total time (s)	optimal histogram entropy method	7.6	8.0	8.0
	MRF method	3.6	3.6	3.2
	algorithm of this book	2.0	2.1	2.0

4.4 Binocular Stereo Vision Obstacle Detection

4.4.1 Application Background

In order to understand the road environment, in addition to detecting and identifying the road navigation signs, it is also necessary to detect the obstacles, so as to improve the accuracy and integrity of road environment understanding. Robot obstacle detection usually adopts two methods: radar ranging method and CCD vision method. This chapter studies the latter.

The typical obstacle vision detection systems of autonomous vehicle are shown in Fig. 4.38.

In DARPA Urban Challenge, the autonomous vehicle Prowler of Princeton Fleet only used three Bumblebee stereo vision cameras and performed road observation and

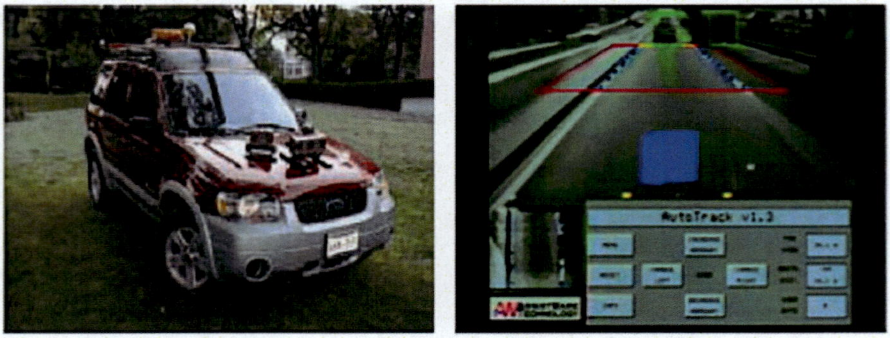

(a) (b)

Fig. 4.38 Typical autonomous vehicle obstacle detection systems **a** autonomous vehicle Prowler relying on stereo vision **b** obstacle detection system of autonomous vehicle Navlab obstacle detection. According to reports, its stereo vision system can detect objects 60 m away, and one set of obstacle detection algorithms is sufficient. Representative stereo vision obstacle detection methods can be summarized as follows.

1. Pixel matching method

Stereo vision calculation is performed for each pixel in the road image that needs to be judged. According to the relationship among image coordinates, camera coordinates and world coordinate system, the height value of each pixel is calculated. If the difference between the value and the ground height value exceeds the set threshold, the pixel is categorized as an obstacle point, otherwise it is passable area. The typical application is with Mars Rover Courage and Opportunity. The method is that, first, the left and right viewpoint matching, peak and small point filtering are performed for each image point, and then whether it is an obstacle is judged according to the height and slope degree of the image point. At the same time, the distance and near layer diagrams of the obstacle is established. At present, only the near obstacles are considered. The issue of this method is that every pixel in the image has to be reconstructed by stereo vision, and the calculation is large, which cannot meet the timeliness requirements of the robot.

2. Projective transformation matrix method

A real-time projective transformation matrix is defined to make the ground points correspond to the image points in the image plane one by one. In the binocular vision mobile robot, the projective transformation matrix is calculated by real-time matching algorithm, and the relationship between the image points on the left and right images can be obtained. If the point (x, y) is located in the passable area, the coordinates of the image point calculated by the mapping coincide with the coordinates of the corresponding point verified by the matching algorithm. Otherwise, it means that the point is higher or lower than the ground, and it is determined as an obstacle.

The disadvantage of this method is that, when the road is uneven, the relative geometric position between the camera imaging plane and the ground is constantly changing, and the mapping relationship between the camera image coordinate system and the ground coordinate system is also changing. Therefore, it is necessary to calibrate the camera in real time, otherwise the detection accuracy will be affected.

3. The method based on image segmentation

A common method is to use color image vision system. Each pixel in the scene is classified into two sets, passable area and suspicious obstacle area, based on color features. That is the preliminary detection of obstacles. Then the stereo vision calculation is carried out, and only the suspicious obstacle area is searched in the calculation to eliminate the influence of false obstacles (such as shadows and leaves) in image segmentation, so as to detect the obstacles more accurately. In order to enhance the adaptability to environmental changes, an adaptive image segmentation method is proposed. The system is trained in advance with color values of passable areas under different lighting conditions, and the adaptive segmentation of robot scene image is achieved by calculating the reference hue, saturation value and threshold value of road surface under current lighting conditions. Some researches use color and texture features for road image segmentation and preliminary obstacle detection, which increases the accuracy of detection, but also increases the amount of calculation.

Although people have done a lot of research on vision based obstacle detection methods, some key technical problems in this field have not been well solved so far. In general, the difficulties in obstacle detection of vehicle vision platform can be summarized as follows.

1. The distinction of obstacles with similar characteristics

Sometimes, the results of feature extraction of some obstacles and passable areas are very similar. For example, the color and texture features may be the same, so it is difficult to identify the obstacles accurately. The multi-feature fusion method can be tried to solve this problem.

2. Pseudo obstacle detection and obstacle definition

Some pseudo obstacles are judged as obstacles according to the definition of system, such as small stones and bricks on the road, flying birds, paper scraps and leaves in the air, as well as dust and pollutant masses. If these pseudo obstacles are often treated as obstacles, it will obviously affect the vehicle speed.

3. Measurement distortion of obstacle information caused by motion vibration. Typical cases, such as autonomous vehicle in the field of complex uneven terrain environment, the vibration during advancing makes the data detected and transmitted by the sensor discontinuous, resulting in inaccurate terrain information, increased false alarm rate of obstacles, which will affect the speed of the vehicle.

In the structured and semi-structured road environment, the obstacles are mainly vehicles and pedestrians. In the unstructured road environment, the obstacles are mainly concave potholes and convex roadblocks such as stones and tree trunks. In general, monocular image segmentation and feature extraction algorithm cannot eliminate the influence of pseudo obstacles such as small obstacles with the same color as the passable area on the detection results, but monocular feature extraction combined with pattern recognition, target tracking and other algorithms can quickly detect and track obstacles. Stereo vision can help the robot further understand the distance and height information of obstacles, and effectively eliminate pseudo obstacles, so as to improve the accuracy of its understanding of the environment. In this chapter, using color binocular stereo vision technology, local invariant feature detection technology and visual saliency theory, two obstacle detection algorithms based on vehicle machine vision are proposed. One is multi-feature fusion obstacle detection algorithm based on depth information, and the other one is dynamic vehicle obstacle detection algorithm based on low level visual saliency feature matching. Due to the use of the calibrated binocular camera, there is no in-depth research on camera calibration, and the research focuses on binocular matching and depth information acquisition.

4.4.2 Steps of Binocular Stereo Perception

The steps of stereo perception based on binocular vision are shown in Fig. 4.39.

Stereo vision calculation includes two main processing modules. The data processing flow is shown in Fig. 4.40.

Parallax is a key concept in stereo vision. Parallax is defined as the difference between the x coordinates of the same point in space in the left and right eye images. Its meaning is explained in Fig. 4.41.

In Fig. 4.41, the parallax of point A is defined as $d_A = x_{\text{Aleft}} - x_{\text{Aright}}$, and the parallax of point B is defined as $d_B = x_{Bleft} - x_{Bright}$, where x_{Aleft} is the x coordinate of A in the left image; x_{Aright} is the x coordinate of A in the right image and so on. After calculation, it will be found that, $d_B > d_A$. This indicates that point B in the scene is closer than point A.

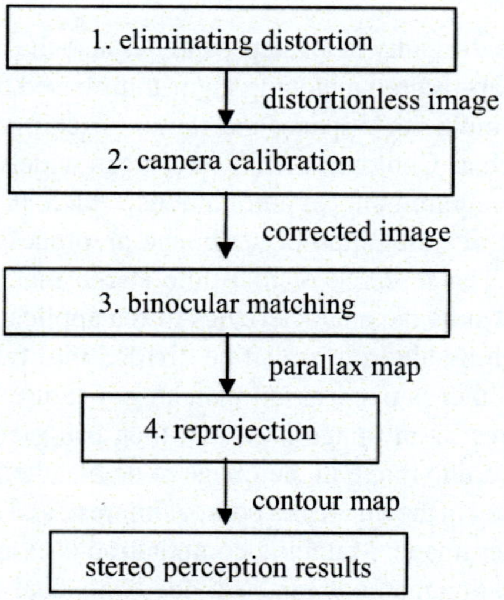

Fig. 4.39 Steps of stereo perception based on binocular vision

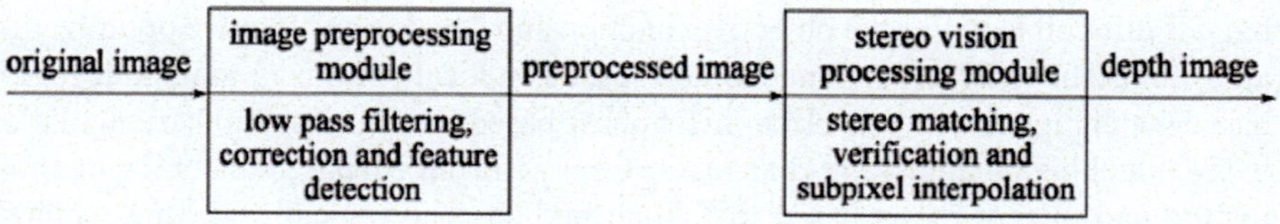

Fig. 4.40 Stereo vision data processing flow

Fig. 4.41 Meaning of parallax

4.4.3 Stereo Matching

Stereo matching is the difficulty in all steps, and it is also the hot spot of stereo vision research. In recent years, Stereo matching algorithms based on image segmentation, image local features, multi-scale space, and stereo matching algorithm and comprehensive operators such as Contourlet-SIFT have been widely studied. According to the visual navigation requirements of autonomous vehicle in different road environments, it is necessary to adopt appropriate image preprocessing method and stereo matching algorithm. The research of real-time stereo matching algorithm can be practical only when it suits the characteristics of the application system.

The existing matching algorithms can be divided into two types: one is feature based matching, the other is pixel based matching. Feature based matching means identifying the features in an image first, and then using constraints to search for similar features in a certain range in the image to be matched. Pixel based matching assumes that every pixel in the image is a point of interest, and the corresponding pixel is searched in the other image. Matching computation is essentially an optimization problem, and optimization methods can be divided into local optimization and global optimization. The local optimization method can only guarantee that the parallax is the optimal solution in a certain region. It cannot optimize the parallax in the whole image range. Global optimization can get the global optimal parallax distribution, but it is difficult to define the objective function and develop an effective optimization algorithm. The typical matching methods based on local optimization include region based matching method, wavelet domain phase based matching method, triangulation based matching method, rank change and Census transform based matching method, feature and mixed region based matching method. The typical matching methods based on global optimization include dynamic programming based matching method, max-flow min-cut method, collaborative iteration method and so on. The commonly used optimization methods include neural network, simulated annealing, genetic algorithm, particle filter, dynamic programming, max-flow min-cut method and so on, but these algorithms are more or less challenged by difficulty of convergence control and low efficiency when facing high-dimensional image data.

The following energy function is generally used to solve the optimization of matching relation:

$$E(f) = E_{\text{data}}(f) + E_{\text{occ}}(f) + E_{\text{smooth}}(f) \tag{4.37}$$

where, E_{data} is the data item, which indicates the consistency degree of the corresponding pixel matching; E_{occ} is the penalty item, which records mismatching; E_{smooth} is the smoothness item, used to constrain adjacent pixels to have consistent parallax. In order to reduce mismatching, the method of adding constraints is usually used. The constraints of stereo matching include uniqueness constraint, similarity constraint, epipolar constraint, smoothness constraint, monotonicity constraint, sequence constraint and obstruction constraint.

In the application environment of autonomous vehicle navigation, the stereo vision algorithm should also meet the following requirements:

1. Timeliness. It should be able to meet the requirements of real-time response control system and navigation decision-making system, and provide necessary data within the specified time limit.
2. Reliability. The algorithm should accurately identify obstacles under different conditions. Once there is an abnormal problem, the algorithm should have the ability of automatic repair and automatic fault tolerance.
3. Environmental adaptability. According to the change of the environment, the algorithm should be able to automatically adjust the collected environment features, learn the feature parameters independently, and finally complete the environmental adaptation through machine learning algorithm.

4.4.4 Depth Information Acquisition and 3D Reconstruction

Under the ideal assumption, in the binocular vision system adopted in this book, the triangulation relationship between depth and parallax is shown in Fig. 4.42. The two images of the left and right cameras are aligned in rows, and the optical axes are strictly parallel. The epipolar distance between the two cameras is T, and the focal lengths f are the same. (c_x^l, c_y^l) is the origin of the left image and (c_x^r, c_y^r) is the origin of the right image. Based on the principle of triangulation, the depth Z can be calculated from the parallax d.

By using similar triangles, it can be deduced that: 1

$$\frac{T - (x_l - x_r)}{Z - f} = \frac{T}{Z} \Rightarrow Z = \frac{fT}{d} \tag{4.38}$$

where x_l and x_r are the abscissas of point P on the imaging planes of left and right cameras respectively. It can be seen from Formula 4.38 that the parallax is inversely proportional to the depth, and the parallax value can be converted into the distance

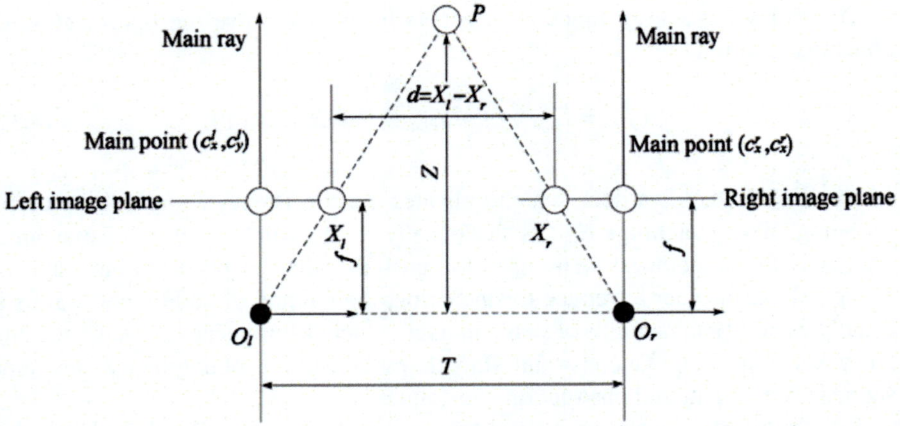

Fig. 4.42 Triangulation relationship between depth and parallax under ideal assumption

value through the principle of triangulation, which is also the reason why stereo vision can be used for distance measurement in reality.

The accuracy of the minimum depth range can be determined by Formula 4.39.

$$\Delta Z = \frac{Z^2}{fT}\Delta d \tag{4.39}$$

where Δd is the minimum parallax increment allowed.

Given the parallax d and coordinates (x, y) of the pixel, the 3D coordinates $(X/W, Y/W, Z/W)$ of the pixel can be calculated by Formula 4.40.

$$Q\begin{bmatrix} x \\ y \\ d \\ 1 \end{bmatrix} = \begin{bmatrix} X \\ Y \\ Z \\ W \end{bmatrix} \tag{4.40}$$

where Q is a 4-by-4 reprojection matrix:

$$Q = \begin{bmatrix} 1\,0 & 0 & -c_x^l \\ 0\,1 & 0 & -c_y^l \\ 0\,0 & 0 & f \\ 0\,0 & -1/T_x & (c_x^l - c_x^r)T_x \end{bmatrix} \tag{4.41}$$

where, T_x is the x direction component of T. Q is determined at the time of stereo calibration. If the main rays intersect at infinity, then $c_x^l = c_x^r$, the lower right item is 0.

Assuming that the coordinates of the target point in the left view are (x, y), the parallax formed in the left and right views is d, and the coordinates of the target point in the world coordinate system with the optical center of the left camera as the origin is (X, Y, Z), then there is a reprojection matrix Q, which makes $Q[x\ y\ d\ 1]^T = [X\ Y\ Z\ W]^T$.

The parameters needed to obtain the distance Z of a point in the world coordinate system are: parallax d, focal length f, camera center distance T_x. If the X and Y coordinates are needed, the offsets C_x and C_y of the original point in the left and right image plane coordinate system and the three-dimensional coordinate system should also be known. The initial values of f, T_x, C_x and C_y can be obtained through stereo calibration. Through stereo calibration optimization, the two cameras are mathematically parallel, and the C_x, C_y and f of the left and right cameras are the same. The essence of stereo matching is to find the parallax d on the basis of the above.

There is also a certain relationship between the distance between the object and the camera and the ranging accuracy. As shown in Fig. 4.43, P_1, P_2 and P_3 are three points in the real world. In the same camera coordinate system, their parallax are d_1, d_2 and d_3, and obviously, $d_1 < d_2 < d_3$.

Based on the analysis of Formula 4.32, when the parallax is small, small parallax changes will cause great depth changes. When the parallax is large, small changes of parallax will hardly cause significant changes of depth. With Fig. 4.43, further analysis shows that the greater the object distance is, the less obvious the ranging depth change caused by parallax change is, which is less favorable to accurate depth

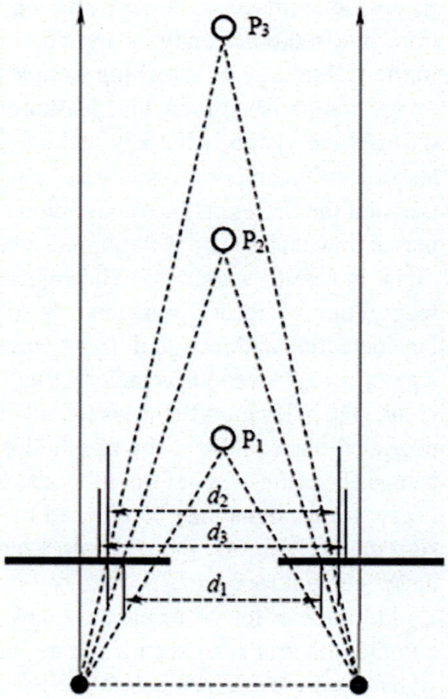

Fig. 4.43 Parallax relationship of different points

measurement. Thus we can get the following conclusion: stereo vision has higher depth measurement accuracy for objects closer to the camera, and the accuracy cannot be guaranteed beyond a certain distance range.

For industrial robot grasping work pieces, 3D modeling and other applications that require accurate 3D world coordinates, 3D reconstruction is needed. However, for the robot stereo visual navigation applications, due to the strong demand for timeliness, complete 3D reconstruction is not required. The visual navigation of mobile robot is essentially the processing, analysis and understanding of video image (image sequence). Other than the basic theory of still image, more attention should be paid to the analysis of image motion information, such as using frame to frame matching to reconstruct the motion trajectory.

4.5 Understanding of Traffic Signs

4.5.1 General Idea

As auxiliary navigation signs, traffic signs play an important role in vehicle autonomous navigation system, and traffic sign understanding is an important part of road environment understanding. Vision based traffic sign understanding includes two processes: traffic sign detection and traffic sign recognition. Traffic sign detection by vision refers to using certain strategies to determine whether there are traffic signs in any video image. If so, the traffic sign area is segmented and its color, shape, position, size and other features are identified. Traffic sign recognition by vision refers to the technology of searching sample signs from the traffic sign database through image comparison or machine learning. Traffic sign detection and recognition based on machine vision is the key technology module of autonomous vehicle, and it is also one of the important research directions of intelligent traffic system (ITS). It can transmit the detection and recognition results to the driving decision-making system in real time, and help the autonomous vehicle understand the road and traffic instructions. In recent years, research projects have been carried out on the detection and recognition of traffic signs, but it is still a relatively immature field compared with lane detection and recognition. At present, for the complex urban road environment, a practical autonomous vehicle traffic sign detection and recognition system is yet to come. The relevant difficulties include: the color characteristics are unstable or even degraded due to the change of outdoor illumination; the change of detection angle brings about the sign deformation; the vehicle vibration makes the traffic sign image fuzzy; traffic signs may be blocked by vehicles, pedestrians and other objects; The sign may be partially damaged. In view of the above difficulties, different solutions have been put forward.

Although a lot of researches have been done, specific work for challenging complex natural road scenes is rare, and projects with results applicable to traffic signs of various shapes colors are not common. Some key technical problems that

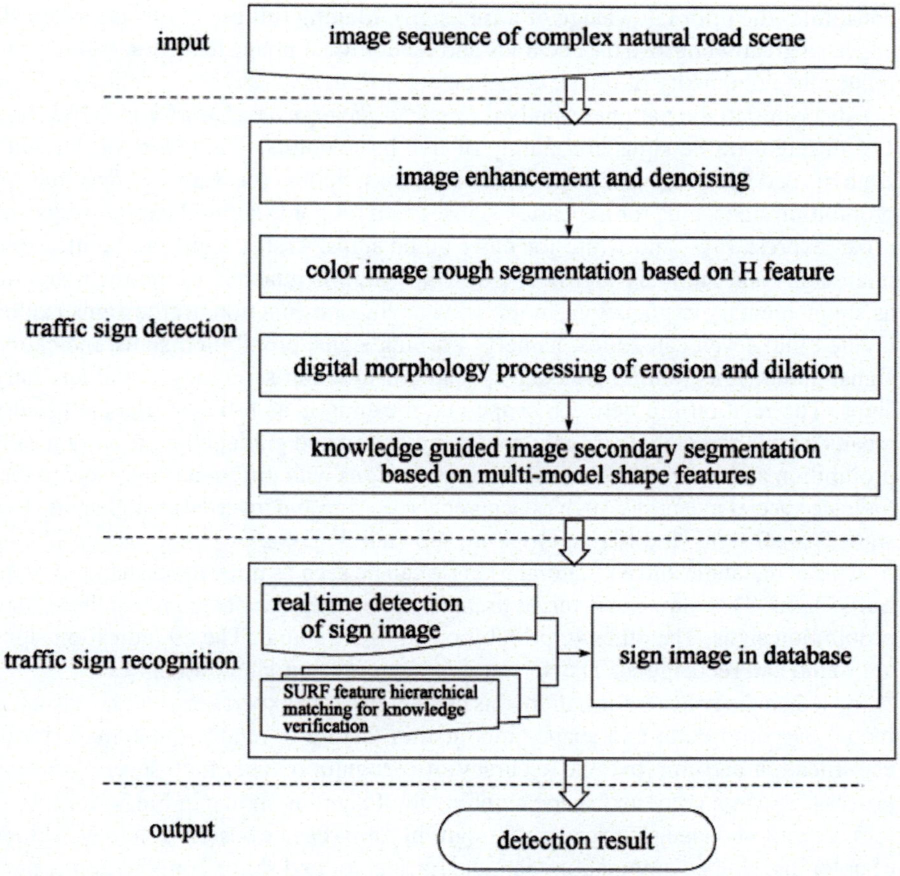

Fig. 4.44 Flow of traffic sign detection and recognition algorithm

prevent the traffic sign detection and recognition system from being practical have not been well solved. For example, how to automatically adapt to light changes? How to detect traffic signs accurately in complex scenes? How to improve the speed and accuracy of online recognition? These are the problems of traffic sign detection and recognition system. In view of the above difficulties, this book proposes the algorithm as shown in Fig. 4.44.

4.5.2 Robust Detection of Traffic Signs

One of the biggest fundamentals of traffic sign detection is that there is a definite relationship between the shape and color of traffic signs. It is this relationship that constitutes the prior knowledge of traffic signs. Making full use of this prior knowledge can greatly improve the accuracy and reliability of image detection system, and reduce the complexity of image detection algorithm.

According to the national standard *Road Traffic Signs and Markings* (GB5768), 116 different traffic signs are mainly of five basic colors. The colors of the outer edge of road traffic signs are specified as follows: yellow for warning signs, red for prohibition signs, blue for indication signs, green for guide signs of expressways and urban expressways, and white for other guide signs. Traffic signs are divided into main signs and auxiliary signs. The auxiliary sign is attached to the main sign to as supplementary explanation. In terms of navigation function, traffic signs can be divided into seven categories, namely: warning signs, prohibition signs, indication signs, guide signs, tourist area signs, road construction safety signs and auxiliary signs. The relationship between shape and meaning is as follows: triangle usually means warning; rectangle usually means indication and guide; circle usually means prohibition and indication. A small number of signs with different shapes and colors indicate special meanings, such as octagonal stop sign and inverted triangle give-way sign. Typical signs of different types are shown in Fig. 4.45.

Circle, rectangle and equilateral triangle can be seen as three main shapes of road traffic signs. There are one inverted triangle and one regular octagon, which are both prohibition signs. The other 40 prohibition signs are round. The 29 indication signs are round and rectangular. The 45 warning signs are equilateral triangles.

Based on the above information, this chapter uses the expert knowledge reasoning model based on color and shape combination feature to reduce the complexity of classification and improve the accuracy of recognition. The multi-level reasoning process based on color and shape combination feature is shown in Fig. 4.46.

The prior knowledge of road traffic signs in China can be described by seven kinds of color and shape combination features. Each color and shape combination feature represents one subclass of traffic signs, and 116 different traffic signs are divided into seven subclasses. Among them, blue indication, red prohibition and yellow warning are the three main types of traffic signs.

Let H_1, H_2, H_3, H_4, H_5 represent the color red, blue, yellow, green and white, then the set of colors is $H = \{H_1, H_2, H_3, H_4, H_5\}$. Let $X_1, X_2, X_3, X_4, X_5, X_6$ represent the shape rectangle, circle, regular triangle, inverted triangle, octagon and others, then the set of shapes is $X = \{X_1, X_2, X_3, X_4, X_5\}$.

According to the above definitions, the categories of traffic signs can be expressed in the form of sets:

Prohibition signs (C1): $JL = \{H_1, X_2\} \cup \{H_1, X_4\} \cup \{H_1, X_5\}$ (only one red octagon sign and one red inverted triangle sign, which are small probability events).

Indication signs (C2): $ZS = \{H_2, X_2\} \cup \{H_2, X_1\}$.

Warning signs (C3): $JG = \{H_3, X_3\}$.

Guide signs (C4): $ZL = \{H_4, X_1\}$.

Road construction safety signs (C5): $ZL = \{H_2, X_1\}$.

Tourist area signs (C6): $LYQ = \{H_1, X_1\}$.

Auxiliary signs (C7): $FZ = \{H_5, X_1\}$.

Fig. 4.45 Typical traffic signs of different types **a** warning signs **b** prohibition signs **c** indication signs **d** guide signs **e** road construction safety signs **f** tourist area signs **g** auxiliary signs

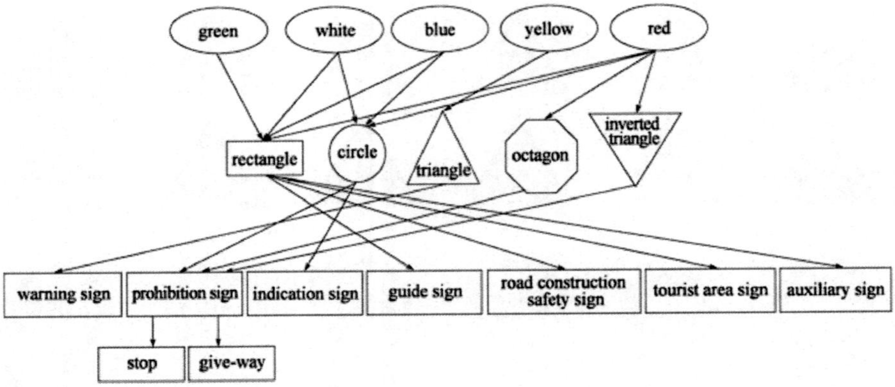

Fig. 4.46 Multi-level reasoning process based on color and shape combination feature

The following individual cases are small probability events: an octagon stop sign and an inverted triangle give-way sign in the red prohibition sign category; a red irregular shape sign in the yellow warning sign category. Apart from these cases, according to the order of color detection and shape detection of the algorithm, taking the main probability as the internal reasoning clue (ignoring the small probability result), the deterministic knowledge among color, shape and sigh categories based on color is summarized, which can be described as the following IF–THEN-THEN statements:

IF red, THEN circle, THEN prohibition sign C1;

IF blue, THEN (circle or rectangle), THEN (indication sign C2 or road construction safety sign C5);

IF yellow, THEN triangle, THEN warning sign C3;

IF green, THEN rectangle, THEN guide sign C4;

IF dark red, THEN rectangle, THEN tourist area sign C6;

IF white, THEN rectangle, THEN auxiliary sign C7.

The significance of the above logic description is that it corresponds to the logical relationship that can be achieved by the actual detection algorithm. It describes the deterministic knowledge relationship, and indicates the combination order of color and shape and two-level reasoning detection logic. For small probability events, the processing method is: if the final result cannot be obtained by using the above logic, then it returns to enter the small probability reasoning process. The small probability reasoning is carried out according to the set representation, which is the same as the main probability event method.

The accurate segmentation method based on shape feature is as follows.

The first step: based on morphological operation, the traffic signs are roughly located.

On the basis of detecting the candidate areas of traffic signs, the appropriate morphological processing algorithm is used to further remove the interference, so as to accurately locate the area of traffic signs. The method based on connected region filling and area threshold can be used to remove small area non-target region. A

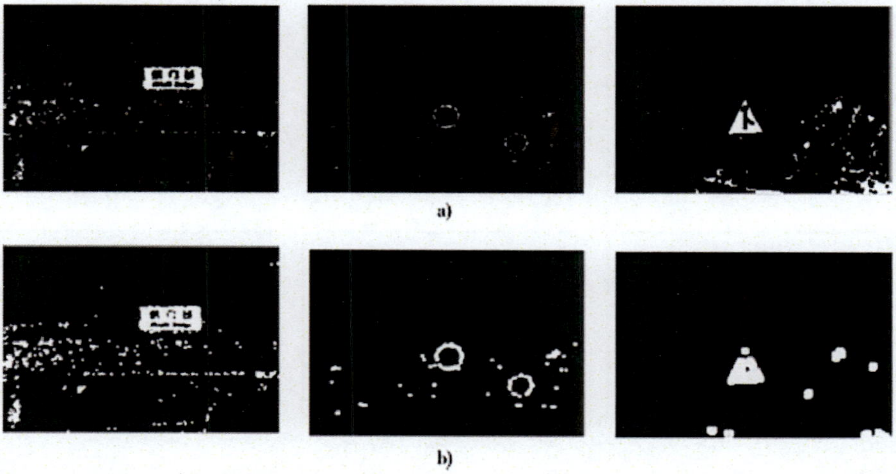

Fig. 4.47 Experimental results of various typical scenes **a** erosion; **b** dilation

seed pixel is determined in the target region. Then starting from the seed pixel, the adjacent pixels are gradually added according to the proximity of the pixel values, and filled with the specified color. When certain conditions are met, regional growth will end. Afterwards the areas of each target region are calculated. If the area of the region is less than the pre-set threshold, it will be the background.

The method adopted in this book is as follows: 3-by-3 square structural element is used on the traffic sign scene image for erosion after the initial segmentation of color features. Then in order to detect the edge of the sign more accurately and reliably, the 3-by-3 square structural element is used twice, and the 3-by-3 cross structural element is used once for dilation. The experimental results of various typical scenes are shown in Fig. 4.47.

Find the largest connected area in the erosion and dilation image, and calculate the center point position. With the point as the center, a square area with large enough to surround the traffic sign is defined as the traffic sign area of interest (sigh ROI). All the regions outside this area are colored white. The sigh ROIs located with largest connected area method are shown in Fig. 4.48.

So far, the detection of traffic signs has been completed. The method based on morphological operation is relatively simple, but its robustness is poor in the case of complex scene image. At this case, the shape feature detection algorithm can be used to locate the traffic sign area more reliably and accurately. Morphological operation steps can also be retained to create better detection conditions for Hough transform. The traffic sign location method based on shape features can obtain shape semantics through detection, so it has the following advantage: it can narrow the feature search scope of traffic sign recognition module, thus reducing the computational burden of recognition system and improving the recognition accuracy.

The second step: traffic signs are detected more accurately based on shape features.

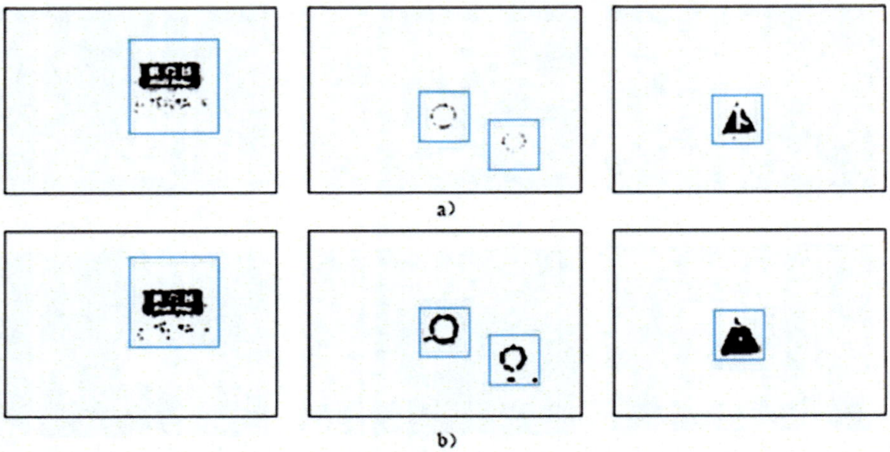

Fig. 4.48 Sigh ROIs located with largest connected area method **a** connection results after erosion; **b** connection results after erosion and dilation

The shape related object features include edge, perimeter, area, corner and so on. These features can be easily extracted without background interference. However, for the complex natural scene of traffic signs, the extraction of target shape feature is not easy. This is due to the great uncertainty of the content of traffic sign scene image and the geometric shape of the object, and the complex interference of the background to the target, which is actually the essential difficulty of the target shape detection in the natural scene. For example, in the absence of background interference, the method of finding the largest target area (connected domain) can be used to determine the target shape smoothly. Yet when there is complex background interference, the area of some objects in the background will even exceed the target, so this method is no longer effective. Typical shape feature extraction methods include: Hough transform, Fourier shape descriptors, shape factor, shape invariant moment, wavelet domain relative moment, etc. The goal of this book is to explore and propose a universal shape detection method for all types of traffic signs. Therefore, Hough transform, which has strong versatility for shape detection, is selected as the breakthrough point, and the algorithm is designed according to the characteristics of the sigh image. Hough transform can effectively detect straight line, circle, ellipse and other foreground targets without or with little background interference. Generalized Hough transform can detect any shape. The harsh requirement of no background interference is very difficult to achieve in actual outdoor scene images. The only thing to do is trying the best to minimize the background interference. In addition, Hough transform cannot directly detect rectangles and triangles. It needs to detect lines first and then use the corresponding algorithm for the detection of rectangles and triangles. Aiming at the automatic detection of various traffic sign shapes (including similar shapes after deformation) in natural scenes, this book proposes a shape feature detection method based on multi-model switching Hough transform.

The detection result of multiple round signs is shown in Fig. 4.49.

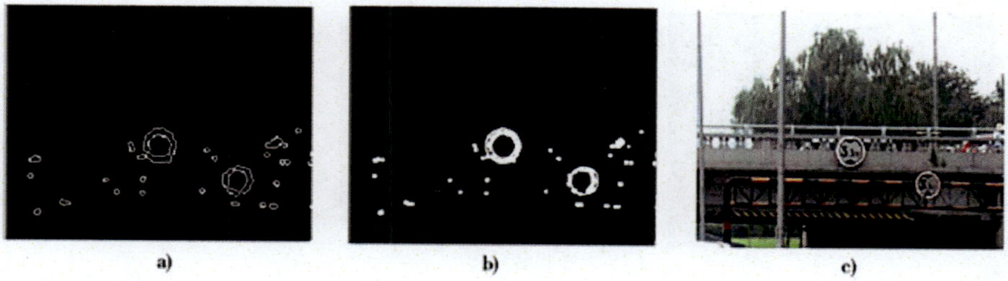

Fig. 4.49 Detection result of multiple round signs **a** edge extraction result; **b** marking the round signs; **c** recovering on the original image

For the round traffic signs with obstruction and deformation, the probability Hough transform circle detection algorithm is adopted. The detection result is shown in Fig. 4.50.

The Hough transform shape detection results of rectangular and triangular traffic signs are shown in Figs. 4.51 and 4.52 respectively.

The third step: in order to enhance the robustness of the algorithm, multi-feature joint estimation is performed.

In order to enhance the robustness of traffic sign detection results in complex natural scenes, on the basis of the feature extraction of color and shape, other auxiliary

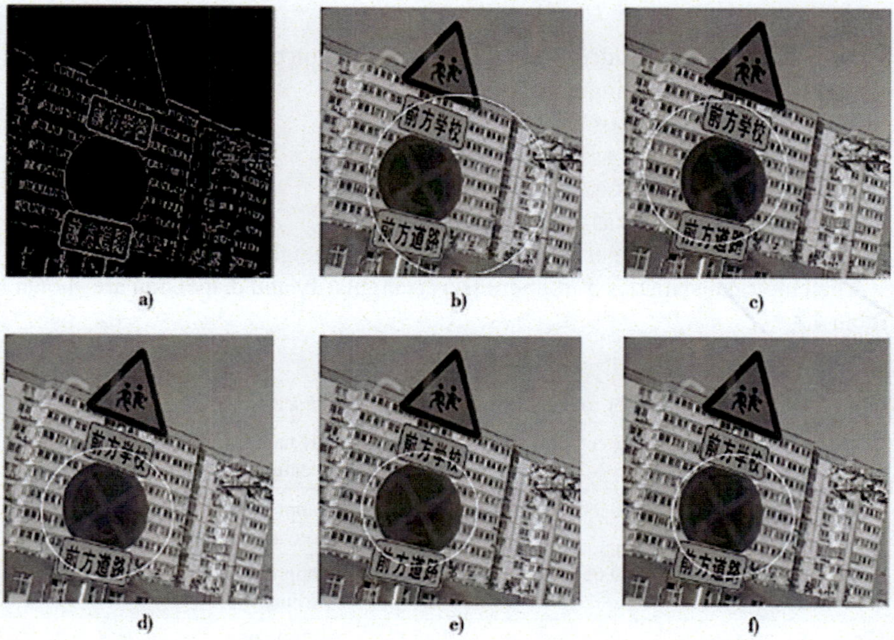

Fig. 4.50 Detection result of a round traffic sign with obstruction and deformation **a** edge extraction result; **b** round region 1; **c** round region 2 **d** round region 3; **e** round region 4; **f** round region 5

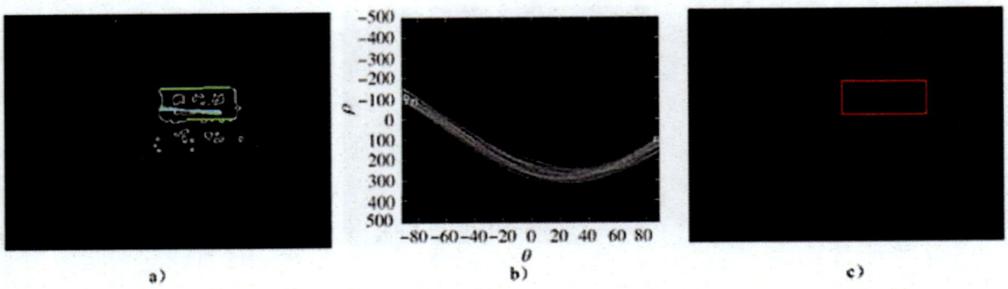

Fig. 4.51 Hough transform shape detection results of a rectangular traffic sign **a** shape detection result; **b** Hough transform parameter space and peak; **c** rectangle fitting result

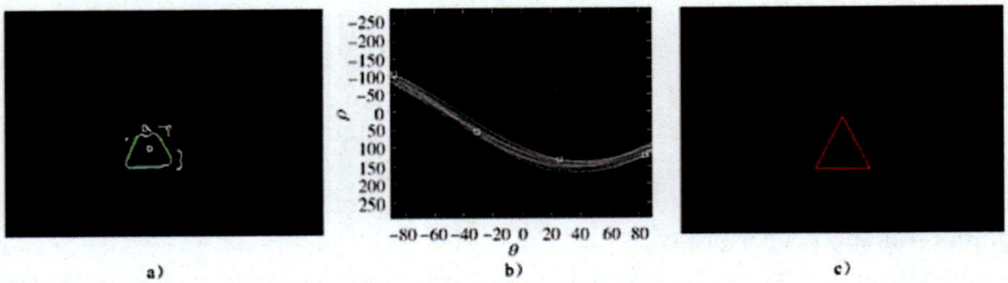

Fig. 4.52 Hough transform shape detection results of a triangular traffic sign **a** shape detection result; **b** Hough transform parameter space and peak; **c** triangle fitting result

features still need to be added. Then the robust detection of traffic signs is completed by multi-feature joint estimation.

The following characteristics are defined:

Roundness, expressed as: $C = 4S/L$;

Rectangularity, expressed as: $R = S/(H.W)$;

Dispersion, expressed as $e = L/S$.

where, L is the perimeter; S is the area; H is the height; W is the width.

The empirical statistics of roundness, rectangularity and dispersion are shown in Table 4.3.

Table 4.3 Empirical values of roundness, rectangularity and dispersion

Shape	Roundness	Rectangularity	Dispersion	Hough fitting feature	Estimation result
Rectangle	0.75	0.99	15.22	approximate rectangle	rectangle
Triangle	0.70	0.65	15.55	approximate triangle	triangle
Circle	1.00	0.71	10.23	circle or ellipse	circle

After the initial judgement of shape, the empirical values of the above features are used for joint estimation to determine the shape. The flow chart of robust detection algorithm after synthesizing the above steps is shown in Fig. 4.53.

In order to verify the above algorithm, the following experiment is carried out.

This book uses the self-developed autonomous vehicle camera to collect traffic sign images of a road section in Beijing, and utilizes the above algorithm for detection. T2T road traffic sign collection results are shown in Fig. 4.54.

The experimental results show that the color features are unstable or even degraded due to the change of outdoor illumination. The sign deformation due to the change of detection perspective, the image blur due to the high-speed movement and vibration of vehicle, the traffic sign being blocked by other vehicles, pedestrians and other

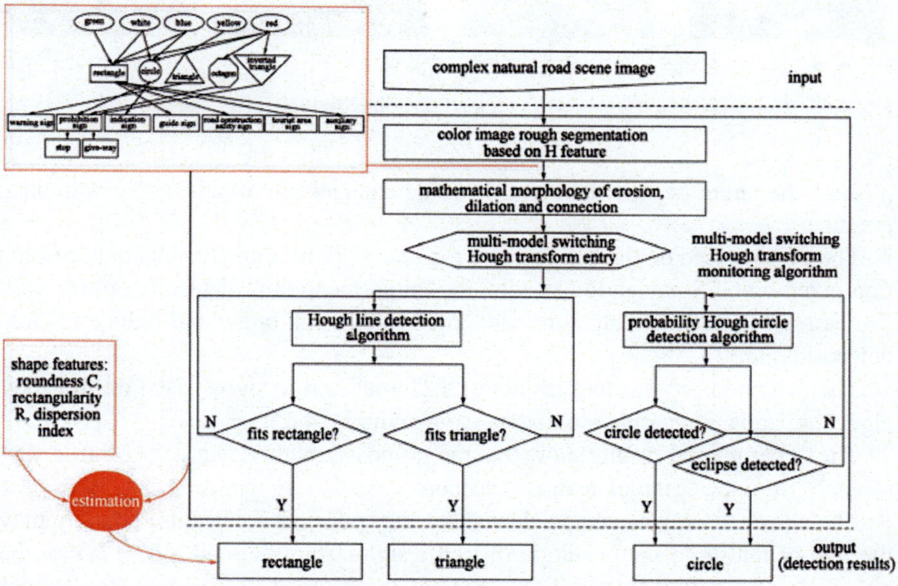

Fig. 4.53 Flow chart of robust detection algorithm for traffic signs

Fig. 4.54 T2T road traffic sign collection results

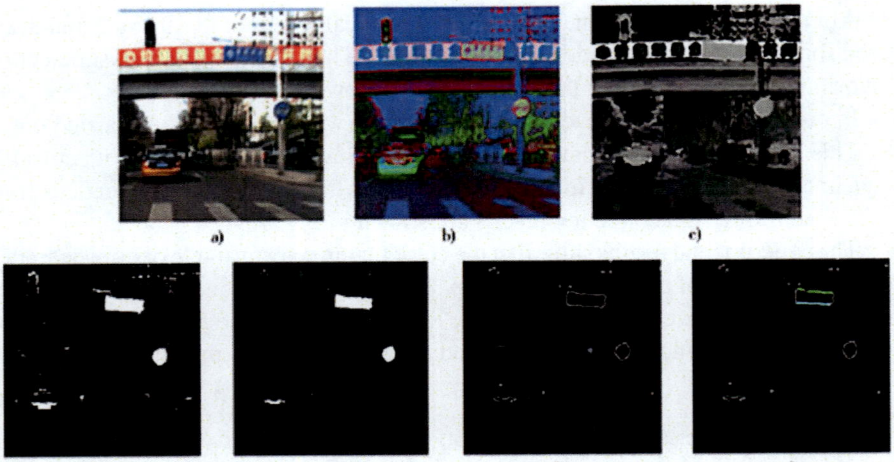

a) b) c)

d) e) f) g)

Fig. 4.55 Traffic sign detection results of a road section in Beijing

objects, the target being small, and the scale changes are all adverse conditions in practice.

Therefore, based on the definition of road network files and the location information given by GPS navigation system, the initial location of detectable traffic signs can be roughly estimated to solve the problem of weak target and reduce the false detection rate.

The detection process and results of T2T road traffic signs based on the above algorithm and engineering method are shown in Fig. 4.55.

The experimental results show that the proposed method can detect traffic signs robustly from the complex real road scene.

The above detection process shows that the traffic sign detection actually plays the part of real-time segmentation of traffic signs from the road video. It becomes practical and significant only if the segmentation is used for the next step, namely traffic sign recognition. In the past, some traffic sign recognition systems only focus on standard traffic sign images, without considering the uncertainty and interference of the actual scene, so the practical value is greatly reduced. The problem which needs to be considered is that: how to automatically select the valid image area. If the area is set too small, the detection algorithm is not practical. If the area is set too large, the scene complexity will increase sharply, and the detection complexity will also increase. When the valid image area is large and the traffic sign is relatively small, how to accurately detect the target? When there is interference with similar color and shape (such as blue rectangular license plate), how to distinguish it effectively?

4.5.3 Traffic Sign Optimized Matching Recognition

Due to the different conditions of collected traffic signs (scene, geometric position, lighting, weather, vehicle speed and camera resolution, etc.), some traffic signs can be separated, some can be partially separated, and some cannot be separated. It is generally believed that only neural network can be used to recognize traffic signs which cannot be separated. For traffic signs which can be separated independently, both neural network and feature matching can be used to recognize them. Real-time processing system usually chooses the feature matching method.

4.5.3.1 Analysis of Image Feature Optimized Matching

There are three major problems associated with image feature matching: feature detection, feature similarity measurement and search strategy. In the application of real-time matching, the matching speed can be improved from three aspects. One is to reduce the total number of matching features, that is, to reduce and optimize the feature space. The second one is to reduce the workload of similarity calculation. The third one is to reduce the number of matching search cycles. This book achieves the above purpose through the improvement of fast SURF feature extraction and hierarchical matching. At the same time, to ensure the matching accuracy and correct recognition of the target, we also need to pay attention to the matching algorithm which reduces the mismatching rate. Through theoretical analysis and experiments, two rules that affect the accuracy of matching can be discovered, namely: first, the larger the matching range and the more similar matching features, the higher the probability of mismatching; second, the more similar positions in the reference map, the higher the probability of mismatching. Therefore, we can improve the reliability of image matching through the following methods:

1. Reduce the size of reference images as much as possible, and reduce the search range through partition constraints and other methods.
2. Choose images with obvious features and rarely affected by the outside world as reference images. Avoid images with many similarities.
3. Choose a robust feature extraction method. Edge feature is easily affected by scale change, rotation, angle change and so on. Therefore, in terms of image matching, edge is not a stable feature. SURF feature is a type of local invariance feature with excellent comprehensive performance, which can effectively overcome the adverse effects of image scale change, rotation, angle change, obstruction and so on. It is very suitable for motion vision occasions with high robustness requirements.
4. Improve the quality of real-time image. Generally, measures can be taken within the hardware and the software. In the aspect of hardware, exposure control system or auxiliary light source can be added. In the aspect of software, the original image can be preprocessed through denoising, enhancing and filtering.

4.5.3.2 Recognition Algorithm Design

Based on the above analysis, in order to reduce the mismatching rate and improve the matching speed, this book takes corresponding measures in the fields of sign image database design, feature extraction and feature matching algorithm to achieve the comprehensive optimization of traffic sign image recognition system.

When the detection conditions are good (for instance in a sunny day), SURF features of two levels lower than the original resolution can be directly extracted, and subsequent algorithms can be performed at this level. In the case of poor detection conditions (hazy or smokey) or high credibility requirements, matching and recognizing is performed on the original image without lowering the resolution. Taking the practicability into consideration, this algorithm provides a framework with strong compatibility and flexibility, which is suitable for a variety of working conditions.

4.5.3.3 Multi-resolution Hierarchical Fast Matching

1. General idea

Although SURF operator is excellent at dealing with rotation, size change, angle change, illumination change, the experiments show that with complex image interference, unreliability still exists. Therefore, on the basis of experiment and theoretical analysis in Sect. 4.4, an optimization method based on multi-resolution image hierarchical matching is proposed. The principle is as follows: when the reliability requirement is moderate, the feature matching recognition is carried out at the lowest resolution level. If the recognition fails, the system switches to the higher resolution level and repeat the above matching recognition steps. In the case of high reliability requirement, multi-recognition method is adopted, that is, matching recognition is carried out in multi-level image space, and then the recognition results at all levels are compared and synthesized. If they are consistent, the recognition results are considered correct. If not, the system runs the recognition algorithm again. For example, in the vehicle autonomous navigation system, the image can be preprocessed into three or two levels. When the image is processed into three levels, the results of two levels are consistent, that is, then the agreed result can be used as the recognition result. When it is processed into two levels, the results will be output only when the results of the two levels are consistent.

Matching recognition at 2 resolution levels does not affect the recognition accuracy, but improves the real-time performance. Multi-level hierarchical matching strategy improves the reliability of matching. Therefore, this method can achieve comprehensive optimization in timeliness and reliability.

2. Fast improvement of SURF operator for real-time image application

In order to achieve fast matching, SURF algorithm has added a new variable in the feature vector—the sign of Laplacian response of feature points, or, the sign of trace of Hessian matrix. In feature matching, only the points with the same sign enter the similarity calculation step. For a system with high timeliness requirements, it is still not sufficient. After analysis, it is found that SURF algorithm consumes a lot of calculation time in the process of determining local extreme points by comparing each detection point with its upper, lower, left and right stereo neighborhood points (image domain and scale domain) one by one. It also needs Hessian matrix to accurately locate feature points. For the application on real-time dynamic image, this method is not practical. This book proposes a method to improve the real-time performance of SURF algorithm by reducing the search space of extreme points. The search space of extreme points is set at $Scale \in (1, 3)$ in order to reduce the complexity of SURF feature extraction and improve the real-time performance of the system.

3. Hierarchical feature extraction real-time image and reference image

The reliability of image matching is inversely proportional to the size of matching region. The larger the matching area is, the greater the probability of mismatching is. However, if the matching area is too small, stable matching point pairs may not be found. Therefore, on the premise of stable matching, we should reduce the size of reference image as much as possible. The specific size depends on the actual application requirements.

Previous research results show that better correlation results can be obtained when the template size is larger than 15 pixels. At the same time, it is proved that a small number of elements in the template can be used for calculation without affecting the results. This is because there is a high correlation between the gray levels of adjacent pixels, so the calculation can be simplified by template subsampling. Template subsampling algorithm takes a pixel value from every n adjacent pixels for calculation, which can reduce the computational complexity to $1/n^2$ of the original, and does not affect the accuracy of evaluation. In this book, each frame of real-time road scene image of traffic signs is modeled as a J-level image pyramid, with the highest resolution level of J and the lowest resolution level of 0. According to the characteristics of practical application, J is set as 2. The reference image and the real-time image of traffic signs are standardized into three resolution levels by wavelet packet analysis. Level 0 is 256 × 256; Level 1 is 128 × 128; and Level 2 is 64 × 64. Sym2 wavelet packet is used to decompose the sign image to get three resolution levels, as shown in Fig. 4.56.

The improved fast SURF algorithm is used to extract SURF features on three levels of approximate images. In order to verify whether the feature position has correspondence, the Level 1 and Level 2 images are matched with the Level 0 image respectively, and the results are shown in Fig. 4.57.

The results of detection and matching in Fig. 4.57 show that the positions of SURF feature points in the lower resolution image are almost the same as those in the highest resolution image. This result is quite useful. It guides us to recognize the traffic signs on the lower resolution image first, and enter the image space of higher resolution for recognition only if that fails. This provides a feasible way to improve the real-time performance of the system.

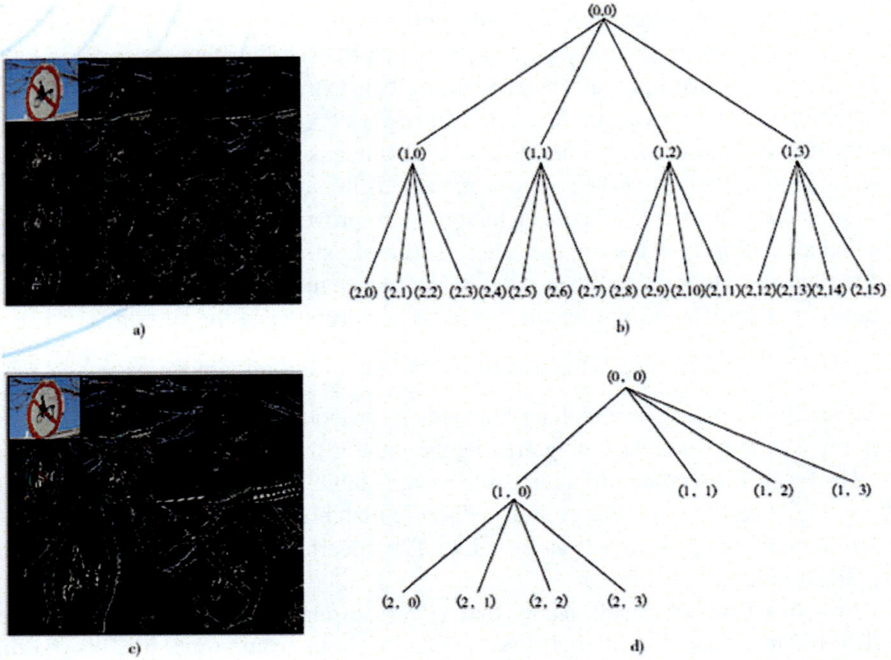

Fig. 4.56 Graphs of resolution levels **a** wavelet packet decomposition node graph **b** wavelet packet decomposition tree structure **c** wavelet packet tree node graph **d** wavelet packet tree structure

4.5.3.4 Sparse Robust Matching Recognition Algorithm with Multiple Constraints to Eliminate Mismatches

1. Analysis of traffic sign mismatch

In real-time traffic sign matching recognition, the causes of image matching error come from three aspects: the quality, size and angle of real-time traffic sign images; the quality and size of reference traffic sign images; and similar features among reference traffic sign images themselves. Compared with aerial image matching recognition, the difficulties of ground traffic scene image matching recognition are as follows. Firstly, because of the smoothness and geometric structure of traffic sign image, there are many similar large regions, which may cause mismatching. Secondly, due to the factors of ground obstacle obstruction, visual angle change, vibration blur and so on, the differences between the real-time images and the reference images of traffic signs are greater, so it is relatively difficult to find the accurate matching features.

In the actual experiment of traffic sign image matching, it is found that when there are local similar regions in the matching image pairs, there may be matching point pairs between signs of the same color and different colors, the same shape and different shapes, as shown in Fig. 4.58. But after careful analysis, we find that there are a lot of mismatches in these results. The mismatching rate is higher when there is interference of noise, vibration, distance change, angle change and so on. Lowe

(a)

(b)

Fig. 4.57 SURF feature extraction and matching results of images with different resolution levels **a** Level 1 (left) and Level 0 (right) sigh images **b** Level 2 (left) and Level 0 (right) sigh images

Fig. 4.58 Matching results of various traffic signs with mismatches

has pointed out that RANSAC algorithm can be used to eliminate mismatches, but the algorithm is not ideal when there are too many similar regions and outliers.

Vector similarity measurement plays an important role in the judgment of mismatching. The basis to verify whether it is a mismatch is to examine the degree of similarity between the matching vectors. There are two methods to measure the similarity between vectors, one is distance measure method, the other is similar function method. Euclidean distance and cosine similarity are the representatives of the two methods, and many other distance measurement methods are evolved from them.

For target recognition using feature matching, the best matching result should be sparse robust matching. Sparsity can reduce the risk of mismatching, but the premise of sparsity is robustness. Only robust matching can ensure the reliability and accuracy of recognition. In this book, we propose a new mismatch elimination method which combines the feature minimum variance distance threshold and matching intensity function to achieve sparse robust matching.

2. Mismatch elimination method based on minimum variance distance threshold

In the original SURF matching algorithm, nearest neighbor (NN) distance is used for matching, and Euclidean distance is used to measure the similarity between two feature vectors. The Euclidean distances between each SURF feature vector in the image to be recognized and all SURF feature vectors in the reference image are calculated iteratively. If the ratio of the minimum Euclidean distance to the second smallest Euclidean distance is less than 0.8, the source feature point and feature point with minimum Euclidean distance are matched. This ratio reflects the similarity difference between the candidate target feature point and a certain source feature point. The larger the ratio is, the smaller the similarity difference between the unmatched candidate target feature point and the source feature point is. In other words, the unmatched candidate target feature point is closer to the matching feature point. In this algorithm, the comparison between images is reduced to the comparison between feature vector sets. The direction information of feature points is not used, so a large number of mismatches are generated. For reliable image matching that requires the elimination of errors, the nearest neighbor distance method is not intuitive and cannot directly reflect the matching status of each feature point. Moreover, the matching algorithm only considers the similarity of feature points within each matching image pairs, and ignores the similarity of the feature points of each image itself.

This book uses the square difference (SD) algorithm to measure the minimum distance between SURF feature vectors.

3. Mismatch elimination method with matching strength constraint

According to the basic principle of image matching, the candidate matching point must satisfy the requirement of uniqueness and continuity. Uniqueness means that only one feature point in the target image can match the source feature point in the source image. Continuity means that there should be a large number of other correct matching points in the neighborhood of this certain matching point.

The matching strength is defined as $E(x_i, y_i)$ to measure the degree of continuity. When there are more candidate matching points in the neighborhood of (x_i, y_i), the value of $E(x_i, y_i)$ is larger, and vice versa.

The uniqueness strength is defined as $P(x_i, y_i)$ to measure the unique correctness of the matching point (x_i, y_i) in its neighborhood. $P(x_i, y_i) = 1 - E'(x_i, y_i)/E(x_i, y_i)$, where $E'(x_i, y_i)$ is the second largest matching strength in the neighborhood of (x_i, y_i), lower than $E(x_i, y_i)$.

Relaxation iteration algorithm is used to implement the principle of continuity and uniqueness. Two array $\vec{A}_E$ and $\vec{A}_P$ are constructed. $\vec{A}_E$ is used to store the q pairs of candidate matching points with the largest $E(x_i, y_i)$ value, while $\vec{A}_P$ is used to store the q pairs of candidate matching points with the largest $P(x_i, y_i)$ value. In each iteration, the first 60% matching point pairs are selected as the outputs, with due consideration of the principle continuity and uniqueness. Through this method, the uncertain pairs in the matching point set can be eliminated, so as to reduce the mismatching rate.

4.5.3.5 Experiment and Analysis

The experimental platform is a PC with Win7 operating system, 2 GHz CPU and 2 GB memory. The programming environment is Visual Studio 2010 with Open CV2.4.1. The algorithm testing and evaluation environment is MATLAB2011. The real-time data of traffic signs are collected from urban roads in Beijing. The traffic sign standard dataset (sign image database) comes from the national standard *Road Traffic Signs and Markings* (GB5768), including 116 signs.

The ideal condition of traffic sign recognition is that the shape, color and visual angle of the sign to be recognized are not changed and are very close to the reference image in the database. Under this circumstance, the data analysis in the process of traffic sign recognition can more clearly reflect the essence of the algorithm. Therefore, firstly, the process of sign recognition under almost ideal condition is studied, and SURF feature matching experiments are carried out by selecting typical positive and negative cases of traffic sign matching.

A typical case of correct matching of front-view sign is shown in Fig. 4.59.

From the above performance test results, it can be concluded that:

Fig. 4.59 Typical case of correct matching of front-view sign (64 × 64)

(a) (b) (c)

Fig. 4.60 Recognition results ($T_{err} = 0.2$) of the "no passing of tricycle" red sign **a** front view, enlarged; **b** side view, enlarged; **c** obstructed

1. The improved algorithm can adapt to all kinds of adverse detection conditions, and shows better comprehensive performance.
2. Compared with the original algorithm, the average mismatch rate of the improved algorithm is reduced by 15–20%.
3. Compared with the original algorithm, the average recognition rate of the improved algorithm is increased by 14–28%.
4. The time consumption of the improved algorithm remains almost unchanged.

The experimental results show that the method proposed in this book can effectively eliminate mismatched points, and can correctly retrieve the target signs in different image database settings, so it has strong universality. At the same time, the recognition accuracy is close to 100%.

After the experiment, we also find that the quality of the real-time images has a great impact on the matching results. So does the quality of the reference images in the database. Therefore, it is very important to prepare decent matching images.

The recognition results ($T_{err} = 0.2$) of the "no passing of tricycle" red sign are shown in Fig. 4.60.

The recognition results of the front-view "non-motor vehicle only" blue sign are shown in Fig. 4.61.

The recognition results of "stop and give way" red sign are shown in Fig. 4.62.

The recognition results of the tilted "watch for children" yellow sign are shown in Fig. 4.63.

The recognition results of the tilted "strictly no parking" blue sign are shown in Fig. 4.64.

The experimental results show that this method can accurately find the position of the sign in the natural complex scene, and the corresponding reference image from the database, to generate semantic interpretation of the sign in real time.

In order to further verify the effectiveness of the matching method, different signs are selected for cross matching experiments, and the results are all not matched. Cross testing of feature matching recognition ($T_{err} = 0.15$) is shown in Fig. 4.65.

In addition, taking multi-view real-time image as the reference template is also an effective method to improve the recognition reliability and reduce the matching error. It is namely the matching recognition method based on multi-view real-time image template.

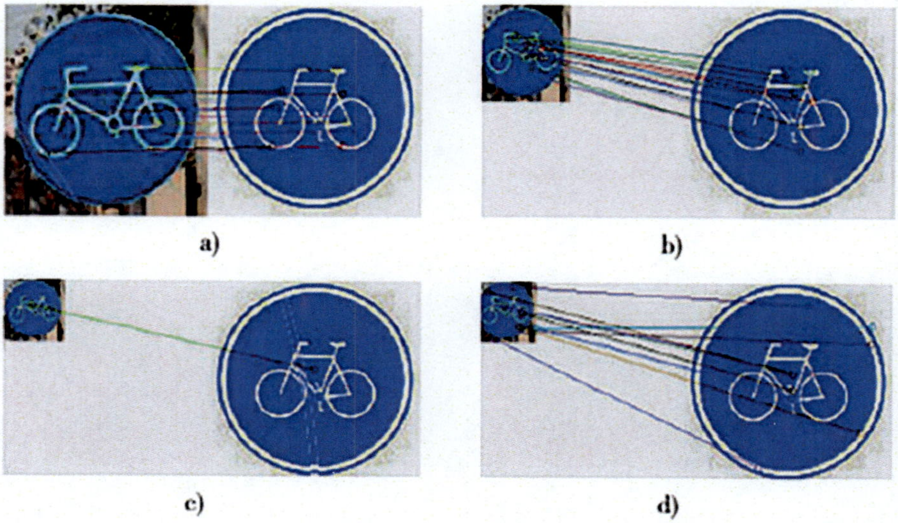

Fig. 4.61 Recognition results of the front-view "non-motor vehicle only" blue sign **a** Size 1, $T_{err} = 0.2$; **b** Size 2, $T_{err} = 0.2$; **c** Size 3, $T_{err} = 0.2$; **d** Size 4, $T_{err} = 0.4$

Fig. 4.62 Recognition results of "stop and give way" red sign **a** front view, $T_{err} = 0.2$; **b** side view, $T_{err} = 0.2$

Fig. 4.63 Recognition results of the tilted "watch for children" yellow sign **a** tilted, $T_{err} = 0.4$; **b** tilted and smaller, $T_{err} = 0.4$

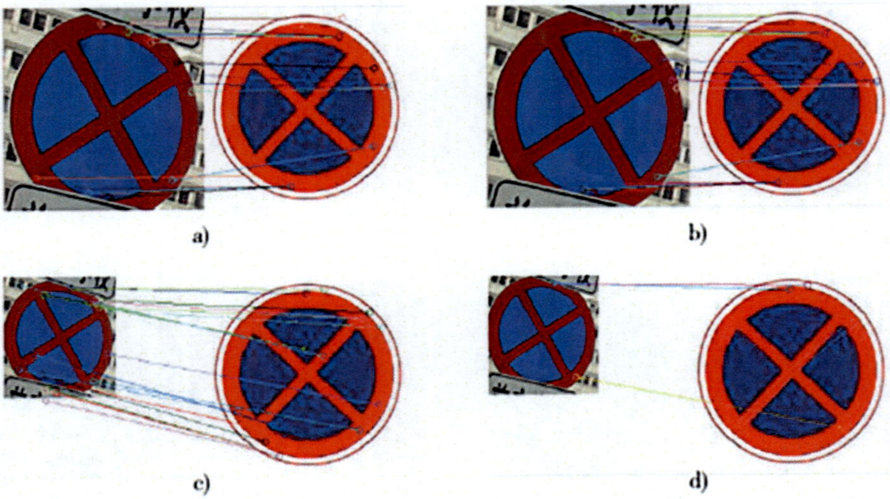

Fig. 4.64 Recognition results of the tilted "strictly no parking" blue sign **a** tilted and backlit, $T_{err} = 0.4$; **b** tilted and backlit, $T_{err} = 0.2$ **c** tilted, backlit and smaller, $T_{err} = 0.4$; **d** tilted, backlit and smaller, $T_{err} = 0.2$

Fig. 4.65 Cross testing of feature matching recognition ($T_{err} = 0.15$)

4.6 Vehicle Detection Based on Convolution Neural Network

Convolution is a special linear operation for two real valued functions. Convolution operation is usually represented by the symbol *. The formula of one-dimensional discrete convolution is as follows:

$$s(i) = (h \times w)(i) = \sum_{j=-\infty}^{\infty} h(j)w(i-j) \tag{4.42}$$

Convolutional neural network is a kind of neural network which uses at least one layer of convolution operation to replace the general matrix multiplication operation.

Where i is the calculated state (time, position), j is the distance to state i (it can be time difference, space distance, etc.). h and w represent two real valued functions respectively. In terms of convolutional neural network, the first function h is called input, the second function w is called kernel function, and the output s is called feature map.

In the application of deep learning, the input is usually a high-dimensional array (such as image), and the kernel function is also a high-dimensional parameter array generated by the algorithm (such as random gradient descent). If two-dimensional image I is input, then the corresponding two-dimensional kernel K is needed, then the two-dimensional convolution can be written as:

$$s(m, n) = (I \times K)(m, n) = \sum_i \sum_j I(i, j)K(m - i, n - j) \tag{4.43}$$

where, (m, n) is the calculated pixel position and (i, j) is the range of consideration.

The input edge in the fully connected layer is actually multiplied by weight and then accumulated, which is essentially a matrix multiplication. Convolution is actually used in the convolution layer to replace the original matrix multiplication in the fully connected layer. The point of convolution is to improve the machine learning system through the following three ideas: sparse interactions, parameter sharing and equivariant representations.

Assuming that the first hidden layer of the network has 40 thousand neurons (40 thousand hidden layer nodes are appropriate for the case of 40-thousand-dimension input samples), then the network has nearly 2 billion parameters. This kind of model training needs a lot of computation and storage space.

The convolution kernel is actually the parameters of convolution network. The convolution kernel slides the window on the input image, which means that the pixels of the input image share this set of parameters. Parameter sharing in convolutional networks means the learning only one parameter set, instead of learning a single parameter set for each pixel, which greatly reduces the storage space required by the model.

Because the whole input image shares a set of parameters, the model is equivariant for the translation of some features in the image. Figure 4.66 is a typical convolution neural network layer (convolution layer). The traditional convolution layer consists of the following three structures: convolution operation, activation function (nonlinear transformation) and pooling. The activation function here plays the same role as the fully connected network, and ReLU is the most commonly used activation function. In general, multiple convolution kernels are used. Different kernels learn different features. Some kernels may learn color features, and some kernels may learn edge and shape features. Figure 4.66 shows the kernel visualization effect of the convolutional neural network that has been trained in the same layer.

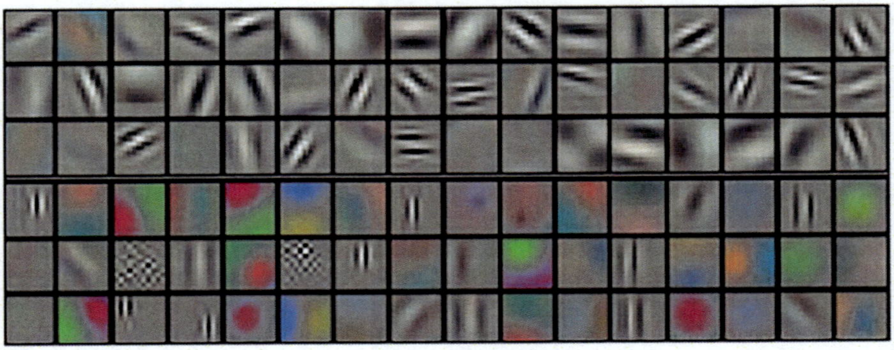

Fig. 4.66 Typical convolutional neural network layer

Since the emergence of deep learning, the research of target detection has made a great breakthrough. The two most noticeable directions are: region proposal based deep learning target detection algorithm represented by RCNN (RCNN, SPP-NET, Fast-RCNN, Faster-RCNN, etc.) and the deep learning object detection algorithm based on regression method represented by YOLO (YOLO, SSD, etc.). Yolo regards target detection as a regression problem. The workflow of a trained network is quite simple, as shown in Fig. 4.67.

In an end-to-end network, the input is the original image, and the output is the location of the target, its category and the corresponding confidence probability. Different from the traditional sliding window detection algorithm, in the training and application stage, YOLO uses the whole picture as the input. The network structure of YOLO is as shown in Fig. 4.68.

The whole network includes 24 convolution layers and 2 fully connected layers. The following is the whole process of YOLO:

1. Pre-training classification network

A classification network is trained on the ImageNet dataset by using the top 20 convolution layers, an average pooling layer and a fully connected layer in Fig. 4.68. The input of this network is 224 × 224. The top 5 accuracy of the model on ImageNet2012 dataset is 88%.

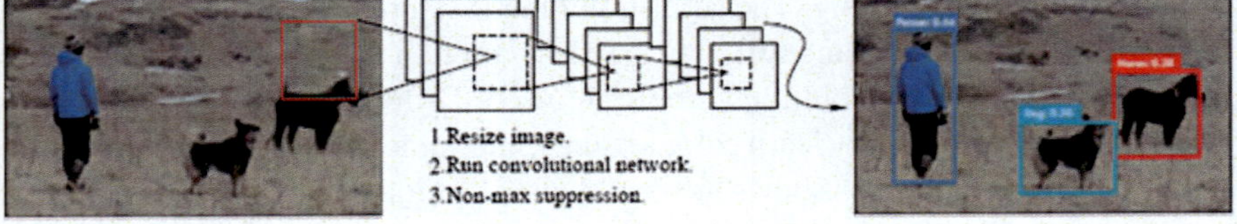

Fig. 4.67 Workflow of a trained network

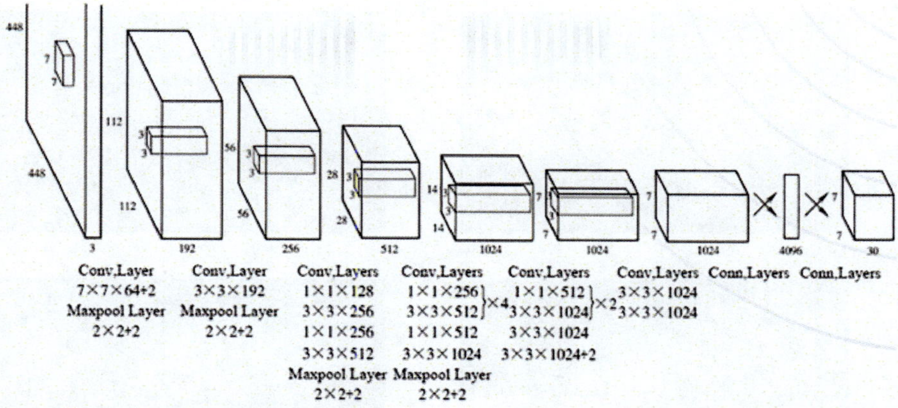

Fig. 4.68 Network structure of YOLO

2. Training detection network

The trained classification network is used for detection. After the 20 pre-trained convolution layers, four convolution layers and two fully connected layers are added (that is, the last four convolution layers and the last two fully connected layers in the structure diagram). The input of the network becomes 448 × 448, the output is a 7 × 7 × 32 tensor. The image input to the detection network is reset to 448 × 448, then divided into 7 × 7 grids. Each grid has a prediction of two bounding boxes, that is, 10 outputs. In addition, there are 20 outputs representing the target category. In order to regress these 30 values well, the design of loss function must achieve a good balance among bounding box coordinates, width, height, confidence and category.

The test results on the test images are shown in Fig. 4.69.

The test result on the test video is shown Fig. 4.70.

Fig. 4.69 Test results on the test images

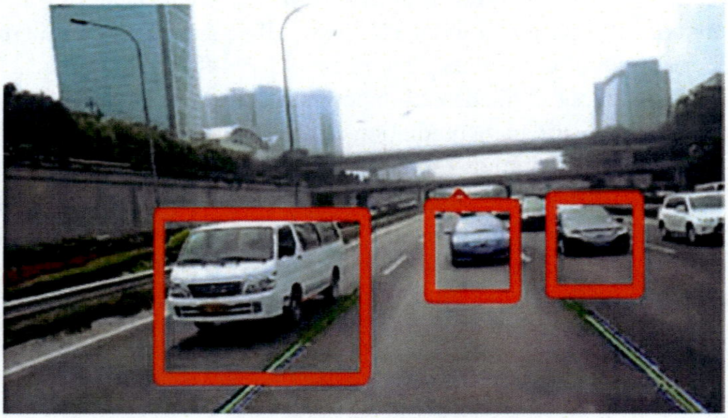

Fig. 4.70 Test result on the test video

Chapter 5
Navigation and Positioning Technology

5.1 High-Precision Electronic Map

5.1.1 Basic Concepts and Characteristics

The traditional electronic map is an abstraction of the road network. The road network is abstracted into the form of a digraph. The vertices in the graph represent intersections, and the edges represent the connections between intersections. Road names, landmarks and road skeleton information can be abstracted as attributes stored in the vertices or edges in the digraph. This abstract form of map representation can well meet the needs of human drivers. The reason is that human beings are born with strong visual recognition and logical analysis ability. In the process of driving, human drivers generally can effectively distinguish roads and road markings, determine their approximate position on the road, find and identify road signs, etc. Referring to the identified information combined with GPS (with a general accuracy of 5–10 m) information in the current electronic map, human drivers can roughly know their position in the actual road network, and plan how to drive next. Based on these abilities of human drivers, the traditional electronic map can be greatly simplified. For example, a curved road can be simplified to a line segment with only a few points. As long as the general outline conforms to the real road network structure, human drivers can locate their current position.

Autonomous driving map plays an irreplaceable role in intelligent network driving scenarios such as high-precision positioning, non-line-of-sight virtual sensing, lane level path planning and navigation, autonomous vehicle decision-making and control. After the advent of high-precision maps, autonomous vehicle can input the slope, curvature, climate, traffic signs and other information of the road into the vehicle controller to ensure safety.

At present, there are multiple ways to obtain remote data, such as V2X technology, where road condition information is sent to the vehicle by vehicles ahead or V2X equipment on both sides of the road. Cloud technology is another solution, which means the cloud sends out information collected by vehicles. But these methods have

M. Du, *Autonomous Vehicle Technology*, Unmanned System Technologies,
https://doi.org/10.1007/978-981-19-4143-6_5

the problems of relatively high construction cost and long construction cycle. With the popularization of navigation software, the electronic maps of Europe, America, Japan and China are basically mature. The accuracy requirement of ADAS for map is not very high, which is similar to that of ordinary navigation electronic map. It required additional ADAS features, such as curvature, slope, heading angle, more accurate number of lanes and so on. The creation cost is relatively low. In ADAS, map is integrated into the system as a special sensor (map sensor) to provide prior knowledge.

Different from the traditional electronic map, the major user of high-precision electronic map is autonomous vehicle, or robot driver. Unlike human drivers, robot drivers lack the innate ability of visual recognition and logical analysis. For example, humans can easily and accurately use images and GPS to locate themselves and identify objects, but this is a very difficult task for the robots. Therefore, high-precision electronic map is an essential part of the current autonomous vehicle technology. High-precision electronic map contains a lot of auxiliary information, among which the most important is the accurate three-dimensional representation of road network (centimeter-level accuracy). For example, the geometric structure of the road, the location of the road markings, the point cloud model of the surrounding road environment, etc. With these high-precision 3D representations, the vehicle robot can accurately confirm its current position by comparing the data of GPS, IMU, lidar or camera. In addition, high-precision map also contains rich semantic information, such as the location and type of traffic lights, the type of road marking and road surface. These can greatly improve the ability of vehicle robot to identify the surrounding environment. In addition, high-precision map can help autonomous vehicle identify vehicles, pedestrians and unknown obstacles. This is because high-precision map generally filters out moving obstacles such as vehicles and pedestrians. If an autonomous vehicle finds objects that are not in the current high-precision map during driving, they are likely to be vehicles, pedestrians or obstacles. Therefore, high-precision map can improve the speed and accuracy of autonomous vehicle to find and identify obstacles.

The characteristics of high-precision map are as follows. Compared with the traditional map serving GPS navigation system, the most significant feature of high-precision map is the accuracy of its representation of road features. Traditional maps only need to achieve meter-level accuracy to achieve GPS navigation, but the high-precision map needs to achieve centimeter-level accuracy to ensure the safety of autonomous vehicle. In addition, high-precision map requires more maintenance than traditional maps. As the road network goes through changes every day, including renovation, road marking line wear and repainting, traffic signs change, etc. These changes need to be reflected on the high-precision map in time to ensure the driving safety of autonomous vehicle. Real time high-precision map is not easy to build, but with more and more autonomous vehicles with multiple sensors driving in the road network, once one or several autonomous vehicles find the change of the road network, through the communication with the cloud, they can upload the updated information for other vehicles, making them more intelligent and safe.

Similar to traditional maps, a high-precision map also has hierarchical data structure. The bottom layer is a precise two-dimensional grid based on infrared radar sensors. Generally, the accuracy of this two-dimensional grid is about 5 × 5 cm. Available road surfaces, obstacles, and the reflective intensity of the road under the lidar are stored in the corresponding grid. In the process of driving, by comparing the data collected by the infrared radar and the high-precision two-dimensional grid in the memory, the specific location of the vehicle on the road can be determined. The characterization of effect of high-precision map two dimensional grid is shown in Fig. 5.1.

Because the grid is very fine, we can clearly identify the road and the location of the road markings from the corresponding radar reflection. The green area indicates the road surface that is not drivable. In addition to the underlying two-dimensional grid representation, high-precision map also contains a lot of semantic information about the road. As shown in Fig. 5.2, based on the two-dimensional grid reference system, high-precision map generally also contains the location and feature information of road markings, as well as the corresponding lane features. Because the

Fig. 5.1 Characterization of effect of high-precision map two dimensional grid

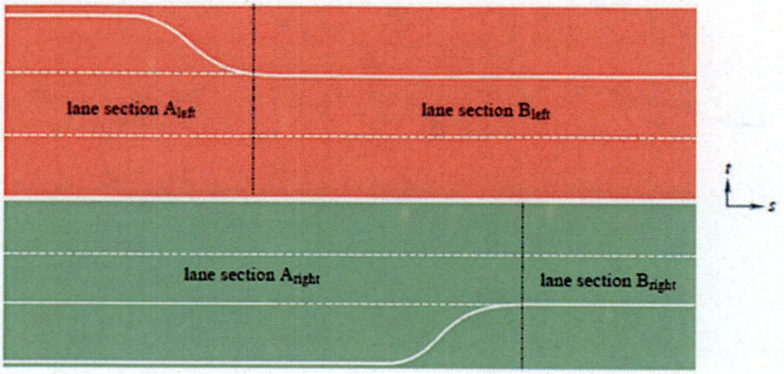

Fig. 5.2 Lane information in high-precision map

vehicle sensors may not be able to reliably analyze the lane information due to bad weather, obstacles and other vehicle's obstruction, the lane information features in high-precision map can help the autonomous vehicle to more accurately and reliably identify road markings, and understand whether it is safe to change lanes.

In addition, the high-precision map will also indicate the location of road signs, traffic signals, etc. relative to the two-dimensional grid. The road markings and traffic signs in a high-precision map are shown in Fig. 5.3.

This information has two functions. One is preparing the autonomous vehicle in advance, instructing it to detect corresponding traffic signs or signal lights at specific locations, improving the detection speed. The other one is that, when the autonomous vehicle fails to detect the traffic signs or signal lights, the driving safety can still be ensured.

The high-precision map used by the autonomous vehicle is a 2D grid, and the data is mainly generated by the lidar. Since the accuracy of lidar is about 5 cm, the highest accuracy of the map can reach 5×5 cm per grid. In such a high precision, how to effectively manage data is a big challenge. First of all, in order to store the map in memory as much as possible, we should try to remove unnecessary data. Generally, the lidar can cover an area of 100 m^2. Assuming that each reflection intensity can be

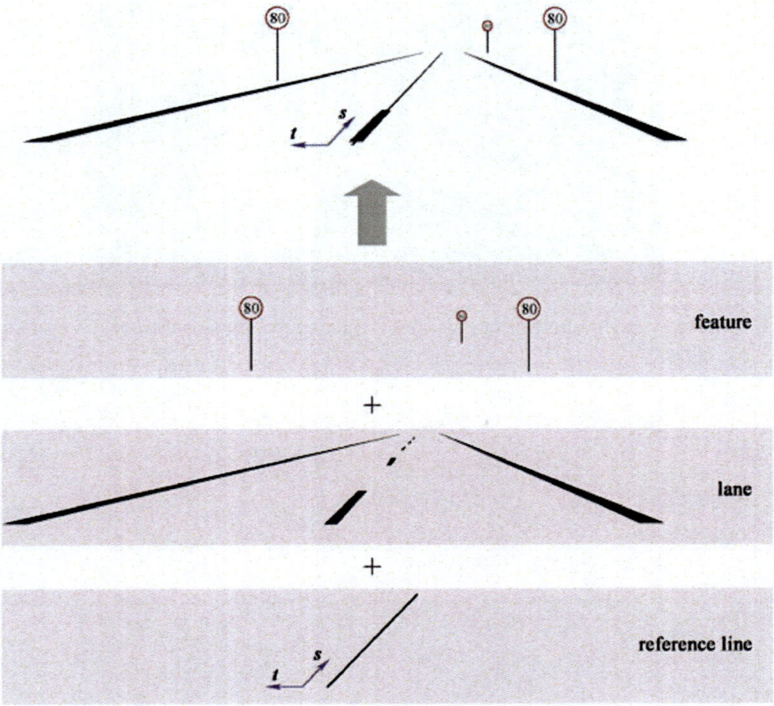

Fig. 5.3 Road markings and traffic signs in a high-precision map

recorded in a byte, each lidar scan can generate 4 MB data. The scanning will include trees and houses along the road, but the autonomous vehicle does not need these data. It only needs to record the data of the road surface. Assuming that the width of the road is 20 m, the data outside the road surface area can be filtered out through data processing, and the data volume of each scan can be reduced to 0.8 MB. On the basis of filtering, lossless compression algorithms such as LASzip can be used to compress map data, which achieves a compression rate of more than 10 times. After these processes, the high-precision map data of more than 100 thousand kilometers of road in China can be stored on a 1 TB hard disk.

5.1.2 *Creation of High-Precision Map*

The traditional electronic map mainly depends on satellite images, and GPS positioning. It can achieve meter-level accuracy. However, the high-precision map needs to achieve centimeter-level accuracy, so relying solely on satellites and GPS is not enough. Therefore, its creation involves a variety of sensors. Due to the huge amount of data generated, data collection vehicles (as shown in Fig. 5.4) are usually used to collect data, and then various data are synthesized to create high-precision maps through offline processing.

The creation of high-precision map is a fusion process of multiple sensors, including the following:

1. Inertial measurement unit (IMU). It generally uses 6-axis motion processing components, including a 3-axis acceleration sensor and a 3-axis gyroscope. Acceleration sensor is a force sensor, which is used to check how much force (including gravity) is exerted on the upper, lower, left, right, front and rear surfaces, and then calculate the acceleration on each axis. The gyroscope is an angular velocity detector, which detects the angular velocity on each axis. Suppose that the autonomous vehicle takes the Z-axis as the pivot and turns

Fig. 5.4 High precision map data collection vehicle here

to 90°, then its angular velocity on the Z-axis is 90°/s. It takes two integrals to calculate the moving distance from the acceleration. Therefore, if there is any error in acceleration measurement, the position error will accumulate and lead to the position prediction error. Therefore, gyroscope alone cannot accurately predict the position of the vehicle.

2. Wheel odometer. We can use the wheel odometer to calculate the position of the autonomous vehicle. The front wheels of a car are usually equipped with wheel rangefinders, which records the total revolutions of the left and right wheels respectively. By analysing the revolutions of the left and right wheels in each time period, we can calculate how far the vehicle has gone and how many degrees it has turned left and right. However, due to the difference of rotation to distance conversion on different ground surfaces (such as ice and cement), the measurement deviation will become larger and larger with time. Therefore, wheel rangefinder alone cannot accurately predict the position of the vehicle.
3. GPS. Its task is to locate four or more satellites, calculate the distance between the itself and each satellite, and then use this information to calculate the its position by trilateral measurement in three-dimensional space. In order to use the distance information for positioning, the GPS receiver must also know the exact position of the satellite. GPS receiver stores an ephemeris, which informs the receiver of the position of each satellite at any time. In the complex dynamic environment, especially in big cities, due to the obstruction of buildings, the problem of GPS multi-path reflection will be more obvious. GPS positioning information obtained in this way may easily have errors of tens of centimetre's or even several meters, so GPS alone cannot create high-precision maps.
4. Lidar. Optical radar first emits a laser beam to the target object, and then determines the actual distance of the target object according to the time of receiving. Then, according to the distance and the angle of laser emission, the position information of the object can be deduced by simple geometric transformation. Lidar system are generally composed of three parts. The first one is the laser transmitter, which emits laser beams with wavelength from 600 to 1000 nm. The second one is the scanning and optical component, which is mainly used to collect the distance, occurrence time and azimuth of the point. The third one is the photosensitive component, which mainly detects the intensity of the return light. Therefore, every point we detect comes with the spatial coordinate information and light intensity information. The light intensity is directly related to the reflectivity of the object, so the object can also be preliminarily identified from the detected light intensity.

Figure 5.5 shows the general high-precision map creation process. First of all, IMU and wheel odometer can predict the current position of the autonomous vehicle with high frequency, but due to its accuracy, the position may have a certain degree of deviation. In order to correct these deviations, we can use sensor fusion meter technology (such as Kalman Filter) combined with GPS and lidar data to calculate the accurate current position of the autonomous vehicle. Then, according to the accurate current position and lidar scanning data, the new data is added to the map.

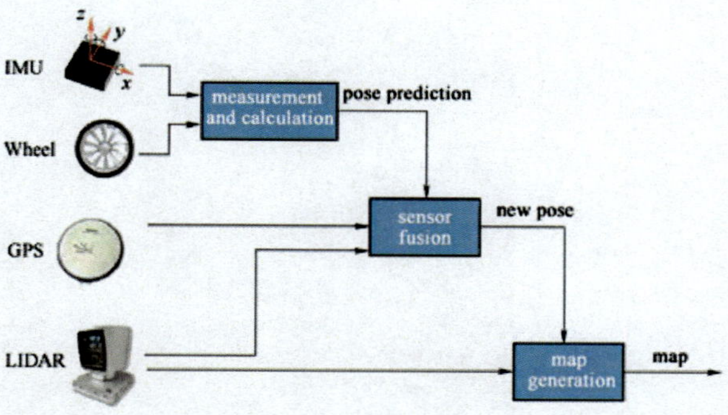

Fig. 5.5 High-precision map creation process

Formula 5.1 is a highly simplified calculation model of high-precision map, where Q represents the optimization equation, Z represents the points scanned by lidar, h is the position and reflectance of the latest scanning point predicted by equation, m is the position of the scanned points in the map, and x represents the current position of the autonomous vehicle. The purpose of this equation is to find out the exact position of the measuring point in the map by minimizing J. In the calculation model, both m and x are unknown at the beginning. We can get x through multi-sensor fusion first, and then get the accurate position m of the measurement point in the map.

$$J = Q[z - h(m, x)] \tag{5.1}$$

AutoNavi plans to complete the creation of national high-speed automatic driving (HAD) level high-precision map of more than 280,000 km, and the data collection of ADAS level high-precision map of national and provincial highway by the end of 2016. At the end of 2017, ADAS level data expand to more than 30 citie's main roads, and HAD level data expand to national and provincial roads and major cities. The accuracy of ADAS high-precision map is about 50 cm. The accuracy of HAD and above level high-precision map is about 10 cm.

In order to obtain these data, AutoNavi currently has two kinds of collection vehicles, which are respectively used to collect high-precision map data required by ADAS and HAD. As shown in Fig. 5.6, Nissan on the left is the HAD level collection vehicle, and Suzuki on the right is the ADAS level collection vehicle.

The vehicle used to collect high-precision map data of ADAS level is shown in Fig. 5.7. Six CCD cameras are installed on the roof of the vehicle. Five of the cameras are arranged in a circle, with a separate camera at the top. Each camera has 5 million pixels, totaling 30 million pixels. There is a display screen for collecting data in the front passenger seat, and the chassis for storing and processing data is in the trunk.

The collection vehicle for HAD level high-precision map data is shown in Fig. 5.8. The top is equipped with two lidars (located in the rear) and four cameras (two in the front and two in the rear) to achieve the required 10 cm level accuracy. The combination of the two solutions can complete the 3D model construction of signs, obstacles, lane lines and other road information. The cost of this car is more

Fig. 5.6 Two kinds of AutoNavi collection vehicles

Fig. 5.7 AutoNavi collection vehicle for high-precision map data of ADAS level

Fig. 5.8 AutoNavi collection vehicle for high-precision map data of HAD level

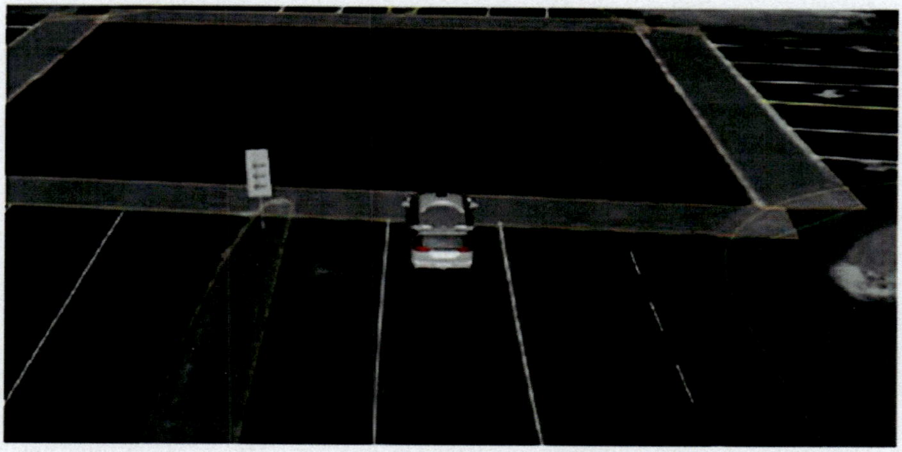

Fig. 5.9 3D map created by Waymo autonomous vehicle

than 8 million yuan, and the RIEGL 3D laser scanning system on it is also the most advanced product at present.

The data collected by these vehicles are only used as the basis. Road information is constantly updated, and with the advancement of automatic driving technology, there will be higher requirements for timeliness. AutoNavi's high-precision map data may be preferentially used in commercial vehicle scenarios with fixed routes such as logistics and sanitation.

The 3D map created by Waymo autonomous vehicle is shown in Fig. 5.9. According to Waymo's report, an autonomous vehicle on the road, like a human driver, usually needs to answer four questions. Where am I? What is around me? What is going to happen next? What should I do? In order to solve the problem of "where am I", Waymo autonomous vehicle does not rely on GPS, but on the detailed 3D map established by its team. The map highlights the road conditions, curbs, sidewalks, lane markings, crosswalks, traffic lights, stop signs and other road feature information, which is often called a high-precision map. By comparing the real-time sensor data with the pre-established high-precision map, the vehicle can determine where it is.

5.1.3 Application in Automatic Driving

High-precision electronic map contains a lot of traffic information, including the geometric structure of the road, the location of road markings, the point cloud model of the surrounding environment and so on. With these high-precision 3D representations, the driverless system can accurately confirm its current position and conduct real-time navigation by comparing the data of vehicle GPS, IMU, lidar or camera.

1. Positioning

Autonomous vehicles have high requirements for reliability and safety, so it is assumed that a map has already been created by high-precision lidar and a multi-sensor fusion. On this premise, this book discusses the tracking and positioning technology of autonomous vehicle.

Autonomous vehicle positioning is mainly carried out by particle filter. The so-called particle filter refers to the process of finding a group of random samples spreading in the state space to approximate the probability density function, using the sample mean instead of the integral operation, and then obtaining the minimum variance estimation of the system state. These samples are called particles, hence the name particle filter. The most common situation (such as in Sebastian Thrun's classic autonomous vehicle paper) is that the particle filter maintains an attitude vector (x, y, yaw), the roll/pitch is assumed relatively accurate, and the motion prediction process obtains acceleration and angular velocity from IMU. Particle filter needs to consider sample impoverishment and other possible catastrophic errors. A small number of particles can be obtained continuously from the current GPS position estimation. As mentioned above, the adaptive control of the number of samples also needs to be effectively adjusted according to the actual situation.

Because there is already a high-precision Lidar point cloud map, it is natural to match the real-time point cloud data with the built map. 3D point cloud matching cannot be discussed without iterative closure point (ICP). The goal of ICP is to calculate the pose between two groups of point clouds, assuming the scene remains unchanged. The earliest ICP principle is that each point of the first group of point clouds finds a nearest match in the second group. Then the mean square error (MSE) is calculated through all the matches, and the estimated pose is adjusted. In this way, multiple iterations are carried out, and finally the relative pose of the two groups of point clouds is calculated. Therefore, with an existing map, the current pose of the autonomous vehicle can be accurately calculated by using the real-time point cloud and a guess about the pose, and a relative pose can also be calculated from two point clouds adjacent in time.

On the other hand, because the autonomous vehicle is a complex multi-system integration, the current standard configuration is lidar, camera and IMU. With the camera and IMU as visual odometry (VO) and visual inertial odometry (VIO), SLAM is the natural choice.

2. Navigation and Control

Tracking and positioning technology is still a passive perception solution. The autonomous vehicle is about fully autonomous driving rather than assisted driving, which requires the vehicle to perform its own intelligent path planning.

Path planning is a topic with a wide range, and several restrictions need to be made first. Firstly, the map should be known. Otherwise, there is no planning to be discussed. If the world is completely unknown to the robot or autonomous vehicle, then the problem is actually SLAM plus exploration. Secondly, for the field of autonomous vehicle, the map is generally 2D or 2.5D. Thirdly, the position of the autonomous vehicle after each step of path planning is assumed accurate. The positioning and path planning are deliberately separated here, but the two are closely related in application. If the positioning is not accurate, the path planning will be affected.

There are two representative and widely used path planning methods. One is the A* algorithm to search for the best path. The core idea is that if there is the best path, it will be found. If the cost per unit path is different, the best path is not necessarily the shortest. A* is to search all the possibilities, choose the best, and use the heuristic algorithm to make a decision.

The other is sampling based path planning. It expands the possible path set by random sampling from the starting point. There is a very important factor that can speed up sampling, which is obstacle detection. If obstacles are encountered, it is meaningless to expand the path in this direction. However, it should be noted that this algorithm focuses on the efficient growth to a large area that has not been searched. In application (especially for autonomous vehicle), with the heuristic algorithm, the real-time path planning focuses on efficiency and needs to be optimized according to the actual situation. Researches in this area include RRT variants and the combination of the two algorithms (such as A*-RRT).

In the implementation of autonomous vehicle engineering, the path planning must perform algorithm selection and make adjustments based on the sensor situation and map quality. No matter how accurate the map is and how good the data quality of the sensor is, the optimization always needs to be carried out under the premise of safety. The information quantity and quality of the high-precision electronic map directly determine the safety, reliability and efficiency of the driverless system. Compared with the traditional electronic map, high-precision map is more accurate (centimeter level), updated faster, and contains more information (semantic information). Due to these characteristics, it is not easy to create high-precision maps, and multiple sensors need to be used to correct each other. After the completion of the initial map, filtering is needed to reduce the amount of data and achieve better real-time performance. After obtaining the high-precision map information, the driverless system can confirm the accurate current position by comparing the vehicle GPS, IMU, lidar or camera data, and conduct real-time navigation. The actual application result of the high-precision map is shown in Fig. 5.10.

The common navigation system provides basic road navigation functions based on electronic map, including path planning from A to B, vehicle and road positioning matching, POI retrieval for Destination query, map display and road guidance, etc. The imaging result of high-precision map is shown in Fig. 5.11.

Human drivers have the ability of extracting, associating and filtering information, as well as making judgment based on vision. With the relatively rough information provided by the navigation system, daily driving and navigation can be completed. Autonomous vehicles are unable to complete such tasks as association, understanding

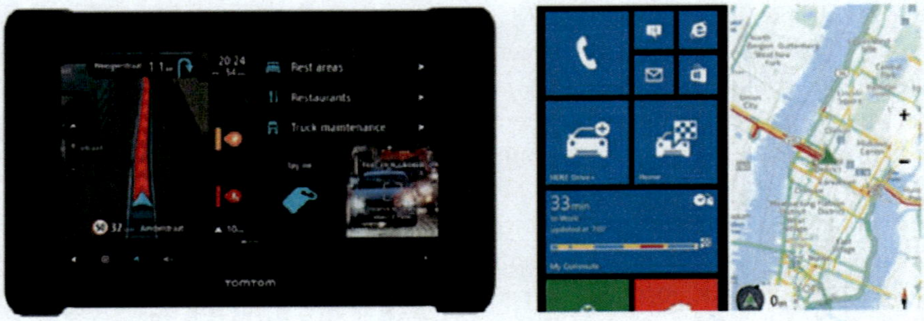

Fig. 5.10 Actual application result of the high-precision map

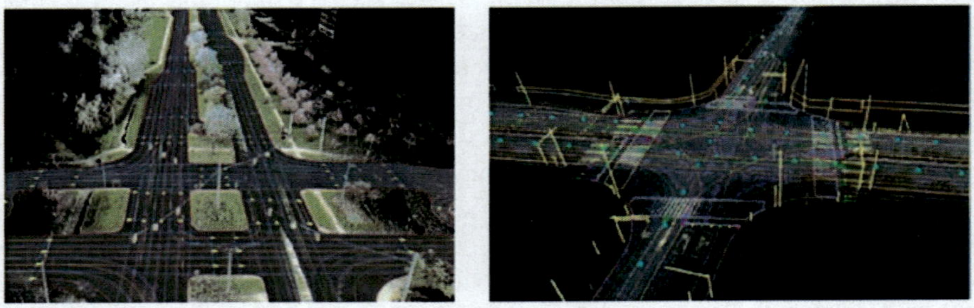

Fig. 5.11 Imaging result of high-precision map

and information integration. The application of artificial intelligence in the future is expected to solve these problems. High-precision map is expected to provide assistance for high precision positioning, road level and lane level planning, and lane level guidance.

In addition, it should be noted that some high-precision maps contain too much information and are not processed through decent information mining analysis. If they are used directly by drivers, they will bring interference. Therefore, a type of map between ordinary navigation electronic map and high-precision map is still widely used in ADAS for active safety scenarios. The accuracy of this type of map is generally 1–5 m. It is expanded on the basis of ordinary navigation electronic map, with additional auxiliary information such as slope, curvature and heading of the road. It also covers the information of the number and width of lanes, and the shape information of roads is more accurate. Yet the accuracy of the above information is significantly inferior to that of the high-precision map. This type of map can be used to support automatic driving when the sensors are sufficient. Most of its application is for active safety scenarios.

5.2 GPS and Beidou Navigation Positioning

GPS positioning information tends to have errors of several meters, which may cause traffic accidents. Due to the low update frequency (10 Hz) of GPS, it can hardly provide accurate real-time positioning when the vehicle is moving fast. Therefore, it is necessary to use other sensors to assist positioning and enhance the accuracy.

The Global Positioning System (GPS), is a medium-range circular orbit satellite navigation system. It provides accurate positioning, velocimetry and high-precision time standards for the vast majority (98%) of the Earth's surface. Developed and maintained by the U.S. Department of Defense, the system meets the needs of military users located anywhere on the globe or in near-Earth space for continuous and accurate determination of 3D position, 3D motion and time. The system consists of 24 GPS satellites in space, a master control station on the ground, three data injection stations, five monitoring stations and GPS receivers on the user side. GPS was developed by the U.S. government in the 1970s and was fully completed in 1994. The basic principle of GPS positioning is shown in Fig. 5.12.

Position differencing is one of the simplest differencing methods, and any GPS receiver can be modified and turned into a part of this differencing system. A GPS receiver installed on a reference station can observe four satellites to perform 3D positioning and solve for the coordinates of the reference station. Due to orbital errors, clock errors, Selective Availability (SA) effects, atmospheric effects, multipath effects and other errors, the calculated coordinates are not the same as the known coordinates of the reference station. The base station sends the correction number using the data chain, which is received by the user station to correct its solved user station coordinates. The final corrected user coordinates have eliminated the common errors of the reference station and the user station, such as satellite orbit errors, SA

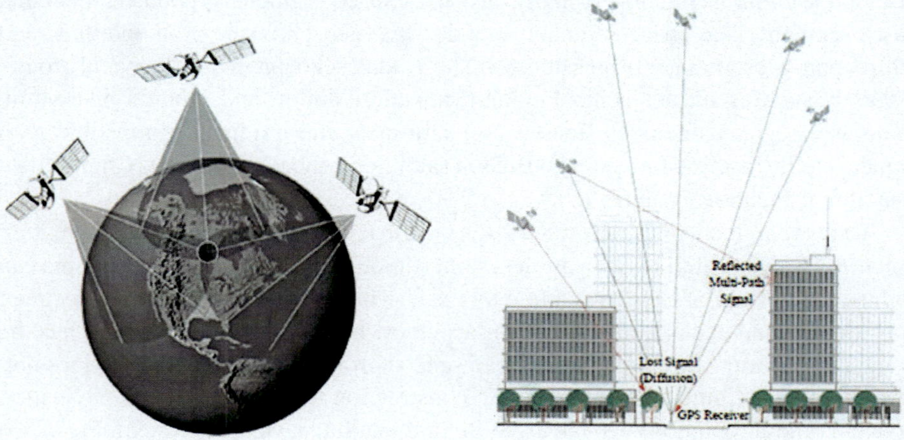

(a) (b)

Fig. 5.12 Basic principle of GPS positioning

effects, atmospheric effects, etc., and improved the positioning accuracy. The precondition for the above situation is that the reference station and the user station observe the same set of satellites. The position differencing method is applicable to the case where the distance between the user and the reference station is within 100 km.

Using the differential technique, errors can be eliminated or reduced. The reference receiver can be placed at a known accurate location, and the reference station can accurately calculate the error of the first receiver. The GPS receiver installed on the reference station can observe four satellites and then perform 3D positioning, solve the measurement coordinates of the reference station, and then calculate the error by comparing the measurement coordinates with the known coordinates. The reference station then sends the error value to the differential GPS receivers within 100 km^2 to correct the measurement data.

The current high-precision military differential GPS can indeed achieve centimeter-level accuracy in static and ideal environments. The ideal environment means that there is no excessive suspended medium in the atmosphere and the GPS has a strong reception signal during the measurement.

Inertial measurement unit (IMU) is a high-frequency (1 kHz) sensor that detects acceleration and rotational motion. The sensor data can be processed to derive vehicle displacement and rotation information in real time. Yet IMU has its own bias and noise problems that affect the results. By using sensor fusion technology based on Kalman filter, GPS and IMU data can be fused to achieve better positioning results.

Beidou satellite navigation system is a self-developed global satellite navigation system being implemented in China. The goal of system construction is to build an independent, open, compatible, technologically advanced, stable and reliable satellite navigation system covering the world, promote the formation of satellite navigation industry chain, form a thorough national satellite navigation application industry support, promotion and guarantee system, and promote the wide application of satellite navigation in various industries of national economy and society. Beidou satellite navigation system consists of space segment, ground segment and user segment. The space segment includes five geostationary orbit satellites and thirty non-geostationary orbit satellites. The ground segment includes several ground stations such as master control station, injection station and monitoring station. The user segment includes Beidou user terminals and terminals compatible with other satellite navigation systems. Beidou satellite navigation system is China's own satellite navigation system.

The development of Beidou navigation system is conducive to China's information security. Beidou satellite navigation system is more accurate and reliable. China can adjust the posture of Beidou satellite by itself, which is conducive to more accurate positioning and can achieve sub meter accuracy. Beidou navigation system has the function of short message, that is, it can send short messages, which GPS does not. Around 2020, China will complete the construction of Beidou satellite navigation system. The autonomous vehicle using Beidou satellite navigation system (Fig. 5.13) has been successfully developed, and Beidou navigation system will be part of the autonomous vehicle industrial chain of China in the future.

Fig. 5.13 Autonomous vehicle with Beidou satellite navigation system

5.3 Vehicle Inertial Navigation System

Inertial measurement unit (IMU) is a kind of sensor which can detect acceleration and rotation (Fig.5.14). A basic IMU sensor include an accelerometer and an angular speedometer. The six-axis IMU based on MEMS is mainly composed of a three-axis acceleration sensor and a three-axis gyroscope. Autonomous vehicles generally use

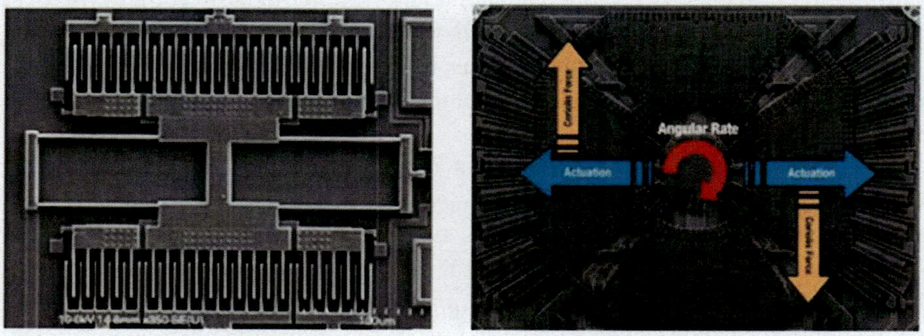

(a) (b)

Fig. 5.14 Accelerometer and angular velocity gyroscope based on MEMS medium and low-level IMU sensors with high update frequency (1 kHz), which can provide real-time position information. However, the fatal disadvantage of IMU sensor is that its error will increase with time.

5.4 Intelligent Integrated Navigation

From the mechanism and error analysis of inertial navigation system, we can see that inertial navigation system has strong autonomy. It can continuously provide all navigation parameters including attitude reference, and has very good short-term accuracy and stability. Its disadvantage is that the navigation positioning error will increase with time, and the speed of navigation error accumulation is mainly determined by the accuracy of initial alignment, the error of inertial sensors used in navigation system and the dynamic characteristics of the vehicle trajectory. There are two ways to solve this problem. One is to improve the accuracy of the inertial navigation system. It mainly depends on the use of new materials, new processes and new technologies to improve the accuracy of inertial devices, or the development of new high-precision inertial devices. Another way is to use integrated navigation technology. It means improving the performance of the inertial system with additional outside information sources, and improving the navigation accuracy through software technology.

During the development process of navigation technology, the integrated navigation system based on inertial navigation was studied first. It has two working modes. The first one is the re-adjustment mode. In the process of inertial navigation, the position measurement information obtained by other devices is used to correct the inertial navigation position. This is a way of using navigation information outside the loop for correction, so there is no change in the response characteristics of the loop. The other one is the damping mode, which adopts the combination of inertial navigation and Doppler radar (or celestial navigation). It uses the velocity (or position) information provided by inertial navigation and Doppler radar to form the velocity (or position) difference, and then uses this velocity difference to correct the inertial navigation system through feedback, so as to reduce the navigation error. This is a damping integrated navigation system, but the damping effect is not ideal in the case of maneuver.

GPS/inertial can constitute an ideal navigation system, which is the main form of current integrated navigation technology. The IMU-GPS integrated navigation system is shown in Fig. 5.15.

The filter designed according to optimal control theory and Kalman filter theory is an important part of integrated navigation system. Its principle is to provide the navigation information collected by all kinds of sensors to the filter, apply Kalman filtering method to process the information, and get the optimal error estimation of the inertial navigation system. Then the controller corrects the inertial navigation system

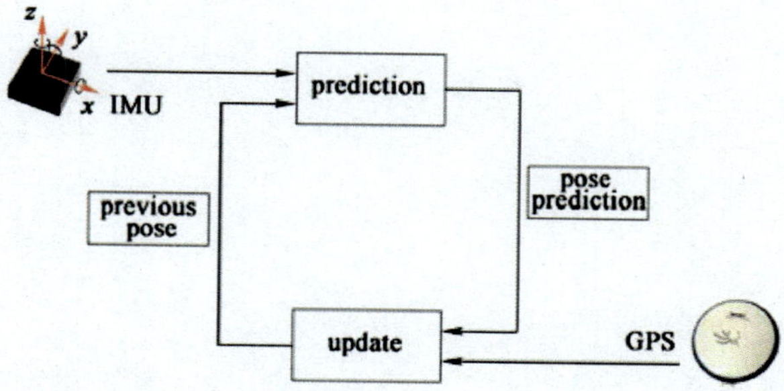

Fig. 5.15 IMU-GPS integrated navigation system

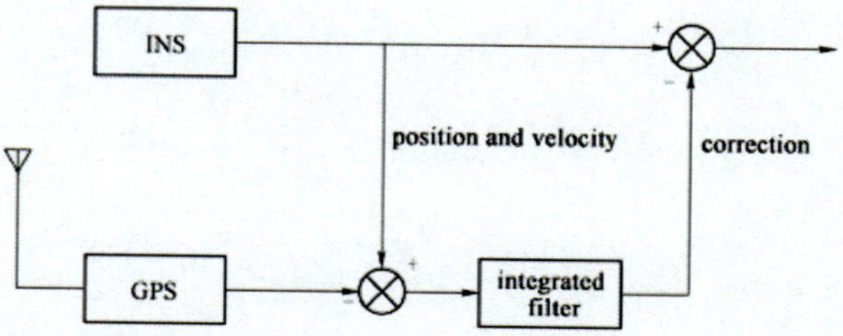

Fig. 5.16 Principle of open-loop correction

to minimize the system error. In order to be distinguished from the general integrated navigation system with re-adjustment and damping, the integrated navigation system with Kalman filter is also called optimal integrated navigation system. In terms of methods of system correction, there are open-loop correction (output correction) Kalman filter and closed-loop correction (feedback correction) Kalman filter. The principle of open-loop correction is shown in Fig. 5.16.

Chapter 6
Motion Control and Planning Decision-Making Technology

6.1 Motion Control of Autonomous Vehicle

6.1.1 PID Control

PID control is one of the earliest control strategies, and its algorithm principle is shown in Fig. 6.1. Because of its simple algorithm, strong robustness and high reliability, it is widely used in various control systems, especially in deterministic control systems with an accurate mathematical model. The commonly used control laws in industry include Proportional (P), Proportional plus Integral (PI), Proportional plus Integral plus Differential (PID). The new PID control methods derived from the classical PID control algorithm include adaptive PID, fuzzy PID, neural network PID, expert PID, self-tuning PID, auto disturbance rejection PID.

PID control is a kind of linear control, which makes up the control deviation $e(t) = r(t) - y(t)$ according to the fixed value $r(t)$ and the actual output value $y(t)$.

1. Proportional Control Law P

The controller with proportional control law is called P controller. The output $p(t)$ of P controller is proportional to the deviation signal $e(t)$ (as shown in Fig. 6.2).

$$p(t)=K_c e(t)$$

The transfer function is as follows

$$G_c(s) = \frac{P(s)}{E(s)} = K_c \tag{6.1}$$

where K_c is the proportional gain or magnification of the controller.

M. Du, *Autonomous Vehicle Technology*, Unmanned System Technologies,
https://doi.org/10.1007/978-981-19-4143-6_6

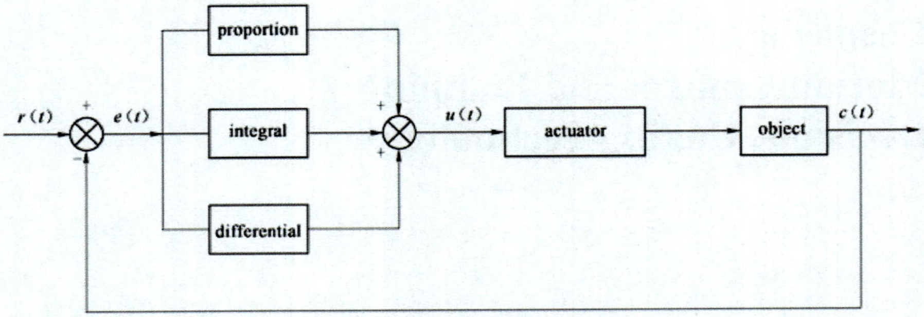

Fig. 6.1 Flow of PID control algorithm

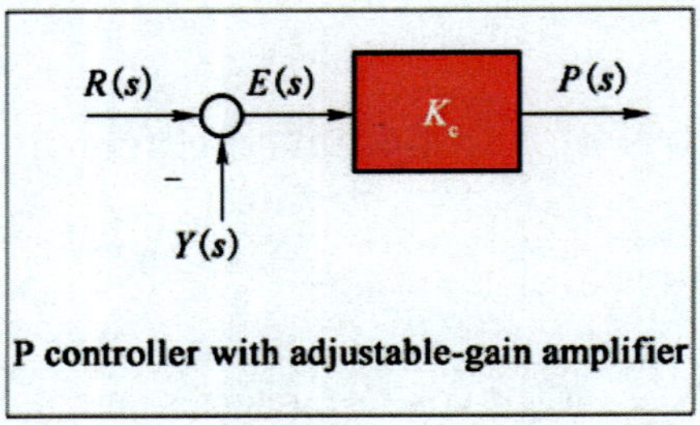

Fig. 6.2 Proportional control law P

The basic principle of P control is shown in Fig. 6.3.

The characteristics of proportional action are:

(a) An amplifier with adjustable gain.

(b) The output waveform is the same as the input waveform, and the amplification is K_c times.

(c) If the generalized object is a type-0 system, the system has steady-state error.

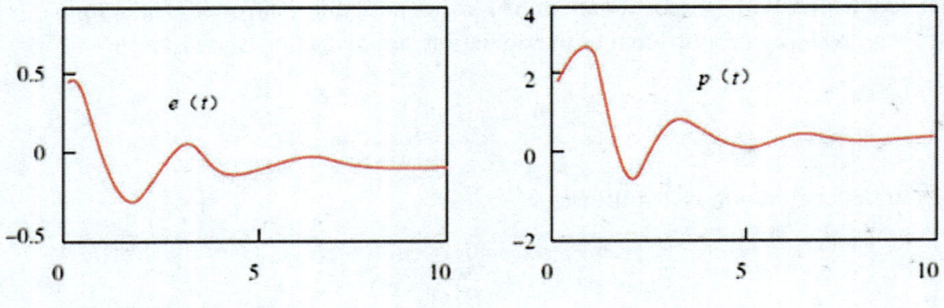

Fig. 6.3 Basic principle of P control

Conclusions:

(a) Proportional control is the most basic control law.
(b) The static error (residual) cannot be eliminated.
(c) The residual is inversely proportional to K_c.
(d) Increasing K_c will reduce the stability of the system, and even cause the instability of the closed-loop system.

2. Proportional Plus Integral Control Law PI

The controller with PI control law is called PI controller.

The relationship between the controller output $p(t)$ and the deviation signal $e(t)$ is as follows

$$p(t) = K_c[e(t) + \frac{1}{T_i}\int_0^t e(t)]dt \tag{6.2}$$

The transfer function of the controller is as follows

$$G_c(s) = \frac{P(s)}{E(s)} = K_c(1 + \frac{1}{T_i s}) \tag{6.3}$$

where K_c is the proportional gain; T_i is the integral time constant.

The basic principle of PI control is shown in Fig. 6.4.

The characteristics of integral action are:

(a) Integration (accumulation) of deviation signal.
(b) As long as the deviation exists, the integral action will continue until it is limited by the actual device (integral saturation).
(c) Only when the deviation is zero, the integral action stops. At this time, the control signal remains at the previous value.

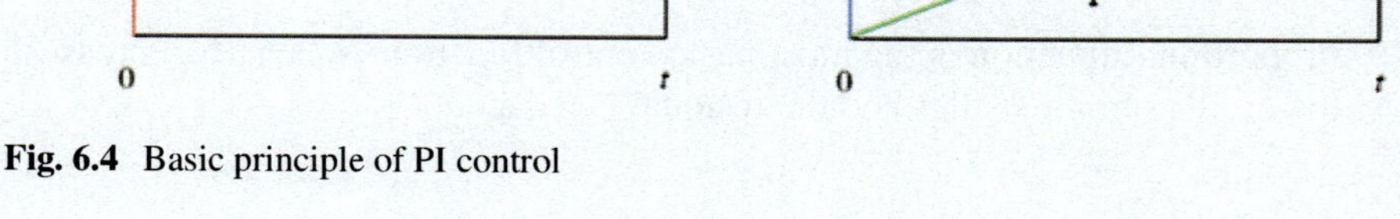

Fig. 6.4 Basic principle of PI control

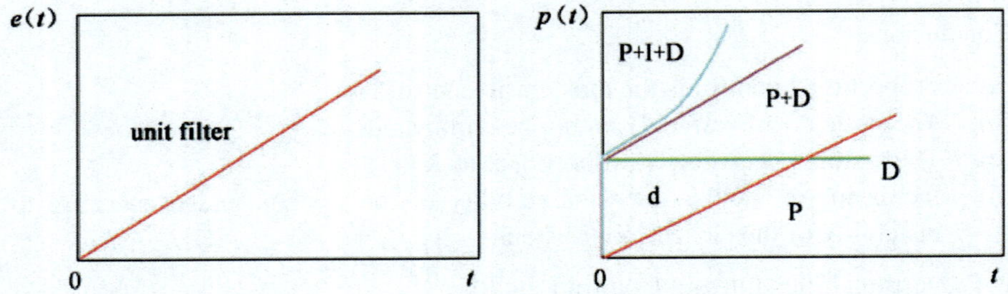

Fig. 6.5 Basic principle of PID control

Conclusions:

(a) Integral action can eliminate the residual (main purpose).
(b) The integral action is inversely proportional to T_i.
(c) It can improve the stability of the system, but it can also produce a 90° phase lag, which has a negative impact on the stability.

3. Proportional Plus Integral Plus Differential Control Law PID (PID Controller)

The transfer function is as follows

$$G_c(s) = \frac{P(s)}{E(s)} = K_c(1 + \frac{1}{T_i s} + T_d s) \tag{6.4}$$

where K_c is the proportional gain; T_i is the integral time constant; T_d is the differential time constant.

The basic principle of PID control is shown in Fig. 6.5.

The characteristics of differential action are:

(a) The differential action is directly proportional to the rate of deviation change, which can improve the sensitivity of the controller response, make the system get the leading effect, and help to increase the stability and control quality of the system.
(b) The differential action is proportional to T_d.
(c) The control function can only work on the dynamic process and is sensitive to noise, so it cannot be used alone. It is often combined with P or PI controller to form PD or PID controller.

PID regulator is a combination of three control functions, which is a relatively mature conventional control law. It can be summarized as follows:

(a) Proportional function is the most basic control function. When K_c increases, the oscillation intensifies and the residual is reduced.

(b) Differential action can increase the stability of the system. When the differential action is increased, K_c can be increased appropriately. The residual cannot be eliminated, and T_d should not be too large.
(c) Integral action can eliminate residual error, also affect the stability of the system. When the integral function is increased, K_c should be reduced.

In control engineering practice, PI controller can be used to improve the steady-state performance of the system. In system calibration, P control is rarely used alone. A single D controller should not be used in series with the controlled object alone.

The PID algorithm formula for continuous case is

$$u(t) = k_p \left[e(t) + \frac{1}{T_1} \int_0^t e(x)dt + T_D \frac{de(t)}{dt} \right] \tag{6.5}$$

Assuming that the sampling time interval is T, at time k

Deviation is $e(k)$

Integral is $e(k) + e(k-1) + e(k-2) + \ldots + e(0)$

Differential is $[e(k) - e(k-1)]/T$

The discrete PID algorithm formula is

$$u(k) = k_p \left\{ e(k) + \frac{T}{T_i} \sum_{n=0}^{k} e(n) + \frac{T_d}{T} [e(k) - e(k-1)] \right\} \tag{6.6}$$

$$u(k-1) = K_p e(k-1) + K_i \sum_{n=0}^{k-1} e(n) + K_d [e(k-1) - e(k-2)] \tag{6.7}$$

Scale factor is K_p.

Integral coefficient is $K_p \times T/T_i$, and can be written as K_i.

Differential coefficient is $K_p \times T_d/T$, and can be written as K_d.

Then the formula can be written as follows.

$$u(k) = K_p e(k) + \sum_{n=0}^{k} e(n) + K_d [e(k) - e(k-1)] \tag{6.8}$$

The incremental PID algorithm formula can be obtained according to the above formula.

$$u(k) = k_p \left\{ e(k) + \frac{1}{T_i} \sum_{j=0}^{1} Te(j) + T_D \frac{e(k) - e(k-1)}{T} \right\}$$

$$\begin{aligned} \Delta u(k) &= u(k) - u(k-1) \\ &= K_p[e(k) - e(k-1)] + K_i e(k) + K_d[e(k) - 2e(k-1) + e(k-2)] \end{aligned} \tag{6.9}$$

6.1.2 *Fuzzy Adaptive PID Control*

Fuzzy adaptive PID control is based on PID algorithm, taking error e and error change rate e_c as input, using fuzzy rules for fuzzy reasoning, querying fuzzy matrix table for parameter adjustment, so as to meet the requirements of e and e_c for PID parameter self-tuning at different times. Fuzzy adaptive PID algorithm is composed of fuzzy controller and PID controller. The fuzzy controller takes error e and error change rate e_c as input, and uses fuzzy rules to adaptively adjust the parameters K_p, K_i and K_d of PID controller, so that the controlled object remains in a good dynamic and static stable state. The functional block diagram of fuzzy adaptive PID algorithm is shown in Fig. 6.6. Compared with the traditional PID control, fuzzy adaptive PID is more flexible and stable, especially for controlled objects with strong time-varying characteristics and nonlinearity.

The following steps are required to design fuzzy adaptive PID control:

1. The determination of fuzzy control rules (fuzzy control rules can be obtained mainly through expert experience and sampling data).
2. e (Error) and e_c (error change rate) are fuzzified by using the fuzzy rule table, and the corresponding degree of membership is obtained.
3. The obtained degree of membership and the abscissa of the corresponding degree (e.g. PB, NS) are brought into the formula for calculation of ΔK_p, ΔK_i, ΔK_d.

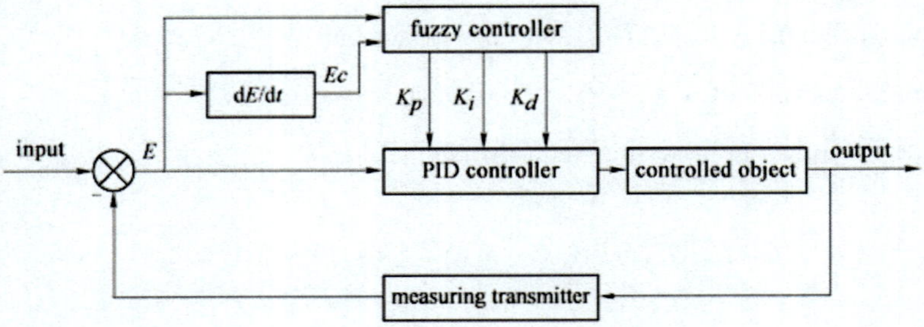

Fig. 6.6 Block diagram of fuzzy adaptive PID algorithm principle

The formula is as follows

$$y = \frac{\sum_{i=1}^{n} \mu_{Ai}(x)\mu_{Bi}(y)z_i}{\sum_{i=1}^{n} \mu_{Ai}(x)\mu_{Bi}(y)}$$

where, $\mu_{Ai}(x)$ and $\mu_{Bi}(y)$ represent the calculated degrees of membership, and z_i represents the abscissa of the corresponding degree of membership (e.g. PB, NS).

4. The tuned parameters K_p, K_i, K_d are obtained from $K_p = K_p + \Delta K_p$ and brought into the PID controller for calculation. The PID formula is as follows.

$$\text{Output} = K_p e(t) + K_I \int e(t)\mathrm{d}t + K_D \frac{d}{\mathrm{d}t} e(t) \tag{6.10}$$

The block diagram of fuzzy adaptive PID algorithm principle is shown in Fig. 6.6.

6.1.3 Adaptive Control

The research object of adaptive control is a system with a certain degree of uncertainty. The so-called uncertainty here means that the mathematical model describing the controlled object and its environment is not completely determined, which contains some unknown factors and random factors.

According to the structure of adaptive control system, the commonly used adaptive control systems include model reference adaptive control system (MRAC) and self-tuning system.

Model reference adaptive control algorithm frameworks can be divided into parallel type, series type and series parallel type, as shown in Fig. 6.7.

(a) parallel model reference adaptive system
(b) series model reference adaptive system
(c) series parallel model reference adaptive system

6.1.4 Neural Network Predictive Control

Predictive control, or model predictive control, adopts online rolling optimization index and feedback self-tuning strategy based on a variety of different predictive models, and strives to effectively overcome the dynamic influence of uncertainty, hysteresis and time-varying factors of the controlled object, so as to achieve the expected control objective of reference trajectory input, and bring robustness and

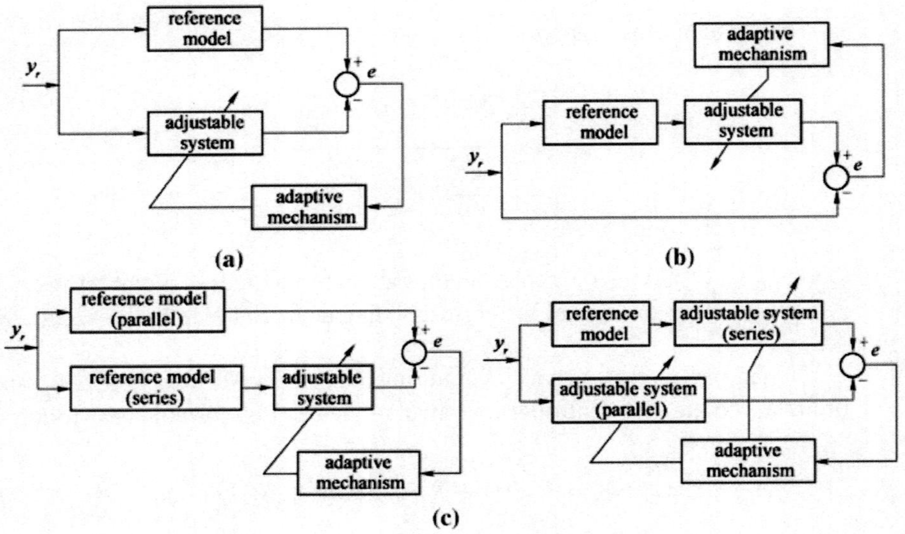

Fig. 6.7 Model reference adaptive control algorithm frameworks

stability to the system. Therefore, the predictive control system roughly consists of reference trajectory, prediction model, rolling optimization and online calibration.

Artificial Neural Network (ANN) can be generally defined as a complex network computing system composed of a large number of simple and highly interconnected neurons. It is one of the main branches of intelligent control technology, which is based on the research results of modern neuroscience. Neural network reflects some basic characteristics of human brain function, and is a very important method to simulate artificial intelligence. A general form of neural network is a machine that models the way the human brain performs specific tasks or functions of interest. Artificial neural network can be realized by hardware or software. It can be seen as a computational model as well as a cognitive model. Therefore, in essence, artificial neural network, parallel distributed processing and neural computer are the same concept. In the field of control, neural network plays an extremely important role. With the continuous improvement of neural network theory, neural network has been used in many aspects of control, such as process control, production control, pattern recognition, decision support and so on.

Neural network has many advantages, such as function approximation ability, self-learning ability, complex classification function, associative memory function, fast optimization calculation ability, strong robustness and fault tolerance brought by highly parallel distributed information storage. The combination of neural network and model predictive control provides a powerful tool for the control of complex industrial processes.

Most of the actual dynamic systems in industrial field are essentially nonlinear systems. The nonlinear system is much more complex than the linear system, because it does not have linearity and superposition. So the predictive control algorithm based on linear system modeling and optimization is difficult to be applied. In addition, the mathematical model used to describe the general nonlinear system and the model based on various kernel functions have some problems, such as specific structure, difficult identification, complex processing and so on. The neural network can fully approach the complex nonlinear mapping, and has the features of learning and adapting to the dynamic characteristics of uncertain systems, strong robustness and fault tolerance. Therefore, neural network has become one of the key technologies to establish the predictive model and optimize the control of nonlinear systems, and a variety of predictive control algorithms based on neural network has been proposed.

In essence, neural network predictive control is still predictive control, and belongs to the category of intelligent predictive control. It combines neural network technology with predictive control, and makes up for the shortcomings of traditional predictive control algorithm, such as low accuracy, suitability for linear system only, lack of self-learning and self-organization function, and weak robustness. It can deal with nonlinearity, multiple objectives, abnormal constraint conditions and other situations.

Neural network can fully approximate any complex nonlinear function, can learn and adapt to the dynamic characteristics of uncertain system, and can use parallel distributed processing algorithm to quickly carry out real-time operation. On this basis, the control law is obtained.

The structure of neural network predictive control system is shown in Fig. 6.8. Where s is the set value; $a = 1, 2, \ldots, n_u$; $b = 1, 2, \ldots, n_y$.

Take quadratic performance index function.

$$J = \min\left\{\sum_{j=1}^{P} e^2(k+j) + \sum_{j=1}^{L} r_j \Delta u^2(k+j-1)\right\} \tag{6.11}$$

where r_j is the coefficient of control right; $\Delta u(k+j-1) = u(k+i-1) - u(k+j-2)$.

From $\frac{\partial y}{\partial \Delta u} = 0$, it can be obtained that.

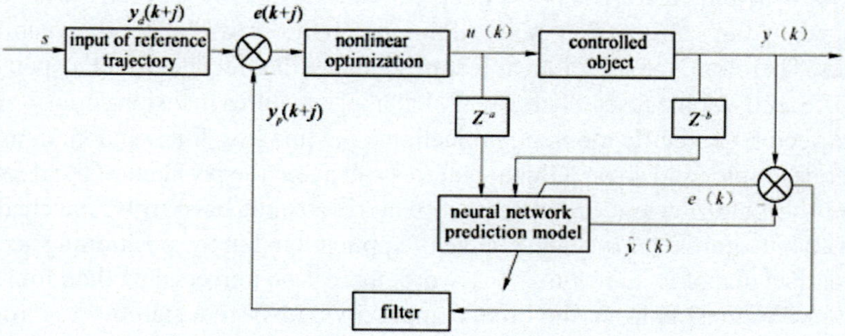

Fig. 6.8 Structure of neural network predictive control system

$$\Delta u(k+j-1) = \frac{e(k+j)}{r_j} \frac{\partial y(k+j)}{\partial \Delta u(k+j-1)} \tag{6.12}$$

$$e(k+j) = y_d(k+i) - y_p(k+j) \tag{6.13}$$

It can be seen from the above that the steps of neural network predictive control algorithm can be summarized as follows ($j = 1, 2, \ldots, P$):

1. Calculate the desired input reference trajectory $y_d(k+j)$.
2. The neural network prediction model outputs $y*(k)$, and the filter generates the prediction output $y_p(k+j)$.
3. Calculate the prediction error $e(k+j) = y_d(k+j) - y_p(k+j)$.
4. Find the quadratic performance function min $J(P, L, r)$; Obtain the optimal control law $\Delta u(k+j-1)$; use $u(k)$ as the first control signal and the input of the controlled object, then go to step 2.

6.1.5 Visual Servo Control

The concept of visual servo, which originated from the field of robot, was proposed by Hill and Park in 1979. Visual servo generally refers to the process during which the image of real object in the world coordinate system is automatically collected and processed by optical devices and non-contact sensors, and the image feedback information is used as the drive, so that the system can further control the machine or make corresponding adaptive adjustment.

The performance of visual servo depends on the image features used in the control loop. Features include geometric features and non-geometric features. Geometric features are commonly used in manipulator visual servo. In the early visual servo, simple local geometric features such as point, line, circle, rectangle, area and their combined features were mostly used, among which point feature was the most widely used. Although local features have been widely used, and can achieve accurate positioning in the case of appropriate feature selection, it is difficult to make accurate operation when the feature is beyond the horizon. Especially for real-world objects, the shape, texture, obstruction, noise, lighting conditions will affect the visibility of features. Therefore, the use of local features alone will affect the robot's operational range. Recently, some researchers use global image features in vision control, such as feature vector, geometric moment, projection from image to line, random transform, Fourier descriptor and so on. Global features can avoid the problem of local features beyond the field of view, and reference features do not have to be matched with observation features. It has a wide range of application, but the positioning accuracy is lower than that of local features. In a word, there is no universal method for feature selection. We must balance the time, complexity and system stability according to the task, environment and system hardware and software performance.

At present, the following classification methods can be used on robot visual servo control system:

1. According to the number of cameras, it can be divided into monocular visual servo system, binocular visual servo system and multi-camera visual servo system.
2. According to the location of the camera, it can be divided into eye-in-hand system and eye-to-hand/stand-alone system.
3. According to the spatial position or image characteristics of robot, it can be divided into position-based visual servo system and image-based visual servo system.

6.2 Planning and Decision-Making

The planning and decision-making mainly involves how to do the planning correctly after getting the fusion data. Planning includes longitudinal control and lateral control. Longitudinal control is speed control, which deals with when to accelerate and when to brake. Lateral control is behavior control, which deals with when to change lanes and when to overtake.

6.2.1 Action Planning

Action planning method based on Frenet coordinate system is one of the most commonly used methods for autonomous vehicle, which was proposed by Moritz Werling of BMW. The core idea of Werling's action planning method is to divide the high-dimensional optimization problem of action planning into two independent optimization problems in lateral and longitudinal directions. For the optimization problem, the vehicle is expected to travel along the direction parallel to the reference line (road centerline). In order to better understand the action planning, the lateral optimal action sequence is defined as: for lateral control, assuming that the vehicle deviates from the expected lane, then the optimal action sequence (or trajectory) is the relatively safe, comfortable, simple and efficient trajectory under the limitation of vehicle braking capacity. The longitudinal optimal trajectory is defined as: if the vehicle is too fast or too close to the vehicle in front, then it must slow down. The comfortable and simple deceleration can be described by the physical quantity jerk, which is the rate of change of acceleration. Generally speaking, high acceleration jerk will cause discomfort to the passengers. Therefore, in terms of riding comfort, the jerk should be optimized. At the same time, the braking cycle T of the trajectory, that is, the operating time of a braking, should be introduced.

$$T = t_{end} - t_{start} \tag{6.14}$$

In the Frenet coordinate system, the centerline of the road is used as the reference line, and the tangent vector $\boldsymbol{t}$ and normal vector $\boldsymbol{n}$ of the reference line are used

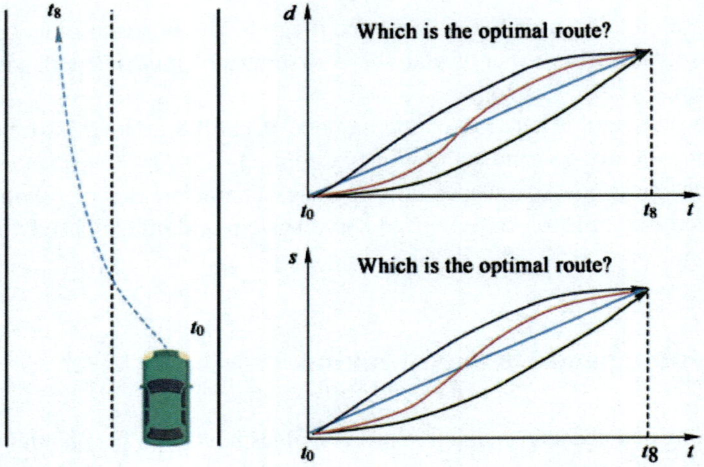

Fig. 6.9 Optimal route expression in Frenet coordinate system

to establish a coordinate system. As shown on the right of Fig. 6.9, this coordinate system is called the Frenet coordinate system. It takes the vehicle itself as the origin, and the coordinate axes are perpendicular to each other, which is divided into the *s* direction (the direction along the reference line, called the longitudinal direction) and the *d* direction (the current normal direction of the reference line, called the lateral direction). Compared with Cartesian coordinate system (the left of Fig. 6.9), Frenet coordinate system obviously simplifies the problem. The position representation based on the reference line can be simply described by the longitudinal distance (i.e. the distance along the road) and the lateral distance (i.e. the distance away from the reference line). Similarly, the speed in both directions can be calculated easily.

Therefore, the configuration space in action planning has three dimensions: (s, d, t), where t is the planned time point of each action. The essential difference between trajectory and path is that trajectory takes time into account. The optimal route expression in Frenet coordinate system is shown in Fig. 6.9.

6.2.2 Behavior Planning

Behavior planning, also known as behavior decision-making, is located between the global task planning and the action planning. It is the middle layer of the three layers (task, behavior, action) of the autonomous vehicle planning module. The function of behavior planning is to determine the current driving state according to the global optimal route information from the upper layer (task planning layer), the understanding of the current traffic scene, the environmental perception information, and to plan reasonable driving behavior under the guidance of traffic rules and driving experience. Figure 6.10 shows the information flow of the behavior decision-making

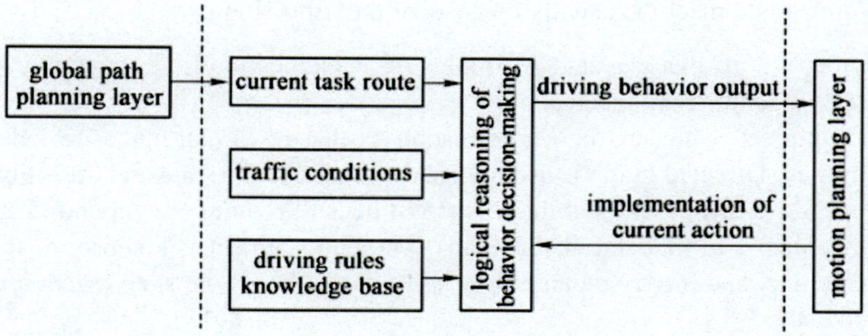

Fig. 6.10 Information flow of the behavior decision-making layer of autonomous vehicle

layer of autonomous vehicle.

When designing the decision-making and planning system of autonomous vehicle, the traffic regulations, driving experience and real-time performance should be considered.

At present, there is no best solution for the behavior planning of autonomous vehicle. Hierarchical Finite-State Machine (HFSM) is generally accepted and adopted. It is also a behavior planning method adopted by many fleets in the early DARPA challenges. Finite-state machine is the foundation of hierarchical finite-state machine.

Finite-state machine (FSM) is a very simple abstract reaction system. It is simple because it only produces a limited number of responses to specific external inputs. In FSM, only a limited number of states can be constructed, and external inputs can only make the state machine jump from one to another in limited states. Figure 6.11 shows a simple FSM.

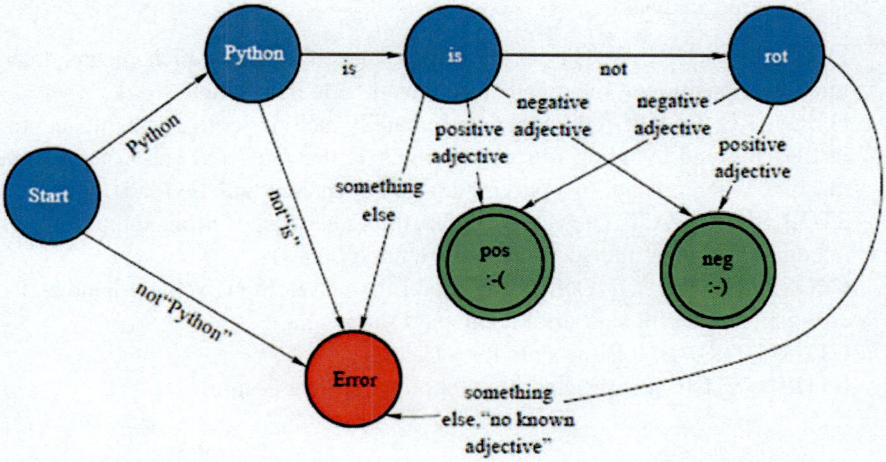

Fig. 6.11 A simple finite-state machine (FSM)

A finite-state machine usually consists of the following parts:

1. Input set, also known as the stimulus set, which contains all the inputs that the state machine may receive.
2. Output set, which is the set of responses that FSM can make, which is also limited. Directed graph is usually used to describe the state and transition logic in FSM. FSM has a fixed initial state (it doesn't require any input and the state machine is in that state by default). The end state set is a subset of the state machine, and it may be an empty set (that is, the whole state machine has no end state).
3. Transition logic, which is the condition for a state machine to transition from one state to another (usually the interaction of current state and input).
4. Acceptors and transducers. According to whether there is output or not, perceptron can be divided into two categories: receivers and transducers. Receiver means that there is no output but there is an end state, while the transducer has an output set. FSM can be further divided into deterministic automata and non-deterministic automata. In deterministic automata, each state has only one exact transition for each possible input. In uncertain automata, a given state may have none or more than one transition to a given possible input.

To some extent, the behavior planning layer of autonomous vehicle is also a reaction system, that is, the decision of the autonomous vehicle (the state of the autonomous vehicle after a period of time) is determined by the current state of the vehicle and the real-time information (input) from the perception module. Stanford University's Junior autonomous vehicle is a typical case of constructing a behavior planning system with hierarchical finite-state machine. It divides the top-level driving behavior into 13 super states (Fig. 6.12), and each driving behavior corresponds to certain sub states to complete this behavior. The top level behavior is managed by an FSM.

State machine analysis:

1. LOCATE_VEHICLE. This is the initial state of Junior, which means determining its position in the map before the autonomous vehicle starts.
2. FORWARD_DRIVE. This super state actually includes going straight, keeping in the lane and avoiding obstacles, which is the preferred state of the state machine when it is not in a parking lot (i.e. open area with no road).
3. STOP_SIGN_WAIT. The vehicle enters this state when waiting at the stop sign (a common sign at intersections in the United States).
4. CROSS_INTERSECTION. In this state, the autonomous vehicle handles the crossing, and it will wait until it can pass safely.

- UTURN_DRIVE. It is the state for a U-turn.
- UTURN_STOP: It is the state for stopping before a U-turn.

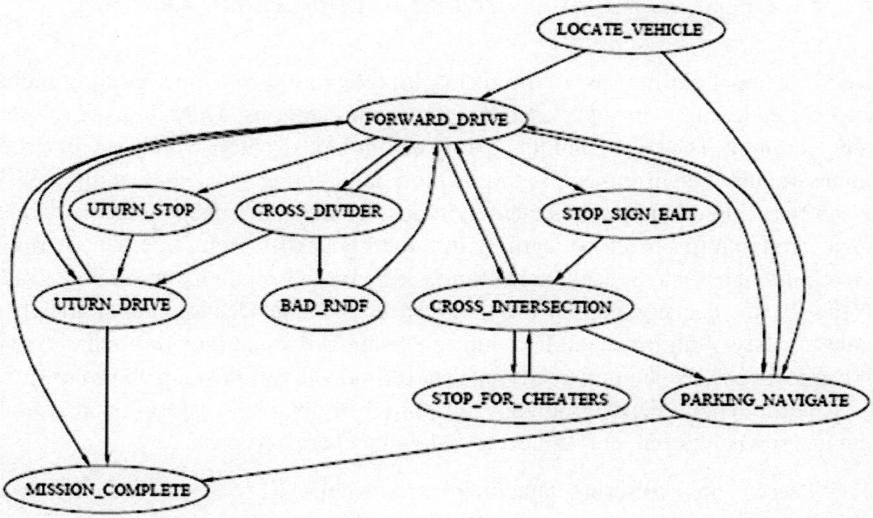

Fig. 6.12 Super states of a hierarchical finite-state machine

- CROSS_DIVIDER. It is the state for driving across yellow lines.
- PARKING_NAVIGATE. It is the normal driving mode in a parking lot.
- TRAFFIC_JAM and ESCAPE. These are the two states for dealing with traffic jam.
- BAD_RNDF. If the current road is different from the road network map created in advance, the vehicle will enter this state. In this state, the autonomous vehicle will use the hybrid A* algorithm to complete path planning.
- MISSION_COMPLETE. When the DARPA Challenge is over, the autonomous vehicle will enter this state, which is the end state of the whole state machine.

The vehicle is usually in normal driving states (FORWARD_DRIVE and PARKING_NAVIGATE) for the most time of regular driving. The system uses stuckness detectors to determine whether to transfer from the normal driving state to other underlying states. After completing the corresponding actions, the behavior module will return to the original normal driving mode. Although some of Junior's strategies are designed for the DARPA Challenge (doing its best to win the competition), its design concept still has reference value. In the actual application of autonomous driving, the state machine to be implemented will be more complex. Although HFSM is more modular than traditional FSM, it still inherits most of the shortcomings of FSM, such as limited reusability.

6.2.3 Decision-Making Based on Reinforcement Learning

Reinforcement learning is a type of machine learning. Machine learning includes supervised learning, unsupervised learning, semi-supervised learning and reinforcement learning. The key to building a truly autonomous vehicle which can drive safely in any required environment, is to pay more attention to the self-learning ability of its software. In other words, the autonomous vehicle is firstly about artificial intelligence and requires machine learning development skills. Reinforcement learning is an important branch of machine learning and an interdisciplinary product. Its essence is to solve the decision making problem, that is, to make decisions automatically and continuously. Reinforcement learning can be applied to autonomous vehicle, so that it can learn to drive automatically in a short time without 3D map and rules.

Reinforcement learning obtains an optimal strategy by exploring and modeling the unknown environment. It has the following characteristics:

1. There is no monitoring data, only reward signal.
2. Reward signals are not necessarily real-time, and are likely to be delayed, sometimes a lot.
3. Time series is an important factor.
4. The current behavior of the intelligent agent affects the subsequent data received.

Supervised learning is to provide a batch of samples in advance, and tell the learner which samples are good and which are bad (the marked information of samples). Through learning these samples, the learner can establish the object model and its strategy. In reinforcement learning, no one tells the learner what to do in certain situations in advance. The learner improves by reflecting on whether the previous action is correct. From this point of view, reinforcement learning can be regarded as supervised learning with time delay labeled information. In many other machine learning algorithms, learners learn what to do. Yet reinforcement learning is to learn which action leads to the greatest reward in a specific situation in the process of trying. Reinforcement learning means learning while obtaining examples and updating the current model to guide the next action. After the next action is rewarded, the model is updated again until it converges.

Reinforcement learning has been widely used in the following fields: UAV motion control, investment management, power station control, robot walking and so on. The British startup company Wayve has recently published an article, *Learning to Drive in A Day*, which is about the application of reinforcement learning in autonomous vehicle. Wayve is building end-to-end machine learning algorithms, which according to itself are different from the conventional ones. Specifically, the company believes that the key to creating a truly autonomous vehicle is the software ability to learn independently. This problem cannot be solved by more sensors. The actual answer is better coordination.

In the decision-making module of autonomous vehicle, the environment information parsed by the perception module supports the automatic control of vehicle behavior. Tesla's automatic driving accident is caused by the failure of sensing module in the strong light environment. Reinforcement learning can help to enhance the robustness of the perception system, so that the vehicle can drive safely even when some modules fail. Successful driving requires a series of correct actions. Minor errors can accumulate quickly and lead to catastrophic consequences. Reinforcement learning can help the vehicle learn to correct errors automatically. Acceleration, deceleration, left turn, right turn, lane change, overtaking are all the outputs of the decision-making module. The decision-making module not only needs to consider the safety and comfort when driving to the destination, but also needs to ensure the safety of passengers in the case of malicious vehicles nearby. Therefore, the decision-making module needs to make long-term planning and predict the behavior of the surrounding vehicles and pedestrians. Moreover, the existing decision-making module is generally built according to the rules. Although the rule-based construction can cope with most of the driving situations, it is impossible to enumerate all kinds of unexpected situations. We need an adaptive system to deal with driving emergencies.

In practice, the model-free deep reinforcement learning algorithms (Fig. 6.13) such as deep deterministic policy gradient (DDPG) can solve the problem of lane tracking. The model input is monocular camera image. The system iterates three processes: exploration, optimization and evaluation.

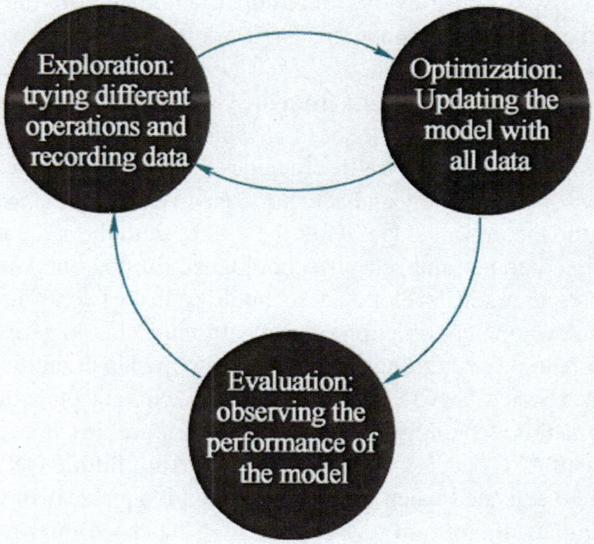

Fig. 6.13 Model-free deep reinforcement learning algorithm

6.3 Application Module: Adaptive Cruise Control (ACC) System

Adaptive cruise control system is an intelligent automatic control system, which is developed on the basis of cruise control technology. In the process of driving, the distance sensor (radar) installed in the front of the vehicle continuously scans the road ahead, while the wheel speed sensor collects the speed signal. When the distance between the vehicle and the front vehicle is too short, ACC unit can coordinate with the anti-lock braking system and engine control system to make the wheels slow down properly and reduce the output power of the engine, so as to restore a safe distance. When the adaptive cruise control system controls the braking, it usually limits the braking deceleration to a level that does not affect the comfortness. When a greater deceleration is required, ACC unit will send out sound and light signals to inform the driver to take the initiative to brake. When the distance to the front vehicle increases and reaches a safe level, ACC unit controls the vehicle to drive according to the set speed. The adaptive cruise control system is based on the traditional constant speed cruise control combined with safe distance keeping control. It monitors the environment (whether there is a vehicle in front, the distance between two vehicles, the relative speed, etc.) through the environmental information perception module. When there is no other vehicle in front or the vehicle is far away, it cruises at the preset speed. When the vehicle in front is within the monitoring range and the speed is lower than the cruising speed of itself, the strategy is to automatically follow the vehicle in front. ACC performs the longitudinal automatic driving under specific conditions, reducing the driver's burden. The working process and basic principle of ACC system is shown in Fig. 6.14.

ACC was first proposed by experts from the University of Michigan in the 1960s. The purpose of ACC is to control the speed of vehicles and the distance between them on expressways, so as to improve traffic efficiency and safety. However, limited by the functions of sensors, controllers and actuators, the research in the next two decades only stayed in the theoretical stage. After the 1980s, with the rapid development of communication, computer and sensor technologies, Europe, the United States and Japan began to research and develop ACC technology. In the twenty-first century, with the continuous development of automotive electronic technology, ACC technology has made a substantial breakthrough and has been applied in the automotive industry. Audi A6, BMW 5 Series, Volvo S60, Cadillac DTS, Lexus and other high-end models are equipped with this technology. At present, joint ventures in China gradually equip some models with ACC, such as Ford Taurus, Audi A6L, Infiniti QX50, etc. At first, ACC can only be activated when the vehicle speed is greater than a certain value. With the continuous progress of technology, ACC has been improved. Now it has the function of stop and go, and its scope of work is extended to full speed range, which means it can deal with multiple signal lights, traffic congestion and other road conditions in the city.

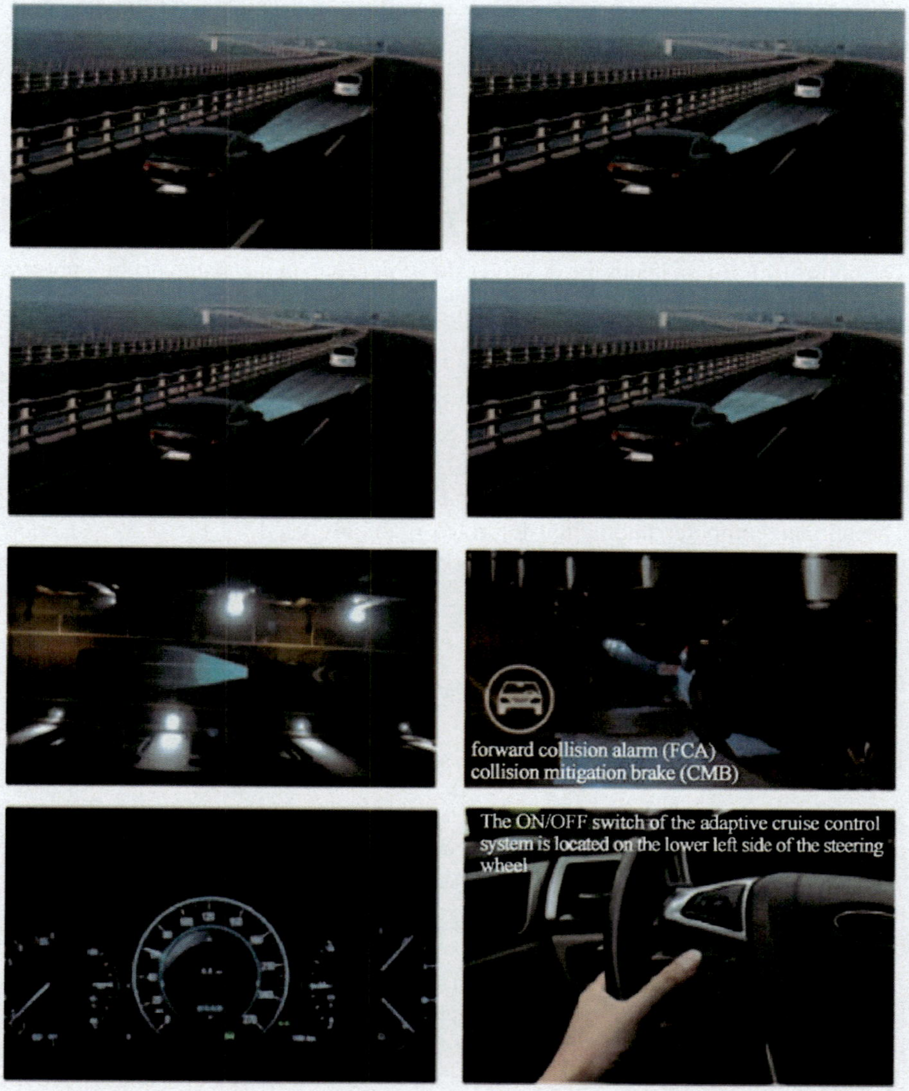

Fig. 6.14 Working process and basic principle of ACC system

6.3.1 Architecture of ACC

ACC system architecture mainly includes four parts: information collection, control system, actuator and human–computer interaction.

1. Information collection

The objective of information collection is to sense the surrounding environment information and monitor the status of the vehicle. Vehicle borne ranging sensor and vision sensor are used to perceive the surrounding environment. Millimeter wave radar is usually used as the vehicle borne ranging sensor, and megapixel camera is usually used as the vision sensor.

2. Control system

The core of the control system is the gap strategy and ACC algorithm. The gap strategy determines the desired safe vehicle gap according to the current driving environment, and provides reference input value for the gap control of the algorithm. The control algorithm keeps the desired safe vehicle gap in the actual driving process by controlling the throttle and brake.

3. Actuator

The actuator mainly includes electronic throttle system, active braking system and automatic shift system applied in low speed control.

4. Human–computer interaction

The human–computer interaction interface is mainly used for the driver to select from the functions and set the parameters of the system.

6.3.2 Technical Characteristics of ACC

1. Typical ACC has activation requirements, while full speed ACC can stop and go.

Typical ACC can only be activated under specific conditions, generally at a speed of more than 30 km/h. However, full speed ACC can perform the function of stop and go, without the speed requirement. According to the traffic rules in China, the maximum cruise speed of ACC should be set at 120 km/h.

2. The following distance is set in time, generally between 1 s and 2.5 s

If the safety distance is too small, the vehicle is often in an unsafe state when it follows the front vehicle, which leads to the driver's mental tension. If the safety distance is too large, it will cause frequent merging of other vehicles and reduce the driver's trust in ACC.

At present, the safety distance control mainly uses the safety headway model, which usually keeps the time of potential collision between 1 s and 2.5 s. In the human–computer interaction interface, there can 3–5 settings, which represent different safety distances, for drivers to choose according to different situations.

3. The automatic braking force is 0.3–0.4 g, which is 30–40% of the maximum braking force of the vehicle

The main purpose of ACC as a comfort configuration is to reduce the driving burden and increase the comfort level. The braking force of ACC automatic braking is the main factor affecting the comfort, which should be determined to ensure safety and comfort as the same time. The automatic braking force of ACC is 0.3–0.4 g, generally 30–40% of the maximum braking force of the vehicle.

4. The sensor is usually millimeter wave radar, while full speed ACC requires millimeter wave radar and camera

Millimeter wave radar is widely used in ACC because of its long detection range and strong environmental adaptability. For full speed ACC, in order to perform the function of stop and go in low-speed and complex urban conditions, the combination of millimeter wave radar and camera is generally used for environmental information monitoring. Most millimeter wave radars operate at 77 GHz, and the cameras are usually one-megapixel.

5. The detection range of sensor varies from 100 to 250 m

77 GHz millimeter wave radars can be divided into long range radars and medium range radars. The detection range of long-range millimeter wave radar is 250 m, while the detection range of medium range millimeter wave radars is 150 m. The detection range of the one-megapixel camera is 100 m.

6.3.3 Technical Difficulties of ACC System

1. Accurately recognizing and following targets

For ACC, it is difficult to detect multiple targets in multiple lanes ahead and judge whether the vehicles are in the main lane according to the environmental information. In addition, it is also difficult to ensure that the target is not lost in special circumstances such as turning.

2. Complex control strategy of vehicle dynamic adjustment

ACC needs to dynamically control the acceleration or braking of the vehicle according to the set distance combined with the speed of the main vehicle and the front vehicle. The control strategy is complex. Only by optimizing the control strategy can ACC be popularized.

3. Optimization of comfortness on the basis of safety

The main purpose of ACC is to increase driving comfort on the premise of ensuring safety. How to optimize the comfortness on the basis of safety is a problem to be solved.

There are more than 20 car models of 11 different Chinese manufacturers equipped with ACC. Except for Hongqi H7, Geely Borui and EC8, the rest are all medium and high-end models of joint ventures. From 2013 to 2015, the number of vehicles manufactured with ACC was 82,100, 111,300 and 152,400 respectively, with an annual growth rate of 35% With the continuous maturity of technology and decline of cost, the ACC market will show a more rapid growth.

Combined with the overall trend of automobile market development, it is estimated that the equipment rate of ACC in the new passenger car market will reach 20% in 2020. The main reasons are as follows:

1. ACC technology continues to mature, showing a strong growth momentum from 2013 to 2015, with a growth rate of more than 30%. It is expected that ACC will usher in a technology growth period in the next five years.
2. ACC and AEB use the same hardware. With national regulations and NCAP requirements strengthening for AEB system, the development of AEB will also stimulate the populariztion of ACC.

Chapter 7
Energy and Power System

7.1 Brief Introduction of Energy and Power System of Autonomous Vehicle

Autonomous vehicles are intelligent and networked vehicles based on traditional power vehicles or new energy vehicles. At present, new energy vehicles mainly include pure electric vehicles, plug-in hybrid vehicles, hydrogen fuel cell vehicles, etc.

The traditional power vehicle mainly uses internal combustion engine as power source. Pure electric vehicle is mainly powered by lithium batteries. Plug-in hybrid electric vehicle uses internal combustion engine and battery as power sources at the same time. Hydrogen fuel cell vehicle is powered by fuel cells. Therefore, according to the source of power, the energy power system of autonomous vehicle mainly includes fuel power system, lithium battery power system, hybrid power system and fuel cell power system.

At present, fuel powered vehicle, pure electric vehicle and plug-in hybrid electric vehicle are the main types of new intelligent networked vehicles on the market. Correspondingly, the energy power system of autonomous vehicle mainly includes fuel system, lithium battery power system and fuel-electric hybrid power system. The representative models of fuel powered autonomous vehicle include Volvo S90, Trumpchi GA4 and Roewe 950. Representative models of pure electric autonomous vehicle include Tesla Model S, NIO ES8, BYTON K-BYTE, etc. The representative models of plug-in hybrid autonomous vehicle include BMW New 5 Series, Mercedes-Benz S-Class PHEV, BYD Song Max, etc.

M. Du, *Autonomous Vehicle Technology*, Unmanned System Technologies,
https://doi.org/10.1007/978-981-19-4143-6_7

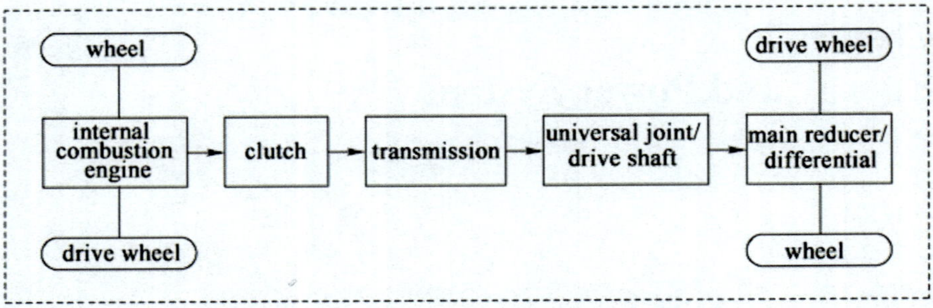

Fig. 7.1 Front-engine rear-drive power system configuration scheme

7.2 Fuel Power System

7.2.1 Dynamical System

Fuel power system uses gasoline or diesel as fuel and internal combustion engine as power source. It is composed of internal combustion engine, clutch, transmission, universal joint, differential and other key parts. The power generated by the internal combustion engine is successively transmitted to the drive wheel through the clutch, transmission, universal joint, transmission shaft and other mechanism to support driving under different conditions. The power transmission sequence is "internal combustion engine—clutch—transmission—universal joint—drive shaft—universal joint—main reducer—differential—drive wheel". According to the position of internal combustion engine and drive wheel, the configuration scheme of automobile power system can be divided into four types: front-engine front-drive, front-engine rear-drive, rear-engine rear-drive and middle-engine rear-drive. Front-engine rear-drive is the typical layout form of fuel vehicle. The front-engine rear-drive power system configuration scheme is shown in Fig. 7.1. The key components and working principles of internal combustion engine power system are shown in Table 7.1.

7.2.2 Fuel System

The energy supply system of transmission power vehicle is fuel system. The fuel system undertakes the fuel filling and transportation to ensure that the fuel reaches the engine smoothly and the engine works well. The fuel system consists of fuel cap, fuel pipe, fuel tank, oil pump, gasoline filter, carbon canister, carbon canister cleaning valve, pipeline and other parts, which can be grouped into the fuel supply system and fuel evaporative emission control system.

The working principle of fuel supply system is as follows. In the working process of automobile engine, the oil pump absorbs the fuel in the fuel tank, increases the pressure, and then filters the fuel through the fuel filter to keep the fuel clean and

Table 7.1 Key components and working principles of internal combustion engine power system

Component	Working principle
Internal combustion engine	It is a thermal engine that directly converts the heat energy released by fuel combustion into power. It is the main device to provide power for vehicles
Clutch	It is responsible for the connection and disconnection of vehicle power and transmission system, which can ensure the stability and smoothness of vehicle in starting and driving shifting
Transmission	It is a mechanism used to change the speed and torque of the engine. It can change the transmission ratio of output shaft and input shaft in fixed or multiple gears
Universal joint	It is a machine part which changes the power transfer angle. It is used where the direction of transmission axis needs to be changed, as the joint of the universal gearing of automobile drive system
Differential	It is mainly composed of left and right half shaft gears, two planetary gears and gear carrier, which can make the left and right (or front and rear) drive wheels rotate at different speeds
Main reducer	It is the machine part which can change the torque and speed in the drive axle. It is basically used to increase the torque from the transmission or universal gearing, reduce the speed and change the transmission direction of the torque
Drive wheel	The drive wheel is the wheel connected with the drive axle. The ground friction on the drive wheel provides driving force for the vehicle

deliver it to the fuel distributor. Under the action of the fuel pressure regulator, it maintains the pressure balance in the distributor, and sends the excess crude oil back to the fuel tank through the oil return pipeline.

The working principle of fuel evaporative emission system is as follows. In the operation of the vehicle, the fuel in the tank will form steam with the increase of temperature, and it is transported to the activated carbon tank through the pipeline, and the activated carbon particles absorb the fuel vapor. When the engine works normally, it is necessary to open the carbon canister solenoid valve. With the vacuum degree of the intake pipe, fresh air enters the carbon canister to wash the steam on the activated carbon particles, and then the desorbed oil gas enters the cylinder for combustion, so as to prevent the fuel vapor from entering the atmosphere to produce pollution, and reduce the waste of fuel.

7.2.3 Market Size

In 2017, the sales volume of internal combustion engines in China was 56.454 million units, with a year-on-year growth of 4.0%. Among them, the sales volume of internal combustion engines for passenger vehicle was 22.054 million, with a year-on-year

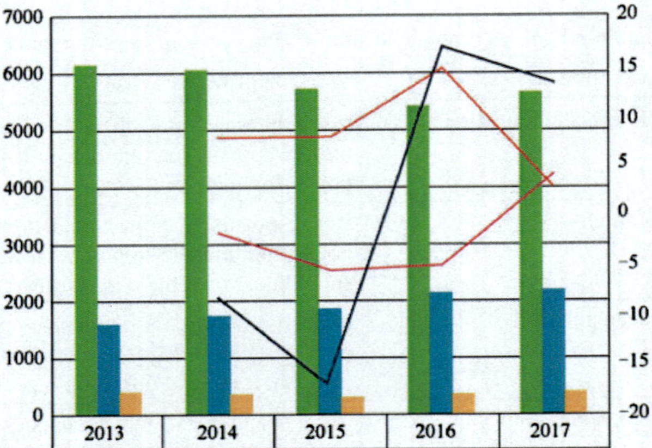

	2013	2014	2015	2016	2017
Total sales volume (million)	6140.89	6038.34	5707.33	5426.46	5645.38
Passenger vehicle sales volume (million)	1602.50	1728.94	1866.33	2142.15	2205.40
Commercial vehicle sales volume (million)	396.61	363.81	301.16	351.92	398.57
Total sales volume year-on-year growth (%)		-1.7	-5.5	-4.9	4.0
Passenger vehicle volume year-on-year growth (%)		7.9	7.9	14.8	3.0
Commercial vehicle sales volume year-on-year growth (%)		-8.3	-17.2	16.9	13.3

Fig. 7.2 Sales volume and growth rate of internal combustion engine in China. *Source* China Internal Combustion Engine Industry Association

growth of 3.0%. The sales volume of internal combustion engines for commercial vehicle was 3.562 million, with a year-on-year growth of 11.7%. The sales volume and growth rate of internal combustion engine in China from 2013 to 2017 are shown in Fig. 7.2.

7.3 Lithium Battery Power System

7.3.1 Dynamical System

Lithium battery power system uses lithium battery as power source. It is composed of electric drive system, power supply system and auxiliary system. The electric drive system is a system that converts the chemical energy of power lithium battery into kinetic energy of the wheel. At the same time, it can convert the kinetic energy of the wheel into electrical energy when the vehicle decelerates or brakes, and charge it into the power battery. It is mainly composed of electronic controller, power converter, motor and mechanical transmission system, and directly affects the dynamic performance and efficiency of the vehicle. Power supply system is the system through which pure electric vehicles provide driving energy to motor. It mainly includes

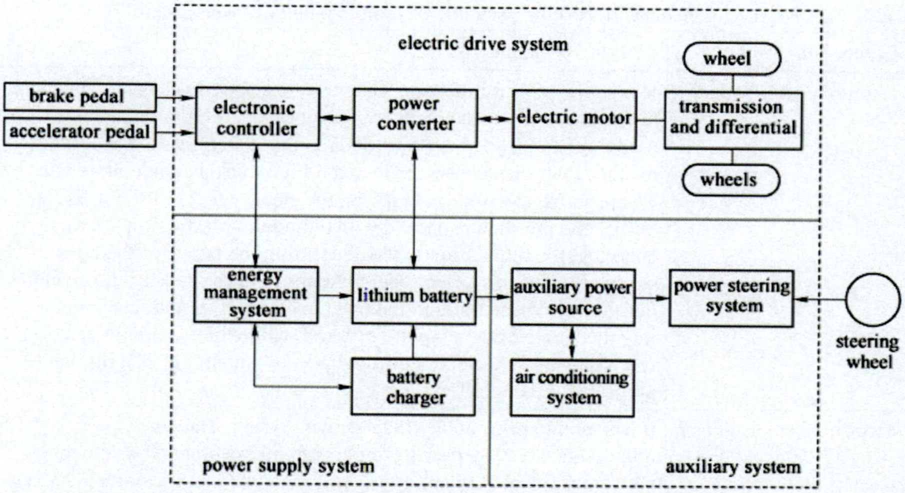

Fig. 7.3 Lithium battery power system configuration scheme of pure electric vehicle

power lithium battery, energy management system and charging device. The auxiliary system is mainly composed of auxiliary power source, air conditioning system and power steering system. The lithium battery power system configuration scheme of pure electric vehicle is shown in Fig. 7.3. The key components and working principle of lithium battery power system are shown in Table 7.2.

7.3.2 Charging System

The energy supply system of pure electric vehicle is the charging system. Charging system provides energy for the operation of electric vehicles, which is an important basic support system. At present, there are three main solutions of charging system, which are wired charging system, ground charging system and wireless charging system.

1. Wired charging system

Also known as the vehicle charging system, it has to work with a dedicated charging pile. There are two different working modes: conventional charging and fast charging. Conventional charging means that the battery should be charged immediately after the discharge is terminated (should not exceed 24 h in special cases), and the charging current is relatively low, about 15 A. This scheme adopts constant voltage or constant current charging with small current, and the charging time is generally 5–8 h, or even up to 10–20 h. Conventional charging mode has the advantages of low charging cost, high charging efficiency and long battery life. Fast charging, also known as

Table 7.2 Key components and working principle of lithium battery power system

Component	Working principle
Lithium power battery	It is a type of power battery. The positive electrode is usually made of binary materials such as lithium cobalt oxide, lithium manganese oxide, lithium iron phosphate, and ternary materials such as lithium nickel cobalt manganese oxide and lithium nickel cobalt aluminate. The negative electrode is graphite, the electrolyte is LiPF6 + EC + DMC, and the shell is packed with aluminum plastic film. During charging, the lithium ion is separated from the lattice of the layered oxide of the positive electrode, migrates to the surface of the layered negative electrode through the organic electrolyte, and then embeds into the lattice of the graphite material, while the remaining electrons reach the negative electrode from the external circuit. The discharge process is the opposite
Electronic controller	It is a miniature computer management center. The core is a microcomputer system with single chip microcomputer as the main part. It takes signal (data) acquisition, calculation processing, analysis and judgment, and decision-making as input, and sends out control instructions to the actuator as output
Power converter	It converts a certain type of current into another, which can be described as DC power conversion or AC power conversion
Motor	It is a piece of equipment which converts electric energy into mechanical energy. It uses the electrified coil (stator winding) to produce the rotating magnetic field and act on the rotor to form the electromagnetic torque. It is divided into DC motor and AC motor
Auxiliary power source	It is a device that supplies power to the power steering system, air conditioning system and other electrical equipment
Power steering system	It is a device that uses the power of the engine to help the driver steer

emergency charging, is to provide short-term charging service for parked electric vehicles with large current within 20 min–2 h. The general charging current is 150–400 A. Fast charging mode has the advantages of short charging time, long battery life, large charging and discharging capacity. However, compared with the conventional charging mode, fast charging has the disadvantages of low charging efficiency and high technology requirements.

2. Ground charging system

It refers to the rapid battery pack replacement system, also known as mechanical charging, or battery change mode. The purpose of charging is achieved by directly replacing the battery pack of electric vehicle. Due to the large weight of the battery pack, the task is quite demanding and requires professional personnel and machinery for replacement, charging and maintenance. This scheme has the advantages of high efficiency, convenience. However, the problems of large battery investment, undefined safety responsibility and lack of unified battery standard system still need to be solved.

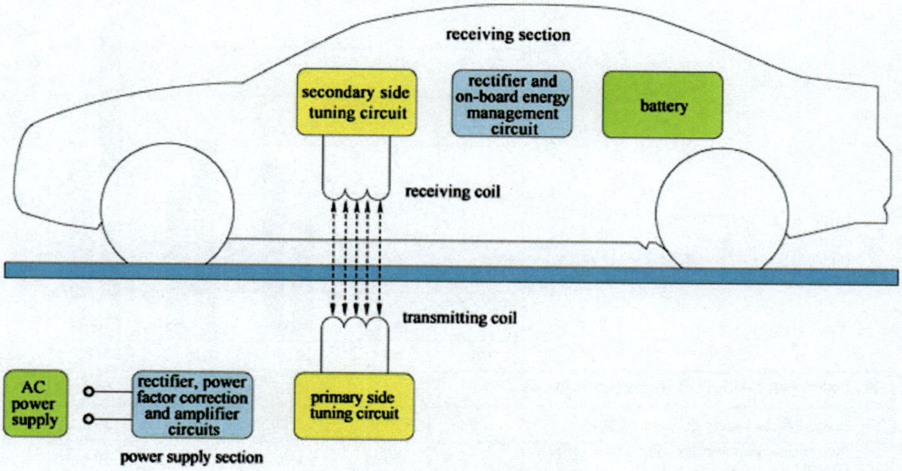

Fig. 7.4 Structure of the wireless charging system for electric vehicle

3. Wireless charging system

Wireless charging system usually includes two sections. The power supply section is composed of AC power supply and ground transmitting device such as transmitting coil. The power receiving section is composed of receiving coil and other on-board receiving devices, storage battery, etc. Wireless charging is to transmit electric energy through the electromagnetic coupling between the transmitting coil embedded on the ground and the receiving coil fixed on the vehicle chassis to charge the power battery. It has a series of advantages such as being safe and environment-friendly, fully automatic, free of maintenance and so on. According to the different distances of wireless power transmission, wireless power transmission can be categorized into far-field transmission and near-field transmission. Near-field power transmission technology is widely used in vehicle wireless charging system. In terms of coupling modes, far-field power transmission technology can be categorized into microwave transmission and laser transmission. Near-field power transmission technology can be categorized into magnetic field coupling transmission and electric field coupling transmission. According to whether resonance occurs, magnetic field coupling includes inductive coupling and magnetic resonance coupling. The structure of the wireless charging system for electric vehicle is shown in Fig. 7.4.

7.3.3 Market Size

In 2017, 39.2 GW·h of power lithium batteries were shipped in China, with a year-on-year increase of more than 30%. In 2017, the sales volume of pure electric vehicles in China was 652,000, which is more than that of 2016. The sales volume of pure

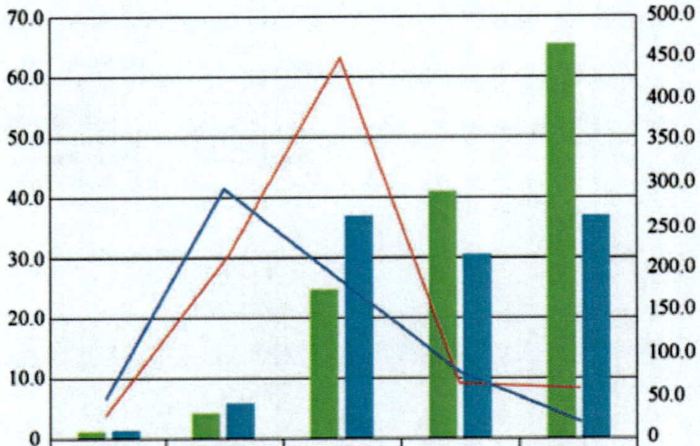

	2013	2014	2015	2016	2017
Pure electric vehicle sales volume (ten thousand)	1.5	4.5	24.7	40.9	65.2
Power lithium battery shipment (GW·h)	1.5	5.9	17.0	30.5	36.9
Pure electric vehicle sales volume year-on-year growth (%)	28.4	208.5	449.4	65.3	59.4
Power lithium battery shipment year-on-year growth (%)	50.0	293.3	188.1	79.4	21.0

Fig. 7.5 Sales statistics of electric vehicle and power lithium battery in China. *Source* Foresight Industry Research Institute; China Automobile Industry Association

electric vehicles, the shipment of power lithium batteries, and the corresponding growth rates in China from 2013 to 2017 are shown in Fig. 7.5.

7.4 Hybrid Power System

7.4.1 Dynamical System

Hybrid power system is a type of power system which uses two power sources of internal combustion engine (motor) and battery to drive the vehicle. The system is composed of internal combustion engine, battery, motor, generator and other key parts. It has the advantages of long working time, decent power performance of fuel engine, while it is also pollution-free and quiet like the electromotor. There are three kinds of hybrid power system, which are series drive system, parallel drive system and hybrid drive system (Figs. 7.6, 7.7 and 7.8). In the series drive system, the engine drives the generator to generate electricity. The converted electric energy can not only charge the battery, but also drive the tire through the same motor and mechanical transmission. In the parallel drive system, the engine is connected with the drive axle through the mechanical transmission device, and the electromotor is also connected with the drive axle through the power combination device. The vehicle can be driven by the engine and the electromotor together or separately. The hybrid drive system is a combination of series and parallel systems. Part of the power from the engine is transmitted to the drive axle through mechanical transmission, and the

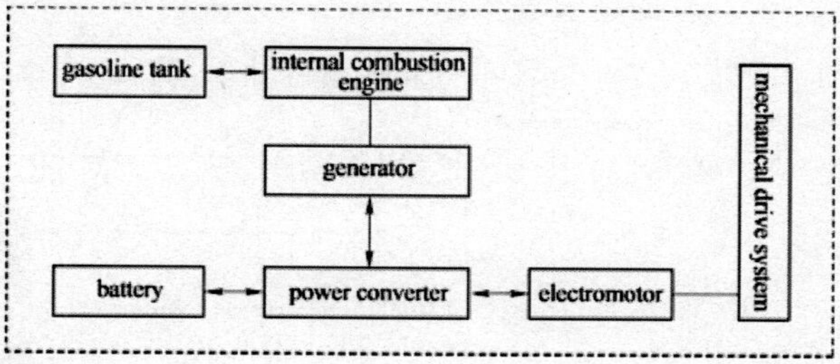

Fig. 7.6 Series drive system

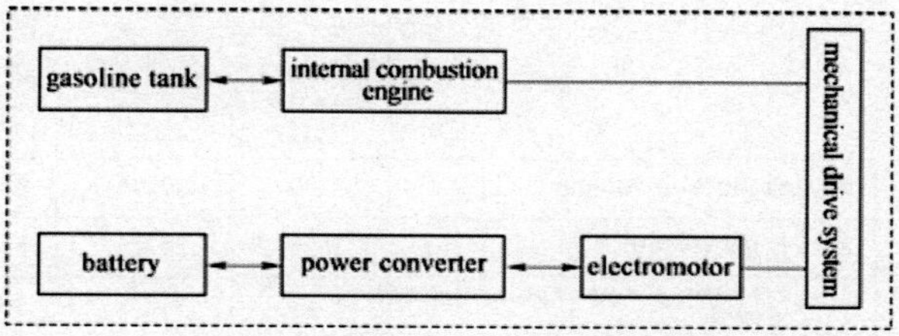

Fig. 7.7 Parallel drive system

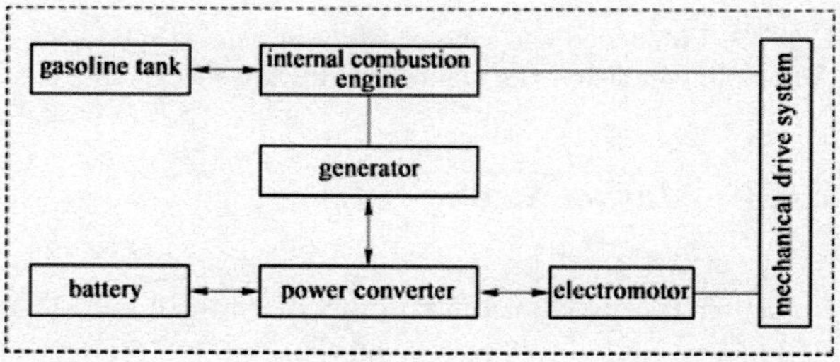

Fig. 7.8 Hybrid drive system

other part drives the generator to generate electricity. The electric energy is controlled by the controller and transmitted to the electromotor or battery. The driving torque generated by the electromotor is transmitted to the drive axle through the power composite device.

7.4.2 Hybrid Power System

The hybrid system requires fueling of the tank and charging of the battery, so the hybrid system is a combination of the fuel system of conventionally powered vehicle, and the charging system of pure electric vehicle. The hybrid power system has two power sources at the same time.

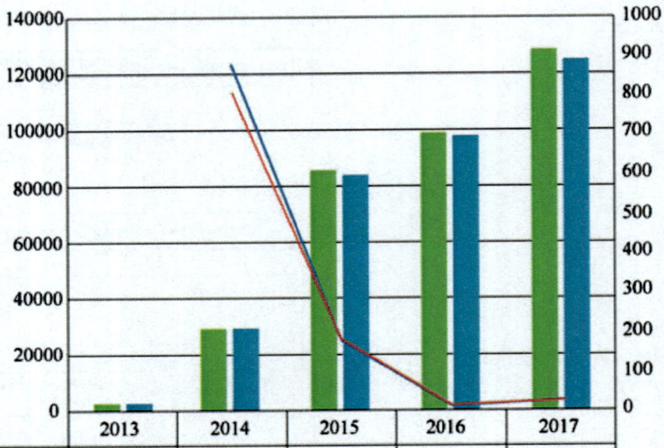

	2013	2014	2015	2016	2017
Production volume	3290	29894	85838	99000	128000
Sales volume	3038	29715	83610	98000	125000
Production volume year-on-year growth (%)		808.6	187.1	15.3	29.3
Sales volume year-on-year growth (%)		878.1	181.4	17.2	27.6

Fig. 7.9 Production and sales statistics of plug-in hybrid electric vehicle in China. *Source* China Automobile Industry Association; www.chinanewauto.org.cn

7.4.3 Market Size

In 2017, the production volume of plug-in hybrid electric vehicle in China was 128,000, with a year-on-year increase of 29.3%. The sales volume was 125,000, with a year-on-year increase of 27.6%. Among them, the production and sales volumes of commercial vehicle were 16,378 and 14,000 respectively, and those of passenger vehicle were 102,073 and 111,000 respectively. The production and sales volume and growth rate of plug-in hybrid electric vehicle in China from 2013 to 2017 are shown in Fig. 7.9.

7.5 Fuel Cell Power System

7.5.1 Dynamical System

Fuel cell power system uses fuel cell as power source, for which the fuel can be hydrogen or methanol. It is composed of fuel cell, rechargeable battery, battery management system, motor and other key parts. Due to the disadvantages of monofuel cell scheme, such as soft output, high cost, starting difficulty, poor transient response, lack of two-way flow of current, and waste of braking energy, this scheme is hardly used in practice. At present, the fuel cell vehicle usually adopts the hybrid drive

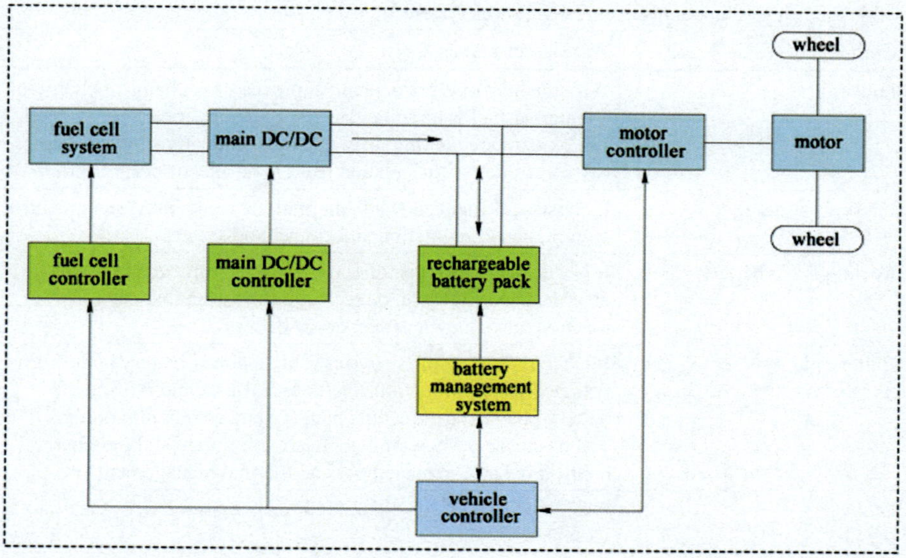

Fig. 7.10 Fuel cell power system configuration scheme

scheme of fuel cell and battery, in which the fuel cell system is the main power source, the battery pack cooperates with the fuel cell system for hybrid drive, and the electric energy is converted into mechanical energy through the motor and transmitted to the transmission system. During acceleration, the battery pack and fuel cell stack output energy together to ensure the acceleration performance. Due to the partial energy provided by the battery pack, the instantaneous acceleration load of the stack is reduced, the oxygen starvation phenomenon of cathode is avoided, and the service life of the stack is prolonged. When braking, the battery pack recovers part of the energy. This process is controlled by the battery management system. In addition, there are other configuration schemes such as fuel cell plus super capacitor, fuel cell plus battery plus super capacitor, fuel cell plus battery plus super high speed flywheel and so on. The fuel cell power system configuration scheme is shown in Fig. 7.10. The key components of fuel cell power system are shown in Table 7.3.

7.5.2 Hydrogen Supply System

Hydrogen is the ideal fuel for fuel cell vehicle, so the main energy supply system of fuel cell vehicle is hydrogen supply system. The hydrogen storage and supply technology involved in on-board hydrogen supply system is the key of large-scale application of hydrogen fuel cell vehicle. In terms of hydrogen storage, there are three major storage methods: compressed gas storage, liquid storage and hydride storage. In terms of hydrogen supply, there are two methods. One is to fill the hydrogen form

Table 7.3 Key components of fuel cell power system

Component	Working principle
Fuel cell	With the help of electrochemical process, the chemical energy of its internal fuel is directly converted into electrical energy. Fuel and oxidant are continuously and independently supplied to the two electrodes of the cell and react at the electrodes
Fuel cell system	It consists of fuel cell stack, thermal management system, control system, power regulation and conversion system
Rechargeable battery	It is a device which can directly convert chemical energy into electrical energy. It can be recharged by reversible chemical reaction, also known as secondary battery
Battery management system	It is an important link between vehicle power battery and electric vehicle. Its main functions include: real-time monitoring of battery physical parameters; battery state estimation; online diagnosis and early warning; charge, discharge and pre-charge control; balance management and thermal management
DC/DC	Also known as DC chopper, it converts a fixed DC voltage into a variable DC voltage

the ground supply station into the on-board hydrogen storage tank. The other one is to directly supply hydrogen for fuel cells through on-board hydrogen production. The first one is currently the major method of hydrogen supply for fuel cells.

1. Compressed gas storage

Storing hydrogen in the form of compressed gas is the most common method. The pressure of standard cylinder is 10–20 MPa. The storage pressure range of fuel cell vehicle is 25–35 MPa. It has been improved to 70 MPa in the United States, Japan and other countries.

2. Liquid storage

This method freezes the hydrogen to s temperature below 20 K. Industrial facilities are needed in the liquefaction process. The energy consumption is at least 15.1 MJ/kg, which is nearly three times of the current refrigeration technology. The liquefaction process requires very clean hydrogen, and involves multiple cycles, including compression, liquid nitrogen or liquid helium cooling and expansion. The liquid is then transferred to the hydrogenation station. Hydrogen filling for a car at a hydrogen station can be completed in a few minutes. The pressure used is usually 0.6 MPa, slightly higher than atmospheric pressure.

3. Hydride storage

In this method, hydrogen molecules are dissociated into atoms near metals and some compounds. If the lattice structure of the metal or alloy is appropriate, the interstitial position can accommodate relatively small hydrogen atoms. As hydrogen enters the lattice, heat is released, and heat must be supplied to drive the hydrogen out of the lattice again.

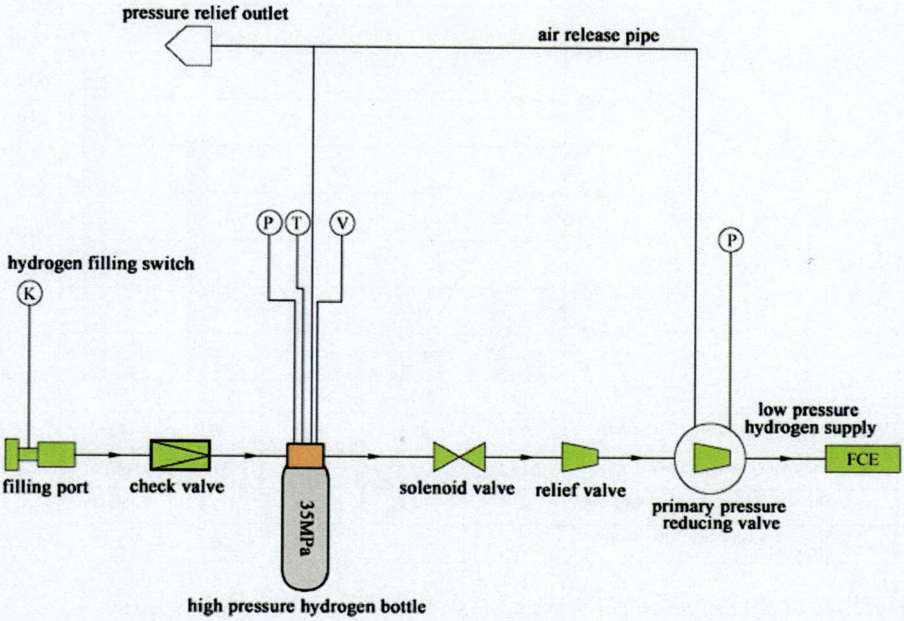

Fig. 7.11 Structure of a typical on-board hydrogen supply system

If the on-board hydrogen supply system is used, it must have the characteristics of large hydrogen storage capacity, good stability and high safety. The system is mainly composed of hydrogen storage device, pressure flow adjusting component, hydrogen leakage sensor, hydrogen supply pipeline and corresponding control system. The structure of a typical on-board hydrogen supply system is shown in Fig. 7.11.

7.5.3 Market Size

In 2017, the production and sales volume of new energy vehicles in China reached more than 700,000. The domestic industry is in a stage of preliminarily mastering the key technologies of fuel cells and electric stacks. Compared with other countries, fuel cell vehicle technology in China is still behindhand.

At present, the fuel cell market around the world is in the initial stage of development and industrialization. In 2017, the fuel cell shipment worldwide was about 73,000 units, with a year-on-year increase of 14.9%. The total capacity shipped was about 669.7 MW, with a year-on-year increase of 29.7%. Global fuel cell shipment quantity, capacity and growth rate from 2013–2017 are shown in Fig. 7.12.

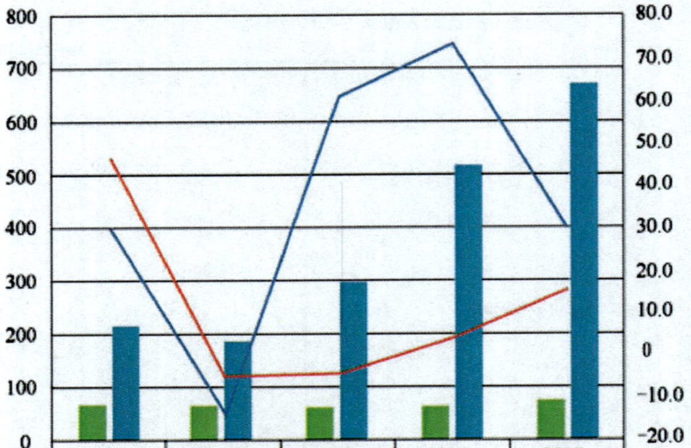

	2013	2014	2015	2016	2017
Shipment quantity (thousand)	66.8	63.6	60.9	63.2	72.6
Shipment capacity (MW)	215.3	182.4	298.1	516.4	669.7
Shipment quantity year-on-year growth (%)	46.2	-4.8	-4.2	3.8	14.9
Shipment capacity year-on-year growth (%)	29.2	-13.9	60.8	73.2	29.7

Fig. 7.12 Global fuel cell shipment quantity, capacity and growth rate. *Source* E4tech

Chapter 8
Internet of Vehicles

8.1 Intelligent Connected Vehicle

The rapid popularization of automobile not only brings convenience to people's life, but also causes traffic accidents, environmental pollution, road congestion and other social problems. Intelligence, electricity and low-carbon emission are the new directions of automobile technology development. Intelligent connected vehicle (ICV) can obtain road traffic information by on-board information sensors. Through V2X technology, it can coordinate people, vehicles, roads and other traffic participants, significantly improve vehicle safety, reduce traffic accident. It has become a main focus of automotive technology research. According to the market data, the annual compound growth rate of the global Internet of Vehicles (IoV) market has reached 25%. According to the statistical data of relevant institutions, since 2014, the market penetration rate of new IoV models has increased year by year. Since 2018, the penetration rate is as high as 31.1%. At present, ICV has become a global trend. The United States, Germany, China, Japan and other countries are actively formulating relevant strategies to be in the lead.

ICV involves many industries, including automobile, information communication, transportation, etc. So far, there is no clear definition of intelligent connected vehicle. Similar concepts include smart car, networked car, IoV, or interconnected autonomous vehicle. The elements of ICV are technology, management, law and society. The first level goal of ICV is to achieve automatic driving, and the higher level goal is to achieve coordinated control between vehicles, and between vehicle and other traffic participants, so as to support the intelligent transportation and smart city. It is a process of driving control or driving responsibility transfer between people and vehicles. With the improvement of the degree of automatic driving, the role of people in driving is becoming weaker and weaker, and the role of vehicles is becoming more and more important. Finally, autonomous driving can be realized. The requirements of building IoV based on ICV mainly include road safety, traffic efficiency and entertainment services.

M. Du, *Autonomous Vehicle Technology*, Unmanned System Technologies,
https://doi.org/10.1007/978-981-19-4143-6_8

The coordination of ICV standards and regulations is becoming the focus of international organizations, and it is carried out in a competitive manner. The United Nations, ISO and IEC are formulating standards and regulations related to ICV. These organizations are all trying to play the leading role. There are other organizations of different industries under the above organizations and they also intend to participate.

SAE in the United States has played a pioneering role with the issuance of two standards. The first one is the automatic driving classification system, which has become a global standard. The second one is the first automobile production standard that has been widely valued and applied around the world. It is a relatively independent organization which is not affected by the rivalry between different countries and institutions.

China sees the development of ICV as part of the national strategies. In 2015, China released *made in China 2025*, which proposed the development direction of low-carbon, informationized and intelligent vehicles, as well as the concept of ICV for the first time. ICV, energy-saving vehicles and new energy vehicles were put forward for the first time as the future development directions of national automobile industry. After that, a series of documents were issued to further emphasize the development of ICV. Some of them also put forward clear requirements for the construction of the standard system of ICV.

8.2 Internet of Vehicles

IOV is a ubiquitous Internet of Things, which is based on Intranet, Internet and mobile Internet. According to the agreed communication protocol and data exchange standard, it carries out wireless communication and information exchange between vehicles and people, vehicles and roads, vehicles and Internet. Through the filtering and cleaning of massive vehicle plus Internet data, and the intelligent processing of data on the IoV platform, intelligent traffic management, intelligent dynamic information service and vehicle intelligent control are completed. The basic composition of the IoV is shown in Fig. 8.1.

The IoV can also be regarded as a vertical field Internet of Things with intelligent vehicles as nodes. It is an important part of urban intelligent infrastructure. The situation of the IoV and urban smart infrastructure is shown in Fig. 8.2

The IoV system makes comprehensive use of artificial intelligence technology, sensing technology, control technology, network technology, computing technology, security technology, etc., so that the system can have a comprehensive perception of traffic conditions and road environment. The vehicle is able to collect holographic information, and upload the information to the big data platform. The uploaded information is summarized, analyzed and processed by the central processor. The system will control each vehicle and each road in real time, so as to provide efficient and safe traffic experience for users.

The IoV is a subdivision of the Internet of Things in the automotive field, an in-depth and practical combination of the mobile Internet and the Internet of Things,

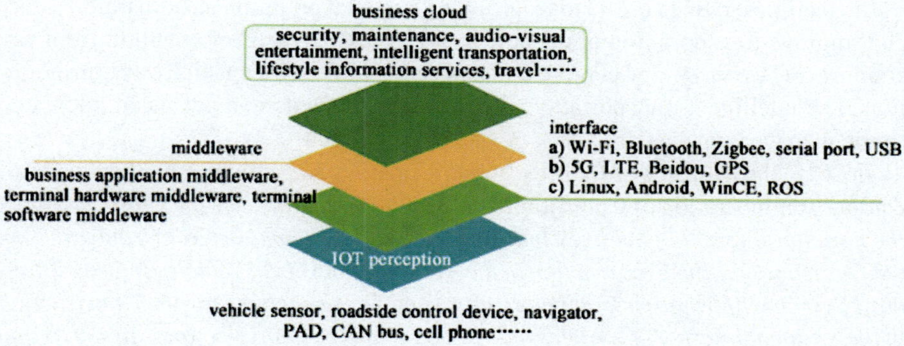

Fig. 8.1 Basic composition of the IoV

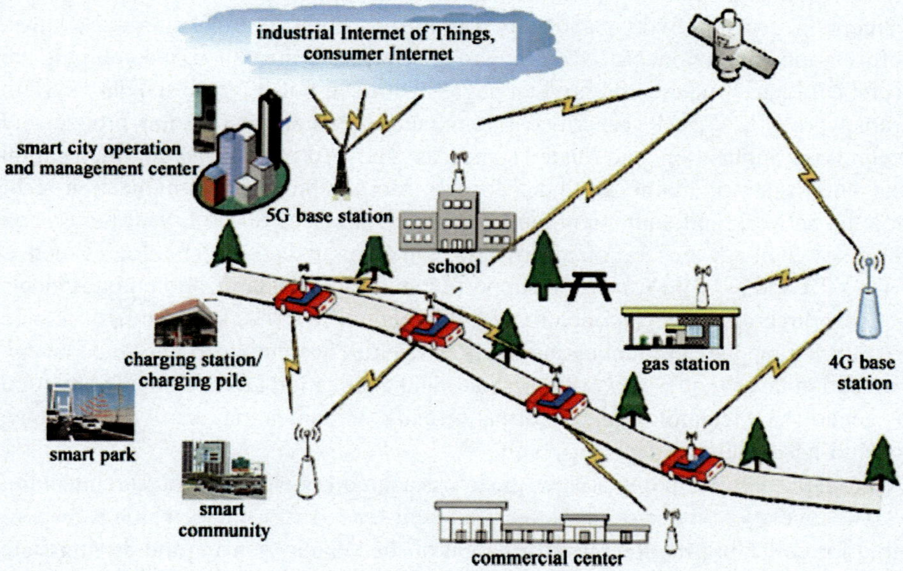

Fig. 8.2 Situation of the IoV and urban smart infrastructure

and a fusion carrier for the application of future transportation, communication, energy, environmental protection, security, city and so on. The vehicle–road cooperative system has always been a key part of intelligent transportation system (ITS). CVIS in Europe, IVHS in the United States, SmartWay in Japan and other systems realize the management and information service of intelligent transportation by establishing effective information communication between vehicles and roads. ITS is the development direction of the future transportation system. It is a real-time, accurate and efficient comprehensive transportation management system that integrates the advanced information technology, data communication transmission technology, electronic sensing technology, control technology and computer technology into the whole ground traffic management system. ITS can make effective use of existing traffic facilities, reduce traffic load and environmental pollution, ensure traffic safety and improve transportation efficiency, so it has been a focus of attention for many countries. The twenty-first century will be a century of intelligent highway transportation. The intelligent transportation system to be adopted is an advanced integrated traffic management system.

In 2018, the Ministry of Industry and Information Technology, and the Standardization Administration of China jointly issued the *Guidelines for the Construction of the national Internet of Vehicles Industry Standard System (General Requirements)*, the *Guidelines for the Construction of the National Internet of Vehicles Industry Standard System (Information Communication)* and the *Guidelines for the Construction of the National Internet of Vehicles Industry Standard System (electronic products and services)*. The standard system of ICV mainly defines the basic framework of definition and classification, the general specification of human–machine interface, functional safety and evaluation, and the application standard related to environmental perception, decision-making and early warning, auxiliary control, automatic control, information interaction, etc. According to the construction method of combining the technical logic structure and product physical structure of ICV, the standard system framework of ICV can be described as foundation, general specification, product and technology application, and related standards. The information and communication standard system is mainly designed for the information and communication technology, network and equipment, and application services of the IoV. It focuses on the research of new generation information and communication technologies such as LTE-V2X and 5G eV2X, so as to support the relevant standardization requirements and key directions of the application development of IoV. The key standards related to information and communication in the IoV industry are divided into three layers: perception layer (end), network layer (net) and application layer (cloud). Supported by generic base technology and information and communication security technology, the end-net-cloud architecture is built.

From the network point of view, the IoV is a three-tier end-net-cloud architecture.

The first layer (end system) is the intelligent sensors of vehicle, which is responsible for collecting intelligent information of the vehicle, sensing the driving state and environment. It is a ubiquitous communication terminal with in-vehicle, vehicle-vehicle and vehicle-network communication. At the same time, it is also a device that enables cars to have the ability of the IOV addressing and network trusted identification.

The second layer (network system) facilitates the interconnection of vehicle to vehicle (V2V), vehicle to road (V2R), vehicle to Internet (V2I), vehicle to human (v2h), etc. It realizes the communication and roaming between vehicle Ad-Hoc network and various heterogeneous networks, guarantees the timeliness, serviceability and network ubiquity in function and performance. It is the unity of public network and private network.

The third layer (cloud system) indicates that the IoV is a vehicle operation information platform based on cloud architecture. Its ecological chain includes ITS, logistics, passenger and freight, specialty vehicles, auto repair and auto parts, car rental, corporate vehicle management, automobile manufacturers, 4S stores, vehicle management, insurance, emergency rescue, mobile Internet, etc. It is a collection of multi-source massive information, so it needs cloud computing functions such as virtualization, security authentication, real-time interaction, mass storage, etc. Its application system is also a composite system of data aggregation, computing, scheduling, monitoring, management and application.

It is worth noting that the current GPS + GPRS combination is not actually the IoV or the Internet of Things, but merely a combination of technologies. At present, most of ITS experiments and IOV concepts in China are based on these technologies. In the future, with the commercialization of 5G technology, the distributed IoV based on 5G vehicle intelligent terminal will be formed. The basic architecture of the IoV is shown in Fig. 8.3.

The overall architecture of the data processing system of the IoV is shown in Fig. 8.4.

The data processing system of the IoV includes data source, data transmission, data preprocessing, data storage and data analysis.

The IoV involves automobile electronics, automobile PC, navigation and positioning system, wireless communication network, vehicle and information security. The elements involved in the industry chain of the IoV are shown in Fig. 8.5.

The IoV systems of A-class cars with outstanding sales including the new generation of eConnect intelligent interconnection technology of Buick Excelle, the intelligent system of Dongfeng Nissan Sylphy, and the CNS 2.0 multimedia system of SAIC Volkswagen Lavida. At present, the interface design, function, third-party app support, remote control and data traffic of these systems are the hot spots of

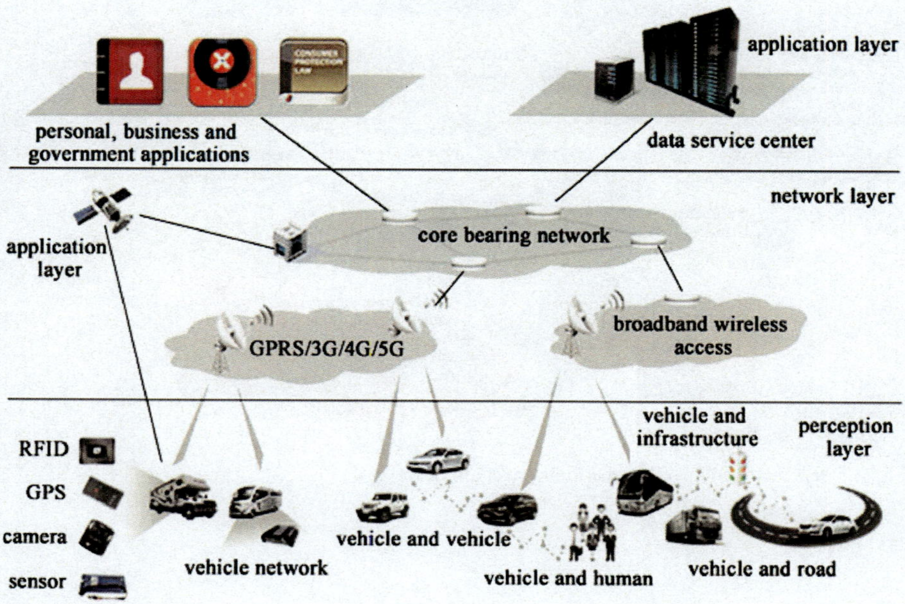

Fig. 8.3 Basic architecture of the IoV

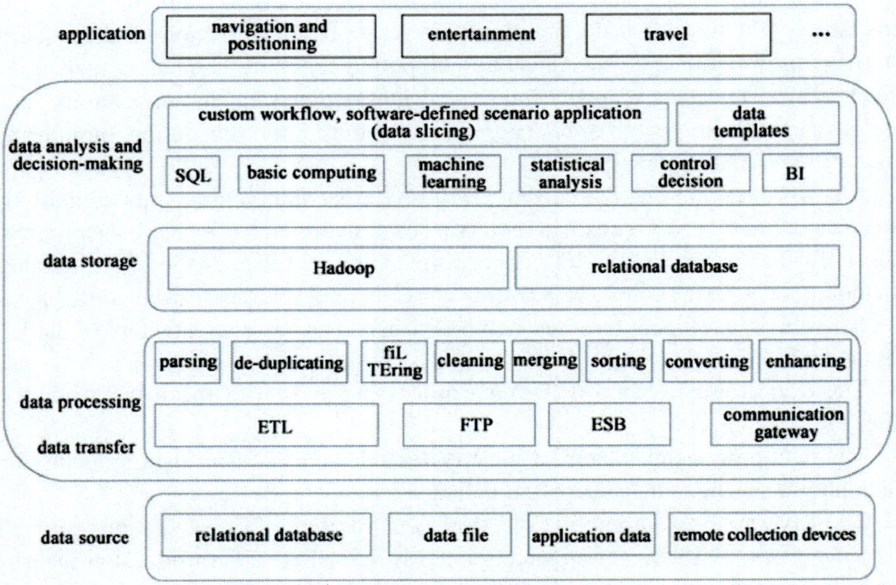

Fig. 8.4 Overall architecture of the data processing system of the IoV

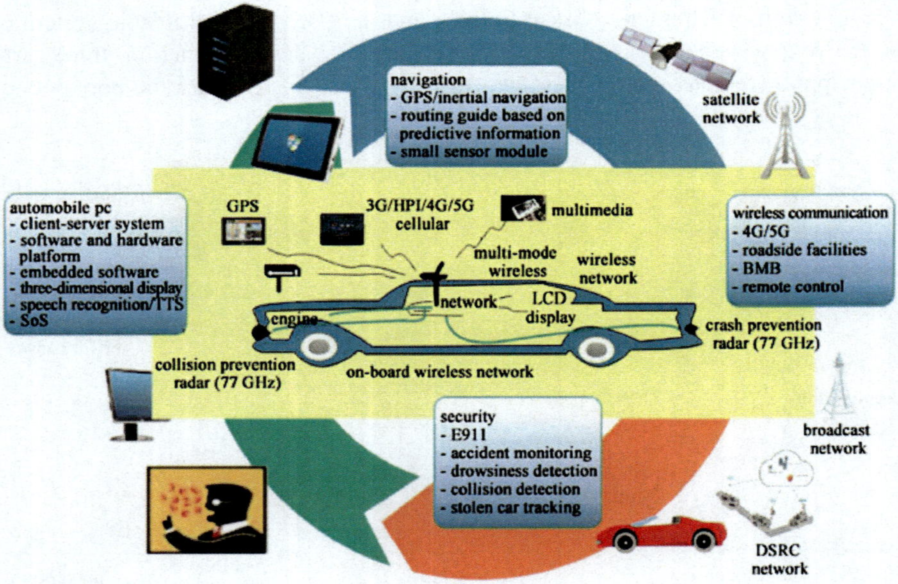

Fig. 8.5 Elements involved in the industry chain of the IoV

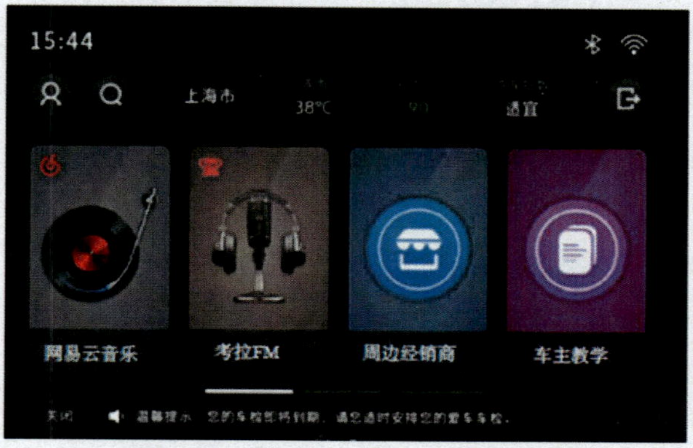

Fig. 8.6 Interface of eConnect intelligent interconnection technology on Buick Excelle

public attention. Figure 8.6 shows the interface homepage of eConnect intelligent interconnection technology equipped on Buick Excelle in 2019.

At present, BMW, Mercedes-Benz, Audi, FCA, GAC, Chang'an, FAW, Geely, Dongfeng, BAIC, SAIC and dozens of other car manufacturers are actively developing their own IoV systems, and establishing cooperative relations with Internet enterprises such as Tencent. Take Tencent as an example. As of March 2019, the enterprise has 45 cooperative models being launched, and has joined hands with more than 300 partners to provide services for car owners. The solutions provided by Tencent include Tencent cloud voice assistant, Tencent My Car mini-program, QQ music, Tencent map, Tencent video and other contents and services. At the same time, exclusive QQ music trial membership, QQ coin discount and other privileges are available. At the beginning of 2019, Tencent and Unicom jointly launched a SIM service plan. Unicom will provide communication services and resources for the IoV. Users can enjoy privileges such as free unlimited targeted data traffic for three years. This will effectively boost the popularization of vehicle networking. While major enterprises are quite active in the IoV field, the formation of the IoV ecology, which integrates the upstream and downstream industrial chain of ICV, is also accelerating. At the beginning of 2019, Shenzhen IoV ecological alliance was established, with members including Tencent, Unicom, Automotive Electronics Industry Association, TP-Link, Z-Link, Topway, Nowada, Imotor, Blackview, Ruihang, Hunydon, Kaiyide, AutoNaviStar, Sndyo, Roadrover, Xinwanhe, Cocolife, Yilu, Changhong, Lelv, Wenwen, Bafangda, Mili and 23 other institutions and enterprises in the upstream and downstream of the industrial chain. In general, IoV has become a hot area which vehicle and Internet enterprises are trying to occupy.

Relying on information and communication technology, the IoV industry provides comprehensive information services through all-round connection and data interaction of within the vehicle, vehicle to vehicle, vehicle to road, vehicle to human, vehicle to service platform. A new industrial form of deep integration of automobile, electronics, information communication, road transportation and other industries is taking shape, which is a global innovation hotspot. The development of IoV is conducive to the innovation of the automobile industry, and beneficial to automobile and transportation services, auxiliary driving and automatic driving, traffic efficiency, reduction of traffic accidents, energy saving, reduction of pollution, and further liberation of productivity.

8.3 3GPP Release and 5G

The development of ICV and IoV communication technology is based on a series of standards formulated by 3GPP, an international organization for standardization. At present, the development of vehicle-vehicle and vehicle–road communication products and technologies largely depends on the progress of 3GPP Release standards and technologies. The timeline of 3GPP Release standards is shown in Fig. 8.7.

At 20:18, June 13, 2018 (11:18, June 14, 2018, Beijing time), in San Diego, the 3GPP plenary session (TSG#80) approved the freezing of the independent networking function of the fifth generation mobile communication technology standard (5G NR). This means that 5G NR has the ability of independent deployment, also brings a new end-to-end architecture. 5G technology can officially enter the commercial stage. The major focuses of 5G includes 5G system and 5G NR (new wireless) access. The former is mainly about complete network slicing, access and mobile management, QoS framework and so on. For the latter, in Rel-15, NR focuses on eMBB and URLLC scenarios. Rel-15 5G access network has two major options, which are non-standalone (NSA) and standalone (SA).

Cellular communication system includes two main parts: radio access network (RAN) and core network. Radio access network is mainly composed of base stations, which provide users with wireless access. The core network mainly provides users with internet access services and corresponding management functions. In 4G LTE system, base station and core network are called eNB (evolved Node B) and EPC (evolved packet core) respectively. In 5G system, the base station is called gNB, the radio access network is called NR (new radio), and the core network is called NGC (next generation core).

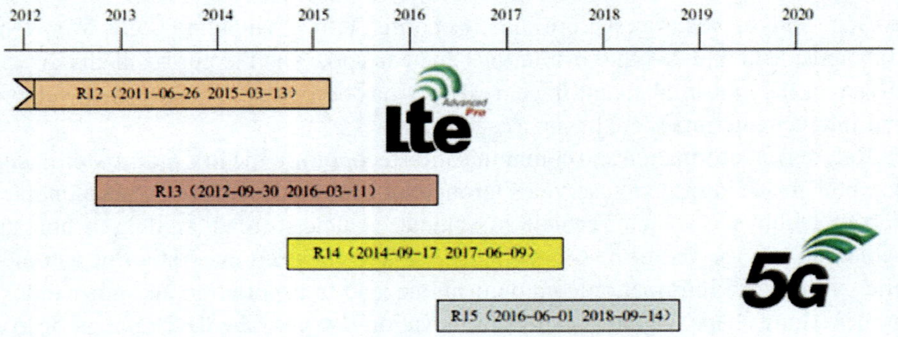

Fig. 8.7 Timeline of 3GPP release standards

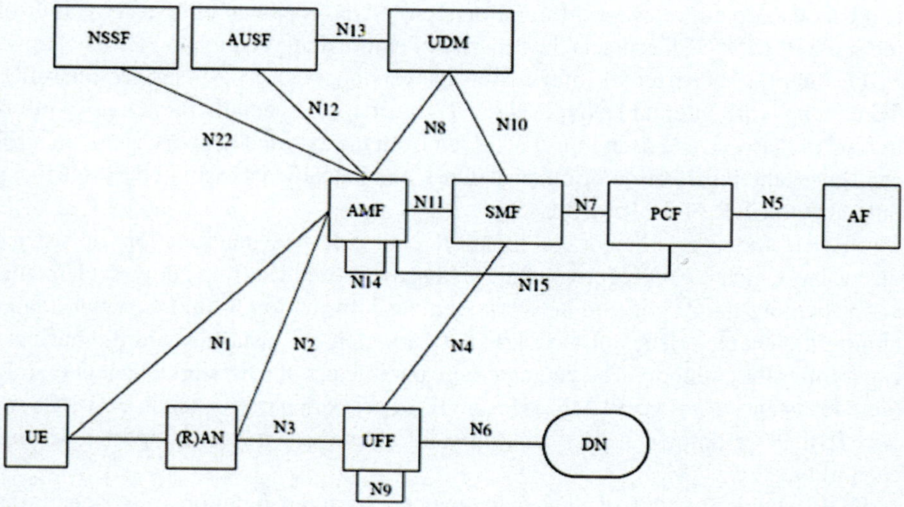

Fig. 8.8 5G system architecture based on reference point

5G system architecture based on reference point is the most common form in communication (traditional point-to-point architecture), which is the architecture mode of non-roaming 5G system, as shown in Fig. 8.8.

(R) AN is access network, which can be 3GPP access network (such as LTE, 5G-NR) or non-3GPP access network (such as common Wi-Fi access). The (R) AN node is the base station in the case of mobile Internet.

AMF is the access and mobility management functional entity. It is the endpoint of (R) AN signaling interface (N2) and NAS (N1) signaling (MM message). It is responsible for the encryption and security of NAS message, registration, access, mobility, authentication, transparent SMS and other functions. In addition, it is also responsible for the allocation of EPS bearer ID when interacting with EPS network. AMF can be compared with 4G MME entity.

SMF is the session management function entity. The main functions of SMF include: the endpoint of SM message of NAS message; the establishment, modification and release of session; UE IP allocation management; DHCP function; ARP proxy or IPv6 neighbor request proxy (in Ethernet PDU scenario); selection and control of UPF for a session; collection of billing data and support for billing interface; decision of the SSC mode of a session; downlink data indication.

UPF is the user interface function entity. It is responsible for packet routing and QoS flow mapping. It is similar to GW (SGW + PGW) in 4G.

PCF is the policy control function entity. It supports the unified policy framework to manage network behavior, provides policy rules to network entities for implementation, and accesses unified data repository (UDR) subscription information. PCF can only access NDR of the same PLMN.

NEF is the network exposure function entity. 3GPP network elements are exposed to other network elements through NEF. NEF stores relevant information in NDR and can also obtain relevant information from NDR. NEF can only access NDR of the same PLMN. NEF protects the security of external application to 3GPP network. 3GPP internal and external information conversion, such as AF-service-identifier, 5G core network internal DNN, S-NSSAI conversion, especially the sensitive information of network and user, must be hidden from the external network element. NEF can access the information of other network elements by accessing NDR. NEF can only access UDR of the same PLMN.

NRF is the network storage function entity. It has the function of service discovery, which means receiving the NF-discovery-request from network elements, and providing the discovered network element information to the requester, maintaining the characteristics of available network element instances and the business capabilities they support. The characteristic parameters of a network element mainly include instance ID, type, PLMN, relevant ID of network partition (such as D-NSSAI, NSI ID), IP or domain name, capability information, name of supported service capability, etc.

UDM stands for unified data management. Its main functions are: generation of 3GPP authentication certificate/authentication parameters; storage and management of subscription permanent identifier (SUPI) of 5G system; management of subscription information; submission of MT-SMS; management of SMS; management of user's service network element registration (such as AMF, SMF, etc., which currently provide services for the terminal).

AUSF is the authentication server network element. It supports authentication of 3GPP access and non-3GPP access.

N3IWF stands for non-3GPP Inter Working Function.

UDR stands for unified data repository. Its functions include UDM storing or reading of subscription data; PCF storing or reading or policy data; storing or reading exposed data. UDR and the accessing NF share the same PLMN, that is, under the same network, the Nudr interface is an internal PLMN interface.

SMSF stands for short message service function.

According to the agreement content of 3GPP 5G NR, the overall architecture and function of NR are as follows.

1. Overall structure

There are two types of NG-RAN nodes:

(a) gNB, which provides NR user plane and control plane protocol and function;
(b) ng-eNB, which provides E-UTRA user plane and control plane protocol and function.

GNB and ng-ENB are connected through Xn interface. GNB/ng-eNB is connected with AMF (access and mobility management function) through NG-C interface, and UPF (user plane function) through NG-U interface.

The overall architecture of 5G is shown in Fig. 8.9. NG-RAN represents wireless access network and 5GC represents core network.

2. Function partitioning

The function partitioning of 5G network is shown in Fig. 8.10. NG-RAN includes gNB or ng eNB nodes, and 5G-C includes three function modules: AMF, UPF and SMF (session management function).

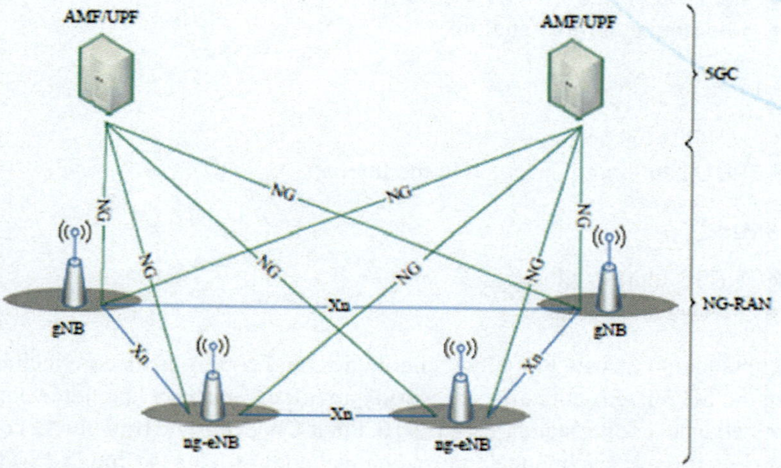

Fig. 8.9 Overall architecture of 5G

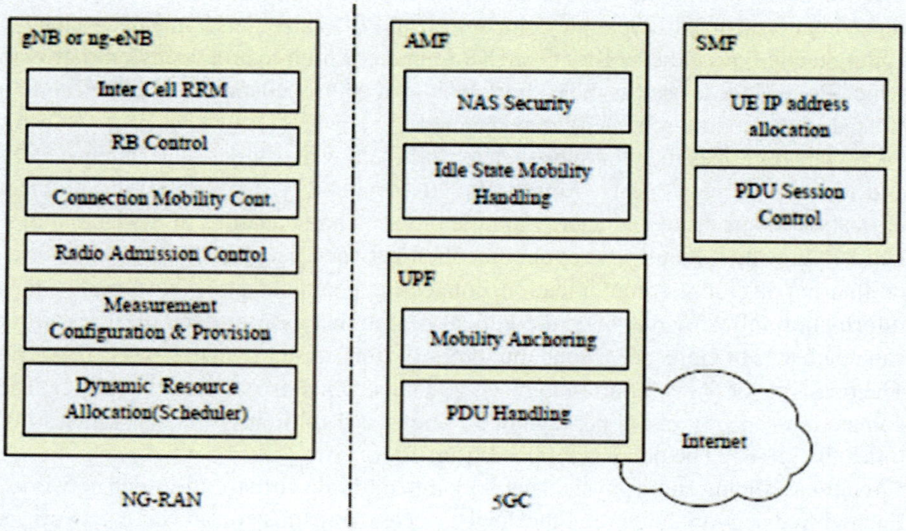

Fig. 8.10 Function partitioning of 5G network

(a) gNB/ng-eNB:

- inter cell radio resource management (RRM)
- radio bear (RB) control
- connection mobility control
- measurement configuration and provision
- dynamic resource allocation

(b) AMF:

- non-access stratum (NAS) security
- idle state mobility handling

(c) UPF:

- mobility anchoring
- PDU handling (connected to the Internet)

(d) SMF:

- UE IP address allocation
- PDU session control

The 5G standard consists of many technologies, and coding is its core technology. Among the 5G related standards, the world's major camps once had fierce competition over channel coding standards. In 2016, Polar Code, pushed by Chinese communication enterprises, became the control channel coding. This is China's first breakthrough in the field of channel coding and lays a foundation for China to win more discourse power in the 5G standard. On November 18, 2016, at the 3GPP RAN1#87 meeting in Reno, Nevada, USA, after several rounds of technical discussions among representatives of participating companies, 3GPP finally determined the channel coding technology scheme for 5G eMBB (enhanced mobile broadband) scenario, in which Polar Code is used as the coding scheme of control channel, and LDPC coding is used as the coding scheme of data channel.

5G integrates key technologies such as millimeter wave, large-scale antenna array and ultra-dense networking. Low delay, high reliability, high data rate, efficient utilization of spectrum and energy are the biggest characteristics of 5G technology. Unlike 4G, which focuses on communication between people, 5G network focuses on Internet of Things communication, connecting people and things, forming a new information infrastructure. The theoretical peak transmission speed of 5G network can reach tens of Gbits per second, hundreds of times faster than that of 4G network. The peak rate of 5G is required to reach 20 Gbps, which is 20 times that of 4G. The connection capacity of 5G per square kilometer is 1 million terminals, which is 10 times that of 4G. The delay is reduced from 10 to 1 ms.

Network slicing cuts a physical network into multiple virtual end-to-end networks. Each virtual network, including the device, access, transmission and core network, is logically independent. The failure of any virtual network will not affect other virtual networks. Network slicing technology is one of the key technologies to improve the flexibility of 5G network architecture to support the needs of various scenarios. According to different business requirements, it can flexibly combine customized network functions into different end-to-end isolated independent networks. Network function virtualization (NFV) is the precondition of network slicing. NFV is to transfer the software and hardware functions of special device in the network (such as MME S/P-GW and PCRF in the core network, digital unit DU in the wireless access network, etc.) to virtual machines (VMs). After the network is virtualized, the wireless access network is called edge cloud, and the core network is called core cloud. VMs in the edge cloud and in the core cloud are interconnected through SDN (software defined network).

8.4 V2X and Its Communication Technology

8.4.1 V2X System

V2X (vehicle to X) includes four parts: vehicle to vehicle (V2V), vehicle to infrastructure (V2I), vehicle to pedestrian (V2P), and vehicle to network (v2N). V2X represents the communication mode between vehicle and other things and is very important for driving safety. For automatic driving, other than the fact that the vehicle needs a perception system to observe the surrounding environment, there are many other requirements. In addition to the sensors, the autonomous vehicle also needs to perform information interaction with all entities that may affect the vehicle, in order to reduce accidents and ease traffic congestion.

1. V2V

V2V is a kind of communication technology not limited to fixed base station, which provides direct end-to-end wireless communication for mobile vehicles. That is, through V2V communication technology, vehicle terminals can directly exchange wireless information with each other without forwarding through the base station. It can be used to monitor the speed, position, and other hidden data of other vehicles on the road, which cannot be disclosed to other drivers. It can automatically predict whether there will be a collision ahead. It is different than the cell phones we use, which rely on a fixed base station to provide signals to interact with other cell phones.

2. V2I

The I of V2I includes base station, street lamp pole, energy supply station, traffic signal lamp, bus station, electric pole, building, overpass, tunnel, roadblock and other facilities. V2I communication function adopts 760 MHz frequency band of on-board intelligent transportation system, which can carry out mutual communication between infrastructure and vehicles without affecting on-board sensors. The vehicle can collect the surrounding environment information through V2I, provide real-time environment model for navigation, control and communication, and help the vehicle better perceive the environment. Frequent change of the base station will cause instability of communication signal. Autonomous vehicle should fully consider this and try to avoid blind areas.

3. V2P

There are many methods for vehicle to perceive pedestrians. In addition to cameras and various sensors, information interconnection is also an effective way. For example, the terminals used by pedestrians, such as cell phones, tablets, wearable devices, can facilitate the interconnection between people and vehicles.

4. V2N

The autonomous vehicle network is designed to obtain abundant information. The basic function is to realize remote network navigation through cloud information, and also to achieve entertainment, anti-theft and other functions. In a broad sense, the Internet of Vehicles can also be considered as a remote sensing technology.

V2X was originally developed on the premise of DSRC (dedicated short range communication) of IEEE 802.11p based on wireless LAN (called DSRC-V2X), which was mainly promoted by Volkswagen, Toyota and GM. Volkswagen announced that all new cars would be equipped with DSRC based V2X technology from 2019. Toyota plans to install DSRC V2X on new cars in the United States in 2021, while GM plans to install DSRC V2X on mainstream cars in 2023 and gradually expand the range of models.

The development process of V2X is as follows. In September 2016, 5GAA was established, and more than 40 automobile enterprises, telecom operators and telecom equipment manufacturers were united. The IoV used to use 802.11p, which is a traditional cellular network system, forming the standard of V2X, also known as C-V2X. With the establishment of 5GAA alliance, many automobile enterprises, manufacturers, telecom operators and telecom equipment manufacturers are supporting the C-V2X industry. Meanwhile, American enterprises such as Qualcomm and Ford also carry out C-V2X tests. The support policy for C-V2X in China is getting clear. In 2016, the Ministry of Industry and Information Technology released the road map of ICV development. In September 2017, Miao Wei, the minister of the Ministry of Industry and Information Technology, proposed to promote the application and deployment of the ICV technology at a relevant conference. In April 2018, the Ministry of Industry and Information Technology, the Ministry of Public Security and the Ministry of Transportation jointly issued the *Road Test Management Practice for Intelligent Connected Vehicles (Trial)*.

LTE-V2X is a V2X communication technology based on mobile cellular network. Two communication modes are defined for vehicle applications: centralized (LTE-V-cell) and distributed (LTE-V-direct). Centralized or cellular mode requires a base station as the control center, while distributed or straight-through mode does not. LTE-V concept and technology was proposed in 2012, and the first prototype was launched in 2014. Its standardization was promoted by 3GPP. In 2017, the LTE-V standard was completed, and its communication module can be integrated into products of other manufacturers. In the past two years, LTE-V has been intensively tested for the frequency RF performance among different manufacturers, ensuring the reliability in practice. In terms of LTE-V standardization, the V2I and V2P standards were completed in March 2017. China Communications Standardization Association completed the wireless connection standard, and China Intelligent Transportation Association and Automotive Engineering Association jointly released the application standard.

The wireless communication technology of IoV includes dedicated short-range communication (DSRC) based on 802.11p and cellular IoV (C-V2X). They both operate on 5.9 GHz ITS spectrum and can work normally in places without cellular network, mobile signal coverage or network operators. With the emerging of 5G commercialization, 5G-V2X will be the main communication mode. Each mode has its own advantages. The former is based on more than ten years of research, with a unified standard and stable technology. The latter has more advantages in coverage, perceived distance, capacity and delay.

In 2019, C-V2X technology flourished. In addition to connecting terminal and device to the network (device to network, D2N), it can also adopt the direct terminal communication mode (device to device, D2D). At present, the frequency band used for V2X in Europe and the United States is 5.8–5.9 GHz. In Japan, the 5.8 GHz band is used for intelligent transportation system (ITS). ITS grows steadily with the popularization of electronic toll collection system (ETC) and vehicle information and communication system (VICS). At present, ETC 2.0 system adopts 5.8 GHz DSRC technology. In other words, C-V2X cannot use this frequency band and will wait for future global arrangement. In November 2018, the Ministry of Industry and Information Technology of China issued the *Management Regulations on the Use of 5905–5925 MHz in Direct Communication of Internet of Vehicles*, which clearly set the 5.9 GHz band as the working band of direct communication of IoV (or ICV) based on LTE C-V2X technology.

LTE-V2X, also known as C-V2X PC5 direct communication, is the core technology of ITS. The national policy is also vigorously supporting the development of this technology. LTE-V2X is a global solution for V2X communication. It has a powerful 5G NR evolution path, aiming at improving vehicle safety, automatic driving performance and traffic efficiency. It is the only V2X communication technology that complies with the global 3GPP specification. The direct connection communication mode based on PC5 means supporting V2V, V2I and V2P communication in the unified 5.9 GHz ITS frequency band, without the assistance of SIM card or cellular network.

8.4.2 DSRC Technology

As early as 1994, CEN/TC278 began to draft DSRC standard in Europe. In 1997, *The Standard of 5.8 GHz DSRC Physical Layer and Data Link Layer* was approved. However, the European standard is different from the American Standard in terms of frequency band and modulation mode. In the past six years, Germany, the

Netherlands and Austria have signed a memorandum of understanding to deploy and test C-ITS on a highway connecting the three countries. The Directorate-General for Mobility and Transport of the European Commission has created a C-ITS platform. In 2016, the ministers of transport of 28 EU Member States signed the Amsterdam Declaration and adopted a European strategic plan on C-ITS. The goal is to achieve large-scale commercial deployment of C-ITS in 2019. DSRC was confirmed as the V2V standard by the U.S. Department of Transportation in February 2014. After years of development and testing, it has been relatively finalized. DSRC is based on IEEE 802.11p standard and supports wireless access of vehicles. It is widely used in expressway toll, non-stop fast lane, parking lot management and other scenarios. The feature of DSRC is that it is suitable for identifying and connecting vehicles with short distance (tens of meters) and high speed. 802.11p specification was established in 2012. It was developed on the basis of the wireless technology 802.11a, which has 20 years of history. The performance test results of 802.11p show that due to the network congestion and the lack of minimum performance guarantee, the coverage of 802.11p is limited, and the performance of large-scale application is unstable. Therefore, it lacks practicability and security.

In terms of manufacturers, Volkswagen, Renault, GM, NXP, Autotalks (an Israeli chip manufacturer) and Kapsch (an American intelligent transportation system provider) have been promoting on-board vehicle Wi-Fi (DSRC) technology. DSRC technology has been sufficiently verified, and fully standardized in European government backed projects. It is more suitable for collision avoidance and target navigation.

8.4.3 C-V2X Technology

Cellular IoV (or C-V2X), also known as LTE-V2X, is a V2X communication technology based on the global 3GPP Release 14 specification. Its direct connection communication mode based on PC5 operates in the 5.9 GHz ITS frequency band, supporting direct V2V, V2I and V2P communication. In addition, C-V2X has a powerful 5G NR evolution path, supports high-speed movement and high-density vehicle scene, and can realize non-line-of-sight sensing and other functions. It is an important part of safe automatic driving in the future. Compared with 802.11p, C-V2X is based on the technology specially designed for high-speed mobile applications, and optimized for automotive applications. The practical experience of 802.11p research and the basic progress of wireless communication are also applied. In addition, it also supports a range of new automotive applications, bringing higher safety and autonomous driving performance.

C-V2X includes direct communication and network-based communication. Direct communication supports V2V, V2I and V2P communication. The vehicle to network (V2N) technology has developed from the original remote information services to the current information and entertainment services, as well as remote operation applications for automatic driving, which are of great importance. The comprehensive C -V2X network is shown in Fig. 8.11.

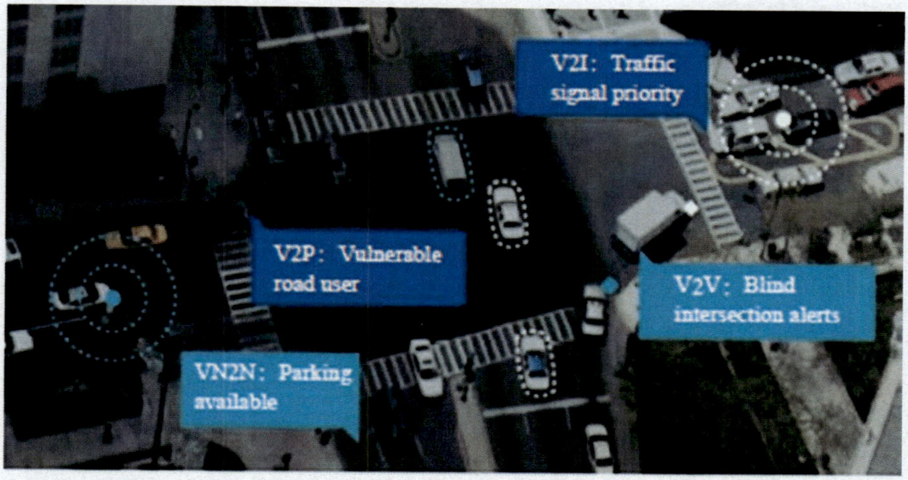

Fig. 8.11 Comprehensive C-V2X network

Compared with the European Union's preference for DSRC, China and the United State have been promoting C-V2X. China has always been the pioneer and leader of C-V2X development, with the continuous expansion of the industry. C-V2X has the following technical advantages: enhanced transmission distance and reliability, support for high-density vehicle scenarios, direct connection autonomous communication with reduced cost and system complexity, collaborative remote information processing platform, reuse of DSRC/C-ITS protocol layer, influence of cellular network ecosystem, support for high-speed vehicle scenarios and smooth evolution to 5G. C-V2X has spawned a large number of new product applications, which can realize V2V, V2I, V2P and other application scenarios. The multi-industry integration and innovation system of C-V2X has also been preliminarily completed, and has its supporters in chip module, platform operation, automobile manufacturing and other fields.

C-V2X realizes V2X communication with 4G (LTE) and 5G technologies. Compared with DSRC technology, C-V2X has the advantages of long communication range, low delay, decent scalability, flexible spectrum bandwidth allocation, wide coverage, higher reliability and large bandwidth support. At present, car manufacturers supporting C-V2X technology include Ford, BMW, Daimler, Audi, Ducati, Peugeot Citroen, SAIC and Tesla. Relevant organizations include Deutsche Telekom, Ericsson, etc., which believe that C-V2X technology can enable cars to connect to other networking devices through cellular network. Sixteen industry giants, including Ford, Volkswagen, Jaguar Land Rover, Intel, Ericsson, Samsung, and Qualcomm, have submitted their opinions to the Federal Communications Commission (FCC) to support the exemption request submitted by 5G Automobile Association (5GAA) to the FCC, requiring an administrative license to deploy C-V2X technology in the 5.9 GHz band. According to the prediction, ICV will account for about 70% of the global automobile market in 2022, and the selection criteria of the European Union may affect the production of some equipment manufacturers.

8.4.4 5G-V2X and Its Future Trend

With the approaching of 5G commercialization, the ceiling of 802.11p has been exposed. C-V2X is evolving to 5G-V2X through the built-in cellular modem in the car. The arrival of 5G is not only a revolutionary storm for mobile phones, but also the most critical event for LTE-V2X. In addition to higher download speed, 5G provides another important feature which is low delay. The powerful communication ability of 5G can make driving more safe and efficient. The evolution route of communication standardization of ICV is as follows. Firstly, the first version of standards is completed according to 3GPP. Then LTE-V (including V2V, V2R, V2P) standardization is achieved. The current standardization is the evolution and enhanced version. The standardization based on 5G new technology is in the research stage, and the whole standardization process ensures the smooth evolution of technology.

Based on 4G, 5G comes with faster network transmission, greater connection density, and lower end-to-end delay. It puts forward higher requirements for fixed network transmission rate and optical device density and speed.

The key performance indicators of 5G are as follows:

1. transmission rate: 100 Mbps–1 Gbps rate for users;
2. system capacity: 1 million connections per square kilometer;
3. NR delay: as low as 1 ms;
4. reliability: close to 100% business reliability guarantee;
5. ultra-low power consumption, ultra-low cost, etc.

The key capabilities of 5G are high speed, large capacity and low delay. Compared with 4G, 5G has a significant improvement in major performance indicators. For example, the peak rate of 5G is required to reach 20 Gbps, which is 20 times that of 4G. The connection capacity of 5G per square kilometer is 1 million terminals, which is 10 times that of 4G. The delay is reduced from 10 to 1 ms.

The driving force for the birth of 5G comes from four aspects, as shown in Fig. 8.12.

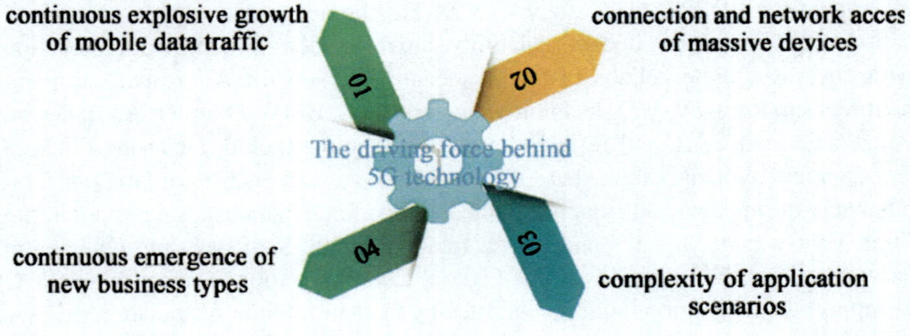

Fig. 8.12 Driving force for the birth of 5G

In the 4G era, the core network mostly adopts the centralized deployment mode. Faced with the requirements of diversified business scenarios, 5G core network will go through cloud evolution. It will achieve a flexible network architecture by centralized deployment or partial sinking according to different business requirements such as low delay ultra-reliable low-latency communication (uRRLC), enhanced mobile broadband (eMBB), massive machine type communications (mMTC). Specific trends include the downward application gateway, coordinated nearby forwarding, local end of traffic, and decentralization. The large bandwidth and low delay of 5G improve the radio access network (RAN) architecture. 4G network is a two-level structure of BBU and RRU, where one base station corresponds to one computer room. 5G evolves into a three-level structure of CU, DU and AAU. The BBU is divided into CU and DU. The CU is placed in the cloud, and DU can be deployed in the far end, achieving centralized management and control. The core network moving down and cloudification is becoming the development trend of 5G, and the corresponding bearer network is also divided into three levels. The 5G RAN deployment architecture is shown in Fig. 8.13.

Unlike 3G and 4G with various standards, 5G will work with a unified global standard. China has deeply participated in the formulation of 5G standards. In November 2016, the Polar Code, represented by Huawei and driven by China, was adopted by 3GPP as the 5G eMBB control channel standard scheme. In June 2017, 3GPP officially confirmed the SBA architecture of 5G core network led by China Mobile as the unified infrastructure. The depth of Chinese enterprises' participation in standardization is unprecedented.

5G will support massive machine communication. Typical application scenarios such as smart city and smart building are deeply integrated with the fifth generation mobile communication. It is expected that 100 billion devices will be connected to 5G network. 5G will also detonate vertical industry applications such as IoV,

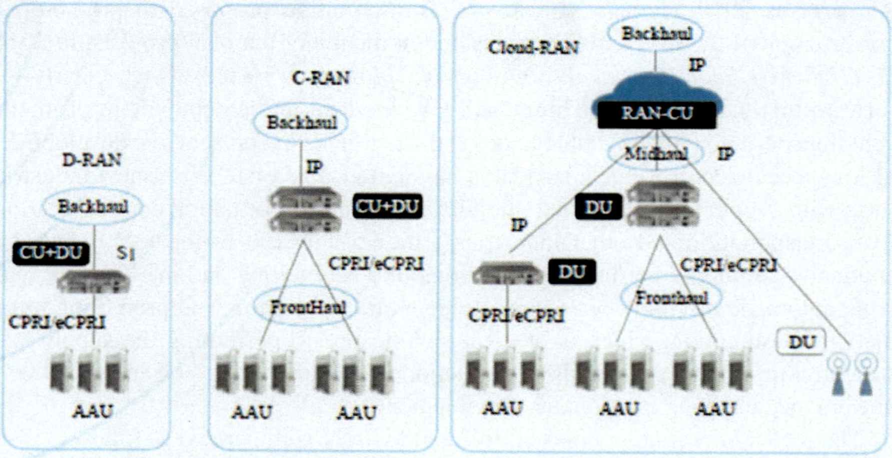

Fig. 8.13 5G RAN deployment architecture

mobile medical, energy Internet, industrial Internet and so on with its excellent performance of ultra-high reliability and ultra-low delay. In the initial stage of 5G network construction, it is difficult to achieve full coverage due to the high frequency band and large propagation loss. There are multiple options of network architecture including NSA (non-standalone) and SA (standalone). The NSA network adopts the joint network structure of LTE and 5G, and uses the existing 4G network with good coverage to realize the rapid introduction of 5G NR. While the SA network can better reflect the advantages of 5G technology to improve the quality of service, it has long introduction cycle and higher requirements for continuous coverage of 5G NR. At present, 3GPP gives priority to the NSA networking mode. It is expected that the NSA networking standard will be completed by the end of 2017, and then the 5G SA networking standard will be completed in July 2018.

In the scenarios of smart city and smart transportation, the evolution of 5G bearer network not only needs to consider the bandwidth, delay and other related network indicators, but also needs to consider the requirements of 5G flexible bearer networking, 4G/5G coexistence bearer and connection with the existing network. The unified bearer of 4G/5G coexistence network is the key problem in the evolution of 5G bearer network.

So far, V2X is based on the combination of DSRC and C-V2X. With the continuous development of application scenarios, communication delay and traffic load will become the new challenges. In the long run, due to the gradual implementation of 5G network construction, which can achieve 1 ms ultra-low delay and a large number of Internet of Things terminal connections, vehicle–road collaboration, autonomous vehicle and the IoV will develop rapidly.

8.5 National Policies, Regulations and Standards

As early as 2010, China's Ministry of Transportation proposed to promote the construction of the IoV, and listed it as one of the major national projects. In 2011, *the 12th Five-Year Plan of the Internet of Things* was issued, which clearly put forward that the Internet of Things will take the lead in the deployment of intelligent transportation, intelligent logistics and other fields. At present, the development of low-speed electric vehicle standards has started. IoV or ICV is a new industrial form with deep integration of automobile, electronics, information communication, road transportation and other industries. The development of the IoV industry is conducive to improving the level of automobile networking and intelligence, realizing automatic driving, developing intelligent transportation, and promoting information consumption, which is of great significance to promoting the supply side structural reform, promoting the construction of manufacturing power and network power, and achieving high-quality development.

The recent important policies on IoV and ICV are summarized as follows.

1. *Action Plan for the Development of Internet of Vehicles Industry*, Ministry of Industry and Information Technology

The Action Plan proposes to take the integration development as the main strategy, give full play to China's industrial advantages, optimize the policy environment, strengthen industry cooperation, break through key technologies, consolidate the cross-industry foundation, and promote the formation of a new ecology of the Internet of Vehicles industry with deep integration, active innovation, safety and reliability, and strong competitiveness.

The Action Plan points out that the goal of high-quality development of the Internet of Vehicles industry will be achieved by stages. In the first stage, by 2020, a breakthrough will be made in the cross-industry integration of the ICV industry. The ICV with high-level automatic driving function will reach scale application in specific scenarios. The penetration rate of users of ICV will reach more than 30%, and the level of intelligent road infrastructure will be significantly improved. In the second stage, after 2020, the technological innovation, standardization system, infrastructure, application service and safety guarantee system will be fully completed. ICV with high-level automatic driving function and 5G-V2X will be gradually put into large-scale commercial application, the human-vehicle–road-cloud structure will achieve high degree of coordination, and the people's growing demand for a better life will be better met.

In accordance with the principle of systematic deployment, overall promotion, innovation leading, application driving, complementary advantages, open cooperation, strengthened management and safety guarantee, five key tasks are proposed on the basis of in-depth research. The first one is to break through key technologies and promote industrial development, make full use of various innovative resources, accelerate the development and application of key parts and systems of ICV, and promote the construction of ICV decision control platform. Relevant authorities will vigorously support the R&D and industrialization of key wireless communication network technologies such as LTE-V2X and 5G-V2X, and build a comprehensive IoV architecture combining communication and computing. The second one is to improve the standardization system and promote test verification and demonstration application, fully implement the guidelines for the construction of national standardization system for the IoV industry, improve and formulate key standards for IoV, issue frequency licenses timely, and build a test and evaluation system for ICV. Relevant authorities will promote the demonstration application of scenarios such as automatic driving and intelligent logistics in airports, ports and parks, build a national pilot area of the IoV, and continuously improve the level of intelligent traffic management and residents' travel service experience. The third one is to jointly build and promote the improvement of the infrastructure of the IoV industry, strengthen the cooperation between departments and provinces, and build the network infrastructure based on LTE-V2X, 5G-V2X and other wireless communication technologies; build a comprehensive big data and cloud platform, promote the informatization and intelligent transformation of road infrastructure, and support the construction of intelligent infrastructure environment integrating sensing, communication, computing and other capabilities. The fourth one is to develop comprehensive application and promote market penetration; vigorously develop IoV users, cultivate innovative applications such as intelligent travel, develop real-time online monitoring system for electric vehicles and big data analysis capabilities, and promote large-scale application of vehicle–road interactive information services; promote the application of pre-accident warning and collaborative control technology, improve the ability of traffic safety and congestion active control, establish a network-based automobile design, manufacturing and service integration system, and achieve the scale application of personalized automobile service based on big data platform. The fifth is to combine technology and management to promote the improvement of the security system; focus on the operation security, data security and network security of the ICV system, improve the safety management system and protection mechanism, build the all-factor safety detection and evaluation system of ICV, IoV data and network, break through the core technology, and strive to improve the level of hidden hazard investigation, risk discovery and emergency response.

2. *Medium and Long Term Development Plan of Automobile Industry* (No. 53, 2017, MIIT LZ), Ministry of Industry and Information Technology, National Development and Reform Commission and Ministry of Science and Technology.

The Action Plan states the idea of "promoting the structural reform of the supply side of the automobile industry, regulating the total amount, optimizing the structure, collaborating for innovation, transforming and upgrading", and the goal of "striving to become one of the world's best automobile manufacturers after ten years of continuous efforts". For the practice of vehicle inspection and test, the objective is as follows: to carry out the demonstration and promotion of ICV; to introduce the test and evaluation system, promote the application demonstration of ICV step by step, and steadily expand the pilot range. In the demonstration area, test and verification environment and corresponding data collection and analysis, management and monitoring platform will be built. Demonstration and evaluation of product performance verification of ICV will be carried out. Information exchange and coordination mechanism among ICV, the Internet, Internet of Things, intelligent transportation network, smart grid and smart city will be established. The innovative development method of ICV suitable for China with cross-field cooperation will be explored.

3. *Action Plan (2015–2018) for Implementing the Guidance of the State Council on Actively Promoting the Internet + Scheme* (no. 440, 2015, MIIT XR), Ministry of Industry and Information Technology

The Action Plan mentions the future development plan of the IoV for the first time. Meanwhile, the *Action Plan for the Development and Innovation of the Internet of Vehicles (2015–2020)* will be issued in due time. In terms of top-level design, it will comprehensively promote the research and development of the IoV technology and the formulation of standards. The analysis points out that promoting the development of the IoV from the strategic level will point out a clear development direction for the whole industry, and further promote the development of future technologies such as automatic driving based on the IoV. It is reported that in the face of the development of new energy vehicles and IoV technology, a number of car companies have launched driverless technology. Baidu, Google and Volvo have conducted road test of relevant models. The Action Plan sets the acceleration of the deep integration of a new generation of information and communication technology and industry as the main line, the implementation of the Internet manufacturing industry as the focus, high-speed broadband network infrastructure and information technology industry as the support, to create a new competitive advantage in the industry under the new situation, combine the decisive role of the market on the allocation of resources and the role of the government, highlight the main position of enterprises, vigorously expand the depth and breadth of the integration of the Internet and manufacturing, actively cultivate new technologies, new products, new business models, deepen the reform of institutional mechanisms, release the development potential and vitality, promote industrial transformation and upgrading, and improve quality and efficiency. Relevant authorities and enterprises should accelerate the construction of the car networking industry, promote the development of car networking technology, standardization development, car networking pilot, 5G technology-based car networking demonstration. In terms of the system solution enhancement, the Action Plan discusses the application testing and pilot demonstration of the cyber physical system (CPS) for aviation, automotive, electronics, petrochemical and other key industries.

The *Guidance of the State Council on Actively Promoting the Internet + Scheme* points out that it is necessary to promote the innovation of intelligent products in key areas, promote the establishment of cross-field innovation platforms between automobile enterprises and Internet enterprises, and accelerate the R&D and application of intelligent assisted driving, complex environment perception, on-board intelligent devices and other technical products. The policies issued by the Ministry of Industry and Information Technology also clarify the future development direction of the IoV, which means that after years of preparing, the IoV will usher in a five-year golden period of development.

4. *Guidelines for the Construction of National Internet of Vehicles Industry Standard System (ICV)* (No. 332, 2017, MIIT LK), Ministry of Industry and Information Technology and Standardization Administration of China

The Guidelines point out that the IoV industry is a new industry with deep integration of automobile, electronics, information communication, road transportation and other industries, and is a global innovation hotspot and commanding height of future development. In order to fully implement the "Made in China 2025" scheme, promote the Internet +, promote the transformation and upgrading of related industries, vigorously cultivate new kinetic energy, and highlight leading role of standardization in the construction of the ecological environment of the IoV industry, the Ministry of Industry and Information Technology and Standardization Administration of China have jointly drafted *Guidelines for the Construction of National Internet of Vehicles Industry Standard System.* According to the main objects of standardization, this series of documents can be divided into general requirements, ICV, information and communication, electronic products and services, etc. See Table 8.1 for the standard system of ICV.

Table 8.1 Standard system of ICV

Standard name and category		Standard type	Standard nature	Status	Corresponding international or foreign standard code
Foundation (100)					
Terms and definitions (101)					
101–1	Intelligent connected vehicle terms and definitions	National standard	Recommended	pre-research	
101–2	Advanced driver assistance systems (ADAS) terms and definitions	National standard	Recommended	Projected 20,171,038-T-339	
101–3	Automotive information security terms and definitions	National standard	Recommended	Pre-research	
Classification and coding (102)					
102–1	Automotive intelligence, networked information classification and code	National standard	Recommended	Pre-research	
102–2	Data structure and transmission format for intelligent and networked automobiles	National standard	Recommended	Pre-research	
102–3	Automotive driving automation rating	National standard	Recommended	Project applied	

(continued)

Table 8.1 (continued)

Standard name and category		Standard type	Standard nature	Status	Corresponding international or foreign standard code
102–4	Automotive networking level classification	National standard	Recommended	Pre-research	
102–5	Automotive information security risk classification and ranking	National standard	Recommended	Pre-research	
102–6	Automotive information security domain and protection hierarchy definition	National standard	Recommended	Pre-research	
Markings and symbols (103)					
103–1	Intelligent networked vehicle signal graphics and signs	National standard	Recommended	Pre-research	
103–2	Smart grid automotive alarm signal	National standard	Recommended	Pre-research	
General specifications (200)					
Functional evaluation (201)					
201–1	General specifications for evaluation of intelligent functions and performance of automobiles	National standard	Recommended	Pre-research	

(continued)

Table 8.1 (continued)

Standard name and category		Standard type	Standard nature	Status	Corresponding international or foreign standard code
201–2	General specifications for the evaluation of the functions and performance of automotive network connectivity	National standard	Recommended	Pre-research	
201–3	Automotive intelligent application working conditions	National standard	Recommended	Pre-research	
201–4	Automotive networking application working conditions	National standard	Recommended	Pre-research	
201–5	Automotive software upgrade technical conditions and functional evaluation specifications	National standard	Recommended	Pre-research	
Human–machine interface (202)					
202–1	Evaluation methods for automotive human–machine interface systems	National standard	Recommended	Pre-research	

(continued)

Table 8.1 (continued)

Standard name and category		Standard type	Standard nature	Status	Corresponding international or foreign standard code
202–2	Automotive alarm signal priority specification	National standard	Recommended	Pre-research	
202–3	General specifications for car alarm signals	National standard	Mandatory	Pre-research	
202–4	Automotive man–machine control conversion system performance requirements and test methods	National standard	Mandatory	Pre-research	
Functional safety (203)					
203–1	Road vehicles functional safety (Part 1–10)	National standard	Recommended	Published GB/T 34,590–2017 (Part 1–10)	ISO 26262
203–2	Failure protection requirements and evaluation methods for human–computer interaction system of intelligent connected vehicles	National standard	Mandatory	Pre-research	

(continued)

Table 8.1 (continued)

Standard name and category		Standard type	Standard nature	Status	Corresponding international or foreign standard code
203–3	Functional safety requirements for automotive interactive interfaces	National standard	Recommended	Pre-research	
203–4	Functional safety requirements for automotive information perception systems	National standard	Recommended	Pre-research	
203–5	Functional safety requirements for automotive decision warning systems	National standard	Recommended	Pre-research	
203–6	Functional safety requirements for automotive auxiliary control systems	National standard	Recommended	Pre-research	
Information security (204)					
204–1	General technical requirements for automotive information security	National standard	Recommended	Project applied	

(continued)

Table 8.1 (continued)

Standard name and category		Standard type	Standard nature	Status	Corresponding international or foreign standard code
204–2	Automotive information security risk assessment guide	National standard	Recommended	Pre-research	
204–3	General requirements for security and privacy protection of automotive data	National standard	Recommended	Pre-research	
204–4	On-board operating system and application software security protection requirements	National standard	Recommended	Pre-research	
204–5	General test and evaluation methods for automotive information security	National standard	Recommended	Pre-research	
204–6	Safety technology requirements for automotive information perception devices	National standard	Recommended	Pre-research	
204–7	Vehicle ECU information security technical requirements	National standard	Recommended	Pre-research	

(continued)

Table 8.1 (continued)

Standard name and category		Standard type	Standard nature	Status	Corresponding international or foreign standard code
204–8	Technical requirements for information security of vehicle bus system	National standard	Recommended	Pre-research	
204–9	Automotive gateway information security technical requirements	National standard	Recommended	Project applied	
204–10	Information security technical requirements for telematics box (TBOX)	National standard	Recommended	Project applied	
204–11	On-board diagnostic interface (OBD) information security technical requirements	National standard	Recommended	Pre-research	
204–12	Driver identification system technical requirements	National standard	Recommended	Pre-research	
204–13	Automotive software upgrade information security protection specification	National standard	Recommended	Pre-research	

(continued)

Table 8.1 (continued)

Standard name and category		Standard type	Standard nature	Status	Corresponding international or foreign standard code
204–14	Technical requirements for information security of electric vehicle telematics service and management system	National standard	Recommended	Project applied	
204–15	Technical requirements for information security of electric vehicle charging system	National standard	Recommended	Project applied	
204–16	Automotive information security breach emergency response guide	National standard	Recommended	Pre-research	
Products and technology applications (300)					
Information perception (301)					
301–1	Automotive reversing view assist performance requirements and test methods	National standard	Recommended	Pre-research	

(continued)

Table 8.1 (continued)

Standard name and category		Standard type	Standard nature	Status	Corresponding international or foreign standard code
301–2	Adaptive front lighting system performance requirements and test methods	National standard	Recommended	Published GB/T 30,036–2013	
301–3	Automotive panoramic image monitoring system performance requirements and test methods	National standard	Recommended	Pre-research	
301–4	Automotive night vision system performance requirements and test methods	National standard	Recommended	Pre-research	
301–5	Automotive distance monitoring system performance requirements and test methods	National standard	Recommended	Pre-research	
301–6	Head-up Display (HUD) system performance requirements and test methods	National standard	Recommended	Pre-research	

(continued)

Table 8.1 (continued)

Standard name and category		Standard type	Standard nature	Status	Corresponding international or foreign standard code
301–7	Vehicle-borne satellite positioning system signal receiving device performance requirements and test methods	National standard	Recommended	Pre-research	
Decision alert (302)					
302–1	Alarm system of commercial vehicles of 3.5 Tons or more	National standard	Recommended	Published GB/T 26,776–2011	
302–2	Automotive blind spot monitoring system performance requirements and test methods	National standard	Recommended	Projected 20,171,039-T-339	
302–3	Pedestrian monitoring system performance requirements and test methods	National standard	Recommended	Pre-research	
302–4	Alcohol detection system performance requirements and test methods	National standard	Recommended	Pre-research	

(continued)

Table 8.1 (continued)

Standard name and category		Standard type	Standard nature	Status	Corresponding international or foreign standard code
302–5	Automotive forward collision warning system (FCW) performance requirements and test methods	National standard	Recommended	Reported for approval	ISO 15623–2013 ECE R131
302–6	Lane departure warning system (LDW) Performance requirements and test methods	National standard	Recommended	Published GB/T26773-2011	ISO 17361–2007 ECE R130
302–7	Door opening blind zone monitoring system performance requirements and test methods	National standard	Recommended	Pre-research	
302–8	Performance requirements and test methods for rear traffic crossing alert system	National standard	Recommended	Pre-research	
302–9	Intelligent speed limit control system performance requirements and test methods	National standard	Recommended	Pre-research	

(continued)

Table 8.1 (continued)

Standard name and category		Standard type	Standard nature	Status	Corresponding international or foreign standard code
302–10	Pre-crash safety system performance requirements and test methods	National standard	Recommended	Pre-research	
302–11	Automotive parking distance measurement warning device performance requirements and test methods	National standard	Recommended	Published GB/T 21,436–2008	
302–12	Low-speed driving assist performance requirements and test methods	National standard	Recommended	Pre-research	ISO 17386
302–13	Extended range reversing assist system performance requirements and test methods	National standard	Recommended	Pre-research	ISO 22840–2010
302–14	Curve speed warning system performance requirements and test methods	National standard	Recommended	Pre-research	ISO 11067

(continued)

Table 8.1 (continued)

Standard name and category		Standard type	Standard nature	Status	Corresponding international or foreign standard code
302–15	Driver attention monitoring system performance requirements and test methods	National standard	Recommended	Pre-research	
Auxiliary control (303)					
303–1	Low-speed following system performance requirements and test methods	National standard	Recommended	Pre-research	ISO 22178–2009
303–2	Adaptive cruise control (ACC) performance requirements and test methods	National standard	Recommended	Published GB/T 20,608–2006	ISO 15622–2010
303–3	Full speed range adaptive cruise control performance requirements and test methods	National standard	Recommended	Pre-research	ISO 22179–2009

(continued)

Table 8.1 (continued)

Standard name and category		Standard type	Standard nature	Status	Corresponding international or foreign standard code
303–4	Passenger vehicle automatic emergency braking system (AEB) performance requirements and test methods	National standard	Recommended	Projected 20,151,489-T-339	
303–5	Commercial vehicle automatic emergency braking system (AEB) performance requirements and test methods	National standard	Recommended	Projected 20,170,405-T-339	ECE R131
303–6	Passenger vehicle lane keeping assist system (LKA) performance requirements and test methods	National standard	Recommended	Projected 20,171,040-T-339	
303–7	Commercial vehicle lane keeping assist system (LKA) performance requirements and test methods	National standard	Recommended	Project applied	

(continued)

Table 8.1 (continued)

Standard name and category		Standard type	Standard nature	Status	Corresponding international or foreign standard code
303–8	Forward collision mitigation system performance requirements and test methods	National standard	Recommended	Pre-research	ISO 22839
303–9	Automotive emergency steering assist system performance requirements and test methods	National standard	Recommended	Pre-research	
303–10	Vehicle lateral and longitudinal control system performance requirements and test methods	National standard	Recommended	Pre-research	
303–11	Parking assist control system performance requirements and test methods	National standard	Recommended	Project applied	
303–12	Automotive driving remote control assist system	National standard	Recommended	Pre-research	

(continued)

Table 8.1 (continued)

Standard name and category		Standard type	Standard nature	Status	Corresponding international or foreign standard code
303–13	Intersection collision avoidance assist system technical requirements and test methods	National standard	Recommended	Pre-research	
Automatic control (304)					
304–1	Automatic parking system functions, performance requirements and evaluation methods	National standard	Recommended	Pre-research	
304–2	Function, performance requirements and evaluation methods of automatic driving system for urban working conditions	National standard	Recommended	Pre-research	
304–3	Automatic highway driving system functions, performance requirements and evaluation methods	National standard	Recommended	Pre-research	

(continued)

Table 8.1 (continued)

Standard name and category		Standard type	Standard nature	Status	Corresponding international or foreign standard code
304–4	Vehicle Line-up automatic following driving system function, performance requirements and evaluation methods	National standard	Recommended	Pre-research	
304–5	Conditional automatic driving system (CA) functions, performance requirements and evaluation methods	National standard	Recommended	Pre-research	
304–6	Highly automatic driving system (HA) functions, performance requirements and evaluation methods	National standard	Recommended	Pre-research	
304–7	Fully automatic driving (FA) functions, performance requirements and evaluation methods	National standard	Recommended	Pre-research	
Information interaction (305)					

(continued)

Table 8.1 (continued)

Standard name and category		Standard type	Standard nature	Status	Corresponding international or foreign standard code
305–1	Automotive event data recording system	National standard	Mandatory	Projected 20,171,835-Q-339	
305–2	Automatic driving record device requirements and evaluation methods	National standard	Mandatory	Pre-research	
305–3	Telematics box (TBOX) technical requirements	National standard	Recommended	Pre-research	
305–4	Intersection signal information and violation warning system performance requirements and evaluation methods	National standard	Recommended	Pre-research	ISO 26684:2015
305–5	Automatic collision alarm system performance requirements and evaluation methods	National standard	Recommended	Pre-research	ISO 24978–2009
305–6	Hazard communication system performance requirements and evaluation methods	National standard	Recommended	Pre-research	

(continued)

Table 8.1 (continued)

Standard name and category		Standard type	Standard nature	Status	Corresponding international or foreign standard code
305–7	Special driving environment warning system performance requirements and evaluation methods	National standard	Recommended	Pre-research	
Related standards (400)					
Communication protocol (401)					
401–1	LTE-V based short and medium range communication protocol	National standard	Recommended	Pre-research	
401–2	5G-based wide-area communication protocols	National standard	Recommended	Pre-research	
Interface (402)					
402–1	LTE-V based short and medium range communication interface	National standard	Recommended	Pre-research	
402–2	5G-based wide-area communication interface	National standard	Recommended	Pre-research	
402–3	Short-range communication interface for automotive safety communication	National standard	Recommended	Pre-research	
402–4	Vehicle positioning and navigation system interface technical requirements	National standard	Recommended	Pre-research	
402–5	Vehicle and external terminal physical interface technical requirements	National standard	Recommended	Pre-research	
402–6	Vehicle and external terminal software interface technical requirements	National standard	Recommended	Pre-research	

5. *National Informatization Plan for the 13th Five-Year Plan* (No. 73, 2016, SC), State Council

The *National Informatization Plan for the 13th Five-Year Plan* proposes to build a modern information technology and industrial ecosystem; strengthen the strategic frontier technology layout; face the world's scientific and technological frontier, major national needs and the main areas of the national economy, focus on strategy, safety and breakthroughs in the field of information technology, general technology, asymmetric technology and disruptive technology; strengthen researches on quantum communication, future network, brain-like computing, artificial intelligence, holographic display, virtual reality, big data cognitive analysis, new non-volatile storage, unmanned transportation, blockchain; accelerate the industry construction of smart wearable devices, advanced robots and smart cars; encourage enterprises to carry out cutting-edge innovation research.

6. *National Scientific and Technological Innovation Plan for the 13th Five-Year Plan* (no. 43, 2016, SC), State Council

The *National Scientific and Technological Innovation Plan for the 13th Five-Year Plan* proposes to develop disruptive technologies; strengthen the early prediction of industrial change trends and major technologies; strengthen the prediction of the inflection point of subversive technologies replacing traditional industries; timely arrange the research and development of cutting-edge technologies in emerging industries; focus on the development of mobile Internet, quantum information, artificial intelligence and other technologies; promote the development of additive manufacturing, intelligent robots, driverless cars and other technologies.

When the Ministry of Industry and Information Technology promotes the development of the IoV industry in an all-round way, all participants in the market have already sensed the trend. As one of the most important application fields of the IoV, autonomous vehicle market has entered the early stage of competition.

On the whole, China's policies to promote intelligent vehicle are in place. With the development of technologies and markets related to autonomous vehicle and vehicle networking, the policy system will be further enriched and improved. The state has paid enough attention to automatic driving technology, creating top-level design and scientific planning for the R&D and industrial application of automatic driving technology. Meanwhile the financial support is also increasing. At present, there is a good trend of collaborative R&D and innovation between automobile enterprises and Internet enterprises. The laws and regulations related to autonomous vehicle need to be revised and improved in order to provide policy support for the R&D, testing and commercial application of autonomous vehicle.

Chapter 9
Autonomous Vehicle Industrialization

9.1 Special Autonomous Vehicle for Various Application Scenarios

With the maturity of autonomous vehicle technology, practical products such as autonomous bus, autonomous mine car, autonomous taxi and autonomous train are emerging and gradually developing towards mass production. At present, the important commercial application scenarios based on the evolution of autonomous vehicle technology are introduced as follows.

9.1.1 Autonomous Mine Car

The mining area environment is relatively closed, and the management and operation personnel can control the vehicles and equipment easily. Therefore, the mining area operation is a very suitable field for the application of automatic driving. In September 2018, the NORINCO Group's 172-ton autonomous electric wheel mining vehicle was tested by Baotou Steel Group. It is reported that after the completion of the test and improvement, the autonomous electric wheel mining vehicle can perform accurate driving and parking, and limit the lateral and longitudinal error to centimeters. At the same time, with the vehicle collaborative operation platform, coordinated operation of the autonomous fleet can be realized. A single vehicle's perception data can be shared with the fleet to facilitate efficient and safe operation in the mining area. An autonomous mine car is shown in Fig. 9.1.

M. Du, *Autonomous Vehicle Technology*, Unmanned System Technologies,
https://doi.org/10.1007/978-981-19-4143-6_9

Fig. 9.1 Autonomous mine car

9.1.2 Autonomous Truck

The development and application of autonomous truck has made great progress in recent years. Some companies are already conducting experiments. Autonomous trucks will be tested in the mining field, and then as port container trucks and long-distance trunk tractors. Autonomous trucks can be widely used in engineering fields, such as construction, railway, energy and so on. They can perform dangerous and repetitive operations and alleviate the shortage of human resources. An autonomous truck is shown in Fig. 9.2.

Fig. 9.2 Autonomous truck

Fig. 9.3 Autonomous electric buses

9.1.3 Autonomous Electric Bus

Autonomous electric bus has been successfully developed. The technical route of linear braking and linear steering system of large bus is different from that of small car. A typical example is the autonomous electric bus demonstrated at the 2018 World Artificial Intelligence Conference. According to statistics, from September 17 to 19, the autonomous bus transported more than 200 passengers per day, with the highest daily volume of 500. This model undertook more than 70% of the passenger transportation of automatic driving experience zone during the whole conference. It is 8.3 m long and equipped with cameras, lidar and millimeter wave radar. Moreover, it is the first 8-m-level autonomous bus for mass production in China, which is highly autonomous based on multiple sensing systems. Autonomous buses designed for operation in a certain area have also been developed. This kind of autonomous bus has no steering wheel, no driver's seat, no accelerator and brake pedal. Two autonomous electric buses are shown in Fig. 9.3.

9.1.4 Autonomous Subway Train

Automatic driving technology enables fully automatic control of the entire operation process of subway trains. The trains will operate according to a near-optimal curve, achieving the purpose of energy saving and environmental protection. At the same time, there is no driver in the train, which saves the labor cost. In addition, the full automation also avoids the operation failure caused by human error. On June 17, 2014, China's first autonomous subway train appeared in the China International Rail Transit Exhibition of 2014. This autonomous subway train is a 3-car train with a maximum capacity of about 1500 passengers. Its appearance is similar to that of an ordinary subway train without an operation cabin. This new type of train does not require driver or crew, and can perform the operation of automatic train wake-up, departure, uphill and downhill driving, accurate parking, door opening and closing,

Fig. 9.4 Autonomous subway trains

etc. In August 2014, Shanghai Metro Line 10 carried out driverless operation. On December 19, 2016, China's first autonomous subway line Nangang Island Line was opened. The autonomous subway trains are shown in Fig. 9.4.

9.1.5 Autonomous Taxi

Autonomous taxis have been launched and are operating in some cities. As long as the destination is set, the autonomous taxi will follow the designated route. Generally, the taxi can carry four people and a trunk, run on batteries and travel at a speed of 25 km/h. At present, South Korea, Spain, Japan and the United States all have autonomous taxis in operation. In some countries, many taxi companies have closed down due to the growing age of drivers and the decrease in the number of people joining the industry. In rural areas, people also have high hopes for autonomous taxis. Therefore, there is a large market space for autonomous taxi in the future. An autonomous taxi is shown in Fig. 9.5.

Fig. 9.5 An autonomous taxi

9.1.6 Autonomous Logistics Robot

Autonomous logistics robot can be categorized as the storage and transportation robot and the delivery robot.

In October 2017, the first autonomous warehouse of JD Logistics was officially launched. It is the first full process autonomous logistics center officially put into use, also the first large-scale green autonomous warehouse in the world. The roof is cover with solar panels, charging during the day and supporting the warehouse at night. In the first half of 2018, the Pegasus intelligent warehouse transportation robot independently developed by JD was applied in its warehouse. JD autonomous warehouse is mainly composed of receiving, storage, order picking, packaging and other operating systems. As a typical example of JD intelligent logistics robot innovation, the Pegasus robot has the function of automatic driving in the storage and logistics scene, which is suitable for the warehousing, picking, merging and handling tasks of small and medium-sized goods. It can greatly improve the picking efficiency by reducing staff movement through human–computer cooperation. Pegasus is jointly built by the JD's R&D teams in Silicon Valley and Beijing. The R&D team in Silicon Valley focuses on artificial intelligence and automatic driving technology, while the team in Beijing is committed to exploring the coordinated use of robots in different logistics scenarios. In addition to the Pegasus robot, many other products are currently in use and constantly upgrading, including the Ground Wolf AGV, Sirius shuttle, sorting robot, intelligent forklift, Vision + manipulator and so on. The handling weight of the latest generation of Ground Wolf AGV has been increased from 300 to 500 kg. Compared with the first generation, the second generation of Sirius robot has half the weight and smaller volume. It increases the storage density by about 15% and the warehousing efficiency by about 25%. At present, JD is continuing to develop the world's most advanced magnetic levitation and direct drive magnetic power technology in the field of warehousing and logistics, providing technical power for the construction of smart city.

The first autonomous delivery robot was launched by JD in 2016, as shown in Fig. 9.6. In 2017, it started the delivery service in universities including Renmin University of China, and appeared in Tianjin Eco City. In 2018, JD announced that its delivery robot had entered the era of mass production. In June 2018, more than twenty JD delivery robots were officially launched in Haidian District of Beijing.

9.1.7 Flying Vehicle

Flying vehicle refers to the vehicle able to fly in the air and drive on the ground. The flying vehicle adopts environment-friendly and fuel-saving dual-purpose engine. The engine uses unleaded gasoline for ordinary vehicles, which can be found in a common gas station. Not only that, the car's internal control system is also dual-purpose. On the ground, the flying vehicle is as comfortable as a luxury car, and more dexterous

Fig. 9.6 JD autonomous delivery robot

because of its balance system. Since the flight altitude of the vehicle is lower than that of a commercial flight, the driver does not have to submit a flight report to relevant authorities in advance. Many countries have loosened the air control, such as the United States, where there are fewer restrictions below 3000 m, and some Western Europe countries. The folding rotor of flying vehicle can ensure the safety of driving and landing. Different from ordinary helicopter, the impeller of flying vehicle can rotate in the air with only aerodynamic force. Even if the engine fails in the air, the impeller at the top can continue to rotate to help the vehicle land smoothly instead of crashing. The vehicle is equipped with GPS, radar, electronic computer automatic control system and satellite control technology, which can also maximize its safety in the air.

As the automatic driving technology progresses, flying car is becoming the next generation of transportation which the automotive industry is closely watching. Compared with autonomous vehicle, flying car may seem fictional. However, the history of flying car can be traced back to 1841, when the brothers William Samuel Henson and John String fellow invented the Henson Aerial Steam Carriage. This was even before the Wright brothers' invention of the airplane. The drawing of Henson Aerial Steam Carriage in an advertisement in 1843 was shown in Fig. 9.7.

In the following 200 years, from sketching on paper to successful take-of, human beings have never stopped exploring flying vehicles. Yet flying vehicle has never made it to mass production. Boeing once launched the Sky Commuter concept car (Fig. 9.8) in 1990, but it has not been successfully put into the market so far.

The world's first flying vehicle (Fig. 9.9) made its first flight in the United States in early March 2009. After landing, the driver only needs to press one button to fold the wings and drive on the highway. On July 6, 2010, American company Terrafugia land and air dual-purpose vehicle was allowed to put into commercial production by the U.S. aviation authority. Inspired by the dream of flying, Slovak engineer and designer Stefan Klein designed and manufactured a vehicle with a maximum speed of 200 km/h in the air and 160 km/h on the ground. On a full tank it can travel 700 km

Fig. 9.7 Drawing of Henson Aerial Steam Carriage in an advertisement in 1843

Fig. 9.8 Boeing's six-million-dollar Sky Commuter concept car

in the air or 500 km on the ground. Typical flying vehicle products include Arrowbile, Airphibian, ConvAirCar, Aerocar, Avrocar, Folding AirCar, X-Hawk, and Folding PAL-V.

On March 6, 2018, the Dutch company launched the mass production version of PAL-V, and began to take orders on its company website. The vehicle will be delivered in 2019, with a price of about $600,000. This flying vehicle is made of carbon fiber, titanium and aluminum. It weighs 680 kg, takes a 165-m runway to take off and a 30-m runway to land. The maximum speed can reach 160 km/h on the ground and 180 km/h in the air. The highest flight altitude is about 3352 m, and the longest flight distance is about 563 km.

Fig. 9.9 World's first flying vehicle

The flying vehicle produced by Xi'an ML Aviation is shown in Fig. 9.10.

The head of the flying vehicle is a flat arc, with LED lights on both sides. Rotors are installed on the top and tail, and four wheels are installed on the chassis. The vehicle can carry two passengers at the same time. The steering wheel and seats used in the flying vehicle are identical to those used in Formula One cars, and the material used to make the seats is fiberglass. The external material is carbon fiber and high-strength titanium alloy. The mass of the whole vehicle is 450 kg. The rotor length is 8.4 m, the length, width and height of the vehicle in traveling mode are

Fig. 9.10 Flying vehicle of Xi'an ML Aviation

Fig. 9.11 Flying vehicle of Geely Terrafugia

4.6 m, 2.0 m and 1.46 m respectively. The maximum takeoff weight is 600 kg. The maximum cruising speed can reach 180 km/h. The two-wheel drive traveling speed is 120 km/h. It is equipped with a 70-L fuel tank for ordinary No. 97 leaded or unleaded gasoline and can fly for 3–4 h on a full tank.

Flying vehicle is well known in China, largely due to the wholly-owned acquisition of Terrafugia flying vehicle company by Geely Group in November 2017.Terrafugia was founded in 2006 by five MIT graduates. In 2009, Terrafugia released its flying vehicle, Transition, which was the first flying vehicle to get a road license in the U.S. Three years later, an updated version of Transition was launched. It is estimated that the first generation of Transition will available for ordering in U.S. market in October 2018, with a pre-sale price of about $190,000. The first batch will be completed in 2019. The flying vehicle of Geely Terrafugia is shown in Fig. 9.11.

China is now one of the countries with the most active innovation and the fastest application of emerging technologies. The mass production of Geely Terrafugia flying vehicle is bound to drive the development of the whole industry chain, promote the integration of automobile industry and aviation industry. It is also expected to bring about a new wave of innovation.

Multiple technology companies and automobile companies have plans in the flying vehicle industry:

1. Tencent participated in the financing of German flying vehicle company Lilium.
2. Google co-founder Larry Page has invested about $100 million in two flying vehicle start-ups, Zee.Aero and KittyHawk.
3. The brand name of Toyota's flying vehicle business is SkyDrive. Toyota is hoping that SkyDrive will make its debut at the Tokyo Olympic Games in 2020 and light the Olympic flame at the Olympic Stadium.

4. Boeing has purchased Aurora Flight Sciences and invested heavily in flying taxis.
5. Uber and Aurora are developing advanced VTOL aircraft and building the future urban air traffic network. They will launch autonomous flying taxi in the next 5–10 years. Uber is even developing a software for flying vehicle booking called Uber Air, which is expected to be launched in 2020.
6. Porsche confirmed that it is considering developing flying vehicles to enter the flying taxi market. Most functions of the vehicle are automatic. The owner can use the car without a pilot's license.
7. The Dutch aircraft manufacturing company PAL-V officially released the mass production version of the PAL-V Liberty flying vehicle at the Geneva Motor Show in 2018, which is expected to be ready for shipment in 2019.

After the idea of flying car was proposed, it took the world nearly 200 years to solve the problem of mass production. Since a flying vehicle has to be able to fly in the air, drive on the ground and park in a garage, the first issue is how to switch between different forms easily with safety and quietness. Secondly, the feasibility and cost of mass production should be considered. Finally, flying vehicles must comply with relevant laws and regulations.

9.2 Design of Autonomous Vehicle Test Field

9.2.1 Industry Analysis

In 2015, the Ministry of Industry and Information Technology issued the *Action Plan (2015–2018) for Implementing the Guidance of the State Council on Actively Promoting the Internet + Scheme*, which defined the future development direction of the Internet of Vehicles. After years of preparing, the IoV will usher in a five-year golden period of development.

Autonomous vehicle test field is a field that reproduces all kinds of road conditions and scenarios for autonomous vehicle. It is used to verify the software algorithm of autonomous vehicle. The test road is like typical roads, including highway, urban road, rural road and tough road that can cause strong bumps. In addition, GPS base station, communication base station, intelligent traffic light and other infrastructure should be arranged in the test field to provide a test environment for the IoV technology. It is more rigorous, more scientific, more rapid and more practical to test automobile in the test field than in the laboratory or under general driving conditions.

9.2.2 Global Development Status of Autonomous Vehicle Test Field

At present, a lot of countries are actively investing in automatic driving technology. Developed countries and regions such as the United States, Europe, Japan are investing in the construction of autonomous vehicle test fields to facilitate the mass application of autonomous vehicle.

At present, a number of test field have been built around the world, such as Mcity in the United States and ITS Corridor in Europe. The world strives to enrich the test scenarios of autonomous vehicle and improve the R&D speed of intelligent connected vehicles on the premise of ensuring safety.

1. Mcity, the world's first professional autonomous vehicle test field

The United States is at the forefront of the world in the field of autonomous vehicle testing. Mcity, a driverless virtual city led by the University of Michigan and supported by the Michigan Department of Transportation, is located in Ann Arbor, Michigan. It covers an area of 129,000 m^2 and is jointly funded by the University of Michigan and the Michigan Department of Transportation with a total investment of $10 million. It is the world's first simulation town specially designed for the test of autonomous vehicle and V2V/V2I IoV technology based on environmental variable control.

Mcity has miles of two-lane, three-lane, and four-lane highways, as well as intersections, traffic lights, signs, and sidewalks with benches and streetlights to separate roads from buildings. On the outside, it looks like a well-designed Hollywood filming location. Mcity is mainly designed by the University of Michigan Mobility Transformation Center, which was established in 2013. At present, the automobile enterprises participating in the project include Ford, GM, Honda, Nissan, Toyota, etc.

Although Michigan allowed autonomous vehicle road testing as early as 2013, Mcity was built for safety reasons. According to the staff of the University of Michigan Mobility Transformation Center, before the automatic driving technology becomes reliable enough, it can be tested in a relatively closed simulation environment to minimize the casualties caused by inevitable accidents.

2. Autonomous vehicle test road in Virginia

Virginia plans to build a 112-km highway in the northern area to test autonomous vehicle. This road will be known as the Virginia automatic driving corridor. Any autonomous vehicle that plans to be tested in Virginia must first be tested on this intelligent road.

The project is headed by Virginia Tech Transportation Institute. Myra Blanco, director of the institute, says that Virginia will provide more convenience than any other state that allows autonomous vehicle testing. On the one hand, Virginia Tech Transportation Institute will provide license plates and insurance for test vehicles, while the Virginia Department of Transportation will be responsible for road maintenance, especially for keeping road markings clearly visible. On the other hand, Here

Maps will develop 3D map technology for the test road, and provide real-time road information for autonomous vehicles, so that vehicles can identify the direction of the road.

The safety issue of the automatic driving test is the focus of attention in the U.S. At present, states such as California and Florida allow the test of autonomous vehicle. After the vehicle passes the field test, the driver has to be in the car for the actual road test, so as to prevent accidents.

3. Ontario with all roads opened for autonomous vehicle testing

Compared with other countries which build test fields and test roads for autonomous vehicles, Canada appoint the roads of a certain province for autonomous research and test. Ontario claims that it will become the first province in Canada to allow automatic driving technology testing because automatic driving and high-tech vehicle technology can improve fuel economy, ease traffic congestion, reduce greenhouse gas emissions and improve driving safety. Mercedes-Benz, Google, Tesla and other companies are expected to start autonomous vehicle test here soon.

As for safety, Ontario claims that during the implementation of the autonomous vehicle pilot project, they will ensure that the roads in the province remain safe as usual. However, although autonomous vehicles can travel on any road in Ontario at any time, participants in the project are limited to automobile manufacturers, technology companies, academic research organizations and automotive parts companies involved in autonomous vehicle industry. In addition, a driver with a valid license has to be in the vehicle, and the relevant company needs to have a minimum US $5 million insurance policy.

At present, the world's leading autonomous vehicle test fields are as follows.

1. Automatic driving demonstration zones in the U.S

The automatic driving demonstration zones in the U.S are divided into two major competing camps: Motor City in Detroit, Michigan and Silicon Valley, which both have two demonstration zones.

In addition to Mcity, the world's first professional autonomous vehicle test field, the University of Michigan is also involved in a project to build a more complex test site in the small town of Ypsilanti, called the American Center for Mobility. This 1.4 km^2 area which is 16.1 km away from Mcity, was used for the production of bombers during World War II. The base has a highway for testing, which will also be equipped with a variety of complex intersections, triple overpasses and off-road surfaces with natural potholes. A panoramic view of the American Center for Mobility test site is shown in Fig. 9.12.

GoMentum Station, located 64.3 km north of Silicon Valley, is an abandoned naval base covering 8.5 km^2. The base has paved 32.2 km of roads, including overpasses, tunnels, railways and other urban infrastructure, as well as barracks and buildings. The environment is similar to that of urban areas. It is an ideal test field for autonomous vehicles. Its unique geographical environment has attracted Internet

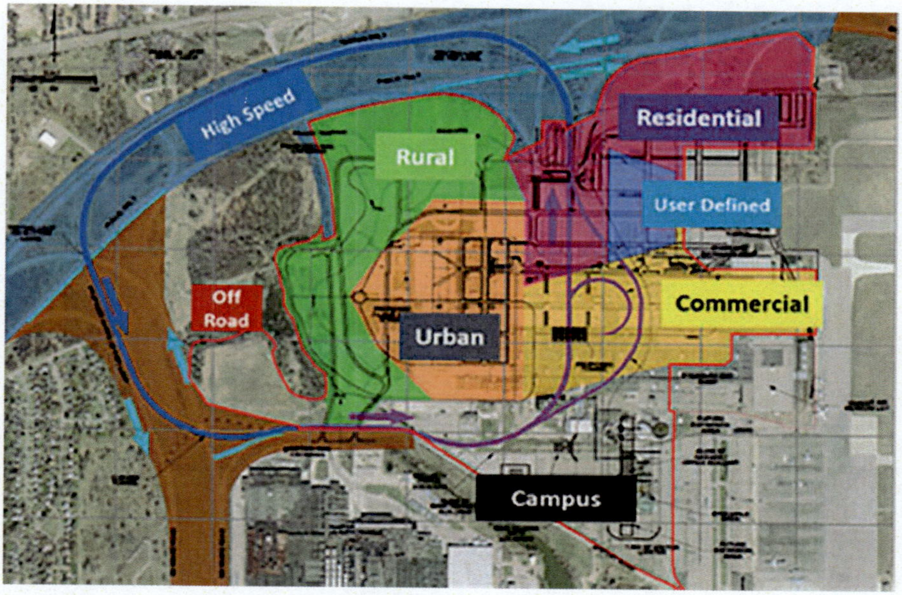

Fig. 9.12 Panoramic view of the American Center for Mobility

companies such as Apple and Google, as well as the local R&D departments of automobile manufacturers including Volkswagen and Nissan. The satellite view of GoMentum Station is shown in Fig. 9.13.

In 2014, Google rented 0.24 km^2 (now about 0.40 km^2) from Castle Air Force Base in California to test its Self-Driving Cars (SDC) and train the drivers. Inside this closed test field, there are streets, branch roads and highway extensions similar

Fig. 9.13 The satellite view of GoMentum Station

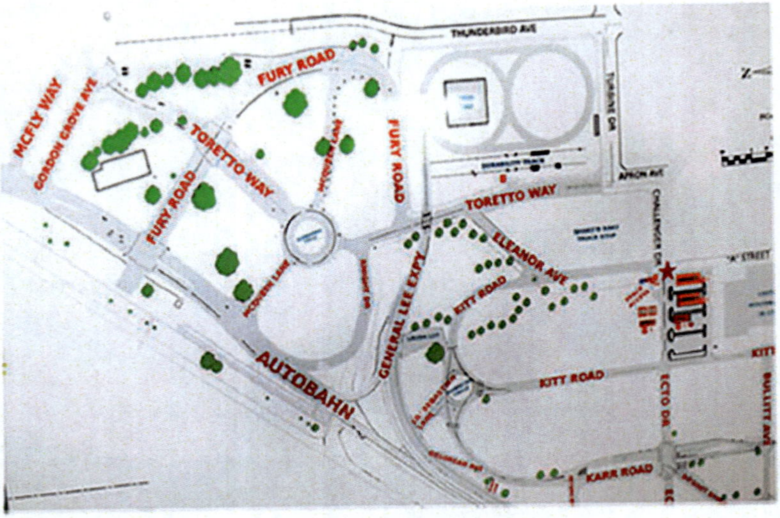

Fig. 9.14 Overall view of Castle

to suburban and urban areas. Meanwhile, there are traffic lights, stop signs, traffic roundabouts, and even a rain simulator. There are 53 autonomous vehicles in the test field. The overall view of Castle is shown in Fig. 9.14.

2. AstaZero test field in Sweden

AstaZero is located in Borås in the southwest of Sweden. This small town is 56 km east of Gothenburg, the second largest city in Sweden, with a population of only 100,000. The designing of AstaZero started in 2011. The construction started in 2012, and by 2013 it had been fully equipped. On August 21, 2014, AstaZero was officially opened. The total investment of AstaZero is only SKR 500 million, about CNY 400 million, since the land in Sweden is quite cheap. The investment of the similar test field in China is at least more than CNY 1 billion.

The primary task of AstaZero test field is to test the active safety system which prevents accidents. The test field has congested urban roads, expressways, multi-lane parallel roads, roundabouts and intersections, which are important for the study of interactions between vehicle and vehicle, vehicle and other traffic participants (pedestrians, bicycles, electric bicycles, motorcycles, animals, etc.). Another important role of AstaZero is the research and development platform of future security technology. Universities and enterprises are working together to achieve zero casualty of traffic. The overall view of AstaZero test field is shown in Fig. 9.15.

The test field covers an area of about 2 km^2, with a total construction area of 0.25 km^2. It is surrounded by a 5.7 km long rural road, with four 40 m × 25 m activity modules to simulate the urban environment. At the same time, there is a 240-m-diameter circular high-speed test area, which is connected with another 700-m multi-lane road through a speed bump. It can simulate various road sections such

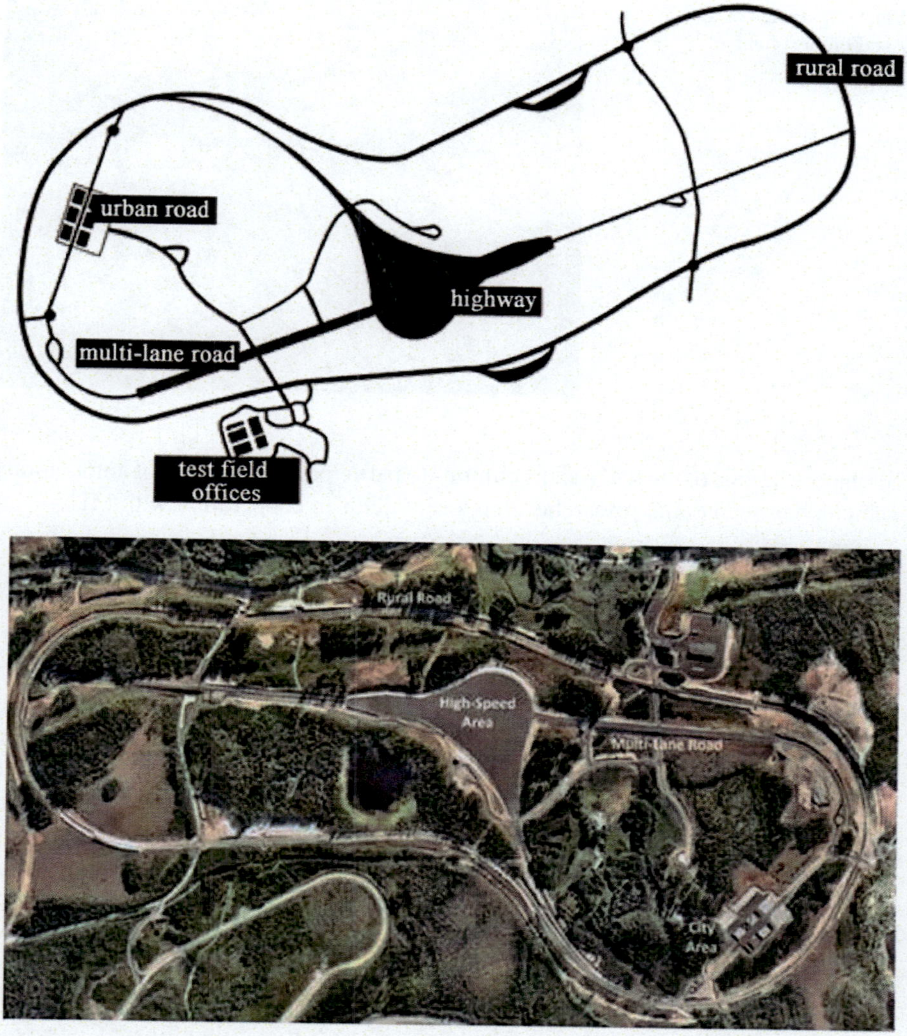

Fig. 9.15 Overall view of AstaZero test field

as urban road, rural road, highway and so on, and is suitable for experiments of autonomous vehicles and robots. An image of the AstaZero test field is shown in Fig. 9.16.

3. One-North tech park in Singapore

In August 2014, Singapore set up the Committee on Autonomous Road Transport to supervise the research and test of autonomous vehicle, and proposed to conduct relevant experiments in the One-North tech park. The experiments are led by the authorities of land and transportation, and are the first public road tests in Singapore.

Fig. 9.16 Image of the AstaZero test field

Singapore, as one of the few Asian countries to participate in the study of autonomous vehicle, has been doing researches in relevant fields in cooperation with MIT.

9.2.3 Development of Autonomous Vehicle Test Field in China

Currently, there are a number of intelligent vehicle and transportation application demonstration zones in China, including the National Intelligent Connected Vehicle (Shanghai) Pilot Demonstration Zone, Chongqing Intelligent Vehicle and Transportation Evaluation and Demonstration Zone, National Autonomous Commercial Vehicle Road Test Base in Tongzhou, Beijing and China Intelligent Vehicle Technology R&D and Test Center in Changshu.

The intelligent Internet demonstration zones approved by the Ministry of Industry and Information Technology are located in Shanghai, Zhejiang, Beijing and Chongqing.

On June 7, 2016, the first closed test field of the National Intelligent Connected Vehicle (Shanghai) Pilot Demonstration Zone (Fig. 9.17) approved by the Ministry of Industry and Information Technology was officially opened. The demonstration zone also known as A Nice City was built by Shanghai International Automobile City (Group) Co., Ltd. The closed test field is designed for test and demonstration of intelligent vehicle and network communication technologies, covering a variety of V2X communication scenarios, as well as technology applications in the fields of safety, efficiency, information and new energy vehicle. According to Shanghai International Automobile City, the demonstration zone will be constructed and put into use in three phases from 2015 to the end of 2017. The areas of the three phases are 5 km^2, 27 km^2 and 90 km^2 respectively. Only about 200 vehicles will be operating in the first phase, and another 10,000 vehicles will be added for the third phase.

Fig. 9.17 National Intelligent Connected Vehicle (Shanghai) Pilot Demonstration Zone

The closed test field (phase I) located on Yining Road, Jiading District, Shanghai, with a road length of 3.6 km. In the test field, one GPS differential base station, 2 LTE-V communication base stations, 16 DSRCs, 4 LTE-V road test units, 6 intelligent traffic lights and 40 cameras of variety have been installed. The whole road area in the field is covered by centimeter-level positioning of Beidou system and Wi-Fi. The simulation traffic scenes can provide 29 scenarios for automatic driving and V2X connected vehicles, including tunnels, boulevards, gas/charging stations, underground parking lots, intersections, T-junctions and roundabouts.

The field will also simulate various natural environments such as rain, fog, ice, ponding, daylight, nighttime and artificial lighting, as well as pedestrians and vehicles, so as to build a realistic scene for evaluation. In terms of testing, the demonstration zone will cover safety, efficiency, transportation, information services, mobile services and other pilot projects, so as to ensure that autonomous vehicles have been put through all kinds of experiments before driving on the road.

On September 11, 2015, the Ministry of Industry and Information Technology and Government of Zhejiang Province signed the *Framework Agreement of Cooperation on the Demonstration Application of Intelligent Vehicle and Transportation Based on Broadband Mobile Internet*. Zhejiang Province will take Yunqi Town, Hangzhou and Wuzhen, Tongxiang City as the core to establish a demonstration zone of intelligent vehicle and transportation. In January 2016, the Ministry of Industry and Information Technology signed similar agreements with Beijing, Hebei and Chongqing. At the same time, a number of demonstration zone projects led by local governments are blooming, while Wuhan and Shenzhen are planning to build driverless towns, and Panjin, Liaoning cooperates with BAIC on an intelligent commercial vehicle project.

To sum up, the well-known automatic driving test fields can provide scenarios including expressway, urban road, rural road and other road types. Some test fields can perform simulations of rain, fog and other weather conditions. Besides the automatic driving technology, the tests also involve V2X and other vehicle networking technologies. At present, most of the test fields are in the initial stage of construction, with future development plans formulated.

In the first stage, automatic driving test fields aims to provide comprehensive test facilities. Only after the technology and safety performance of autonomous vehicles have been verified, can they proceed to public roads. The interconnection of things is the future trend, but the premise of networking is safety. In the second stage, with the cooperation of enterprises, governments and individuals on technology and capital, the commercial value of the autonomous vehicle test field can be maximized.

9.2.4 Market Analysis

Internet of Vehicles is currently one of the hottest emerging industries in the world, and its most important application area is smart car and intelligent transportation. At present, the IoV industry in China is still in the early stage of development and the market scale is small. In 2018, the market scale of IoV in China reached CNY 200 billion, accounting for about 60% of the global market.

The *Chinese Autonomous Vehicle Industry Market Development Analysis and Investment Trend Research Report (2016–2020)* shows that the current global market size of automatic driving technology is about USD 1.29 trillion. The market space of pan-intelligent automatic driving in China is expected to reach CNY 230 billion by 2020.

Global Internet and automotive companies including Google, Toyota, Honda, Nissan, Daimler, GM, Ford, Volvo, BMW are actively developing automatic driving technology. The IoV market in China has become a focus of foreign car giants. With technical advantages, foreign car companies have equipped traditional cars with some auxiliary automatic driving functions, such as automatic parking, automatic following, automatic braking, etc.

Major Chinese car companies have already set their objectives in the field of automatic driving. A brief overview is as follows.

FAW: it will produce small quantities of L4 Hongqi autonomous vehicles in 2019. Mass production will take place in 2020.

GAC: it will mass-produce L3 autonomous vehicles in early 2020, becoming the first Chinese car company to fully launch autonomous vehicles. Mass production of L4 vehicles will take place in 2022.

BAIC: L3 automatic driving in congested conditions is expected to be realized in 2020. L4 automatic driving under restricted road conditions will not be in place until 2025 or later.

Great Wall: mass production of L3-class autonomous commercial vehicles will take place in 2020. Mass production of L4 commercial vehicles is planned for 2023. L5 commercial vehicles will be launched in 2025.

Chang'an: It is expected to mass-produce L3 autonomous vehicles in 2020 and L4-level autonomous vehicles in 2025.

Chery: It aims to achieve mass production of L3-level autonomous vehicles in 2020. It plans to achieve L4/L5 or higher level automatic driving in 2025.

Geely: It plans to achieve mass production of L2 models in 2019 and release L3 platform (GPILOT3.0) in 2020.

JAC: L3 technology development is to be completed in 2019.

NIO: It plans to launch L4 autonomous models in 2020.

Byton: It plans to achieve L4 level automatic driving after 2020.

Chehejia: Prototype of L4 autonomous vehicle is to be completed in 2019. Mass production will take place in 2023–2025.

Peng: massive shipments of L3 models are to be made at the end of 2018 or in early 2019.

Singulato: It plans to achieve L3 automatic driving in 2019.

Leapmotor: L3 software upgrade is planned to take place before 2020.

Internet companies also have taken actions in the field of automatic driving. Baidu and the Institute of Highway Science, Ministry of Transport have joined hands to work in intelligent driving, traffic safety, vehicle–road coordination, policy and regulation research, technology standardization and other aspects, in order to promote the development of autonomous vehicle industry in China. SAIC and Alibaba have also reached cooperation. Google, Tesla, BMW, Audi, Mahindra and other well-known manufacturers have announced R&D projects of autonomous vehicle.

9.2.5 *Typical Cases*

1. Shanghai Anting Town Driverless Demonstration Area (Fig. 9.18)

Shanghai Anting Town Driverless Demonstration Zone covers an area of 90 km^2, where overall testing of intelligent connected vehicle and demonstration of intelligent transportation will be carried out. In the closed test area, 29 test scenarios will be constructed in the first phase, and nearly 100 test scenarios will be formed in three years. Vehicle–road cooperation applications will be carried out on a total of 20-km road sections Demonstration applications such as vehicle-vehicle communication

Fig. 9.18 Driverless Demonstration Zone in Anting Town, Shanghai (the lower right area)

and warning, bus priority and automatic parking will be gradually experimented on open roads with smart lighting transformation. These tests will stimulate the industry chain of vehicle chip, vehicle software, auto parts, and vehicle integration, thus accelerating the development of Shanghai's automotive electronics industry.

The first partners include SAIC, Volvo, GM, Ford, Chang'an in vehicle manufacturing, Huawei in chip design, SEARI in intelligent transportation, Beidou and Aerospace 802 Institute in navigation, and universities such as Tsinghua and Tongji. The zone will construct new scenes on the basis of Phase I to reach a total of 15–20 scenes. In the core area of the Auto City, a demonstration model of more than 1,000 intelligent connected vehicles with the applications of information communication, safety warning and control, energy saving will be formed to promote the integration of vehicle network, location network and road traffic network, improve driving safety and road traffic capacity.

Illegal parking is one of the reasons for road congestion. Regulations and drivers' needs should both be considered for parking management. At present, the Shanghai Parking smartphone APP is mainly for large parking lots, while the information of roadside parking is not accurate. According to authorities, the trial application of intelligent roadside parking is being carried out and the relevant information will be incorporated into the Shanghai Parking APP in due time. In the future, drivers can obtain real-time information of roadside parking, and pay fees via WeChat, Alipay and so on. With intelligent parking management, if a vehicle occupies two parking spaces, the parking attendants will be required to intervene.

2. Beijing Intelligent Vehicle and Transportation Based on Broadband Mobile Internet Demonstration Zone

In January 2016, the Ministry of Industry and Information Technology, the Beijing Municipal People's Government and the Hebei Provincial People's Government signed the *Framework Agreement of Cooperation on the Demonstration Application of Intelligent Vehicle and Transportation Based on Broadband Mobile Internet*. Meanwhile, China Transinfo, LeTV, Baidu, BAIC New Energy, E-town Capital and BUAA signed the *Cooperation Agreement on Beijing Intelligent Vehicle and Smart Transportation Industry Innovation Center*. Yizhuang Pilot Zone then became the Intelligent Vehicle and Smart Transportation Demonstration Zone in Beijing.

Baidu has been carrying out automatic driving tests smoothly within the jurisdiction of Yizhuang. The research topics in the demonstration zone include green car, smart road network, smart driving, convenient parking, happy car lifestyle, and smart management. The most important feature of the demonstration zone is the synchronization of intelligent vehicles and smart transportation. The site of the closed test field in the demonstration zone has been determined, which covers a total area of 914,000 m^2, and the construction is in progress.

9.3 Design of Driverless Town

9.3.1 Strategic Objective

This section discusses the construction project of the Driverless Town, which is located in a city in southern China. The strategic objective of this project is to build China's first town of autonomous vehicle and intelligent networked driving, and to implant an autonomous vehicle industry ecosystem and a smart community in the town. The autonomous vehicle industry ecosystem includes five major industry clusters: smart car, robotics and artificial intelligence, smart transportation, smart energy, and new generation information technology. The project should achieve the following goals.

1. Global demonstration

The Driverless Town is going to be the world's first autonomous vehicle and intelligent network driving test area supporting the smart city research.

2. National demonstration

(a) The Driverless Town is going to be China's first town of autonomous vehicle and intelligent networked driving.
(b) The Driverless Town is a pilot demonstration zone of smart city and smart transportation.
(c) The Driverless Town is also a demonstration zone of green ecological civilization.

As a comprehensive industrial ecosystem which is unprecedented, the Driverless Town aims to bring together researchers and manufacturers around the world, stimulate the cooperation on R&D, production and marketing. With the technological development of autonomous vehicle, intelligent networked driving and smart transportation, it is expected that an advanced business model will be formed and introduced to the world.

9.3.2 *Technical Proposal*

9.3.2.1 Platform Composition

The social laboratory platform for autonomous vehicle, which is located in the Driverless Town, is composed of four major parts: closed site, public community, social scene and future city (Fig. 9.19).

The closed site refers to the closed test field for autonomous vehicle technology, with a planned area of about 2 km^2 and a total construction area of 0.25 km^2. It is surrounded by a 5.7 km long rural road, with four 40 m × 25 m activity modules to simulate the urban environment. At the same time, there is a 240-m-diameter circular high-speed test area, which is connected with another 700-m multi-lane road through a speed bump.

The public community refers to the urban and suburban living environment, including residential communities, roads, boulevards, gas stations and gardens, etc.

Fig. 9.19 Composition of social laboratory platform for autonomous vehicle

Fig. 9.20 Expected panorama of autonomous vehicle social laboratory platform

Fig. 9.21 National Economic and Technological Development Zone

The social laboratory will build the facades of 40 buildings, right-angled intersections, roundabouts, bridges, tunnels, gravel roads, building barriers and many other obstacles.

The social scene refers to real scenes for urban residents, including children on the street, bus stops, schools, hospitals, shopping malls and parking lots.

The future city is mainly composed of intelligent medical system, intelligent transportation system, intelligent environment management, open Internet application and intelligent automobile industry cluster, etc. The expected panorama of autonomous vehicle social laboratory platform is shown in Fig. 9.20.

9.3.2.2 Spatial Distribution

The overall construction area of the Driverless Town is about 56 km^2. The National Economic and Technological Development Zone, on which the town will be based, is shown in Fig. 9.21.

1. Planned area

The Development Zone area reaches to Dapan Shoal in the east, Industry Road in the south, the mountain in the west and Merchants Avenue in the north, with a total area of 16.51 km^2.

2. Planning scale

(a) Population. The area has a residential population of 163,000, excluding Shuangyu Island (20,000) and the faculty and students in Zhangzhou campus of Xiamen University (30,000).
(b) Land. The total area is 16.51 km^2, including 14.42 km^2 of urban construction land and 2.08 km^2 of water and other areas.

3. Land layout

(a) Residential land. The planned residential land is 5.85 km^2, accounting for 35.41% of the whole area.
(b) Land for public facilities. The planned area for public facilities is 3.69 km^2, accounting for 22.39% of the whole area.
(c) Land for municipal facilities. The planned area for municipal facilities is 0.25 km^2, accounting for 1.49% of the whole area.
(d) Green space. The total area of planned green space is 2.57 km^2, including 1.52 km^2 of public green space. The green space accounts for 15.59% of the whole area.

The planned layout of the National Economic and Technological Development Zone is shown in Fig. 9.22. Based on the layout, how to integrate automatic driving into people's lives should be considered.

Fig. 9.22 Layout of the National Economic and Technological Development Zone

Fig. 9.23 Traffic planning of the National Economic and Technological Development Zone

4. Road traffic

The planned road and plaza area is 1.78 km^2, accounting for 10.77% of the whole area, of which 1.72 km^2 is road area and 0.06 km^2 is plaza area. According to the urban road classification standard, the road system in this area is divided into three levels: trunk road, secondary trunk road and branch road. The planned traffic facilities mainly include public parking lot, plaza and other facilities.

The traffic planning of the National Economic and Technological Development Zone is shown in Fig. 9.23. Based on the planning, how to arrange the roads of the test field should be considered.

9.3.2.3 Function and Facility Planning

1. Function description

Autonomous vehicle social laboratory is an intelligent ecosystem with real-time interaction among human, vehicle, road and environment. It can support tests of multiple vehicles and scenarios at the same time, including urban road test, expressway test, rural road test and so on.

2. Scene design and distribution

High-speed test area for simulating expressway environment includes four-lane highways, intersections, three-layer overpasses, ramps, toll stations traffic lights, traffic signs and markings. The pavement materials include cement and asphalt, and there are natural potholes on the pavement.

Low-speed test area for simulating urban and suburban environment includes two-lane and three-lane roads, intersections, overpasses, traffic lights, traffic signs and markings. There is asphalt pavement, soil pavement, brick pavement, pavement covered with leaves. There are also streets, branch roads and highway extensions like urban and suburban areas.

The road scenes in the low-speed test area includes the following elements: autonomous cars, complex terrains, moveable models of building, wireless road devices, pedestrians and bicycles in fixed positions, road signs and signals, traffic management devices, obstacles, etc. Problems that drivers may encounter in cities and suburbs are also present, such as vandalized road signs are faded lane markings.

3. Facility and equipment

Wi-Fi and sensors are equipped on roads, street lights, traffic lights and other facilities. The whole road area in the field is covered by centimeter-level positioning of Beidou system and Wi-Fi. In the test field, one GPS differential base station, 2 LTE-V communication base stations, 16 DSRCs, 4 LTE-V road test units, 6 intelligent traffic lights and 40 cameras of variety will be installed.

4. Construction plan

The construction project is divided into three phases.

Phase I completes the construction of closed test site. A variety of simulated traffic scenes such as expressways, tunnels, gas stations, charging stations, crossroads, T-junctions, roundabouts, overpasses will be built. After the completion of the phase I, the test field can hold 200 test vehicles and 15 km closed roads within a 5 km^2 area. Experiment of high-speed autonomous vehicles will lay the foundation for the application of automatic driving in future smart city.

Phase II completes the construction of public communities and social scenes. There will be facades of 40 buildings, including schools, hospitals, shopping malls, etc. Gardens, avenues, parking lots, cobbled roads, together with expressways, trunk roads, secondary trunk roads, branch roads, bus stops and gas stations will be built. After the completion of Phase II, the tests of autonomous vehicles on open roads are expected to be carries out.

In Phase III, the construction of the whole zone will be finished and the area of the social laboratory will be expanded. Wireless charging devices will be built. The smart medical system, intelligent transportation system, intelligent environment management, open Internet applications and intelligent automobile industry cluster will be formed.

9.3.2.4 Test Content and Method

The test contents and methods are as follows:

In the social laboratory, closed road tests and partially open road tests are carries out on autonomous vehicles, in order to verify their cognitive abilities in regard of road objects, traffic sign objects, vehicle objects and pedestrian objects. The tested abilities include basic cognitive ability, advanced cognitive ability and comprehensive cognitive ability.

1. Basic cognitive ability test mainly verifies the cognitive ability in terms of audio-visual information such as traffic signals, vehicle light signals, horn sounds, etc. According to the degree of difficulty, the test can be divided into two levels, including the recognition of standard audio-visual information and the recognition of fuzzy audio-visual information.

(a) For the recognition of standard information, the vehicle is required to output images, words, symbols or sounds representing the corresponding audio-visual information. The main performance indicators are the number of correct recognition results and the recognition time. Standard audio-visual information includes traffic signals (signs, markings, lights and police commands), vehicle lights and horn sounds. Traffic signals mainly refer to the signals in line with national standards. Vehicle lights mainly refer to the lights used by vehicles in traffic, such as steering lights, headlamps, brake lights, etc. Horn sounds mainly refer to the sound of horns, or alarms on special vehicles, such as police cars, fire engines, ambulances, etc.

(b) The recognition of fuzzy information mainly includes: the cognition of non-standard audio-visual information, such as information with the interference of shadow, stain, obstruction, and tilted, faded, illuminated, deformed information; the cognition of audio-visual information under the condition of feature binding; the cognition of multi-dimensional audio-visual information. The vehicle is also required to output images, words, symbols or sounds representing the corresponding audio-visual information.

2. Advanced cognitive ability test is about the cognitive ability of autonomous vehicles for environment and behaviors in certain traffic scenes. The tests can be divided into two categories: moving audio-visual information for stationary vehicle, and stationary audio-visual information for moving vehicle.

The moving audio-visual information mainly refers to the situation that the information source has obvious longitudinal movement, but limited lateral movement, which is consistent with the actual road scene. The moving vehicle refers to a vehicle driving on the road.

The traffic scenes include intersections, roundabouts, urban fast ramps, parking areas, tunnels, overpasses, interchanges, sharp curves, uphill and downhill roads, roads with lateral inclinations, etc. The traffic behaviors include turning, lane changing, overtaking, meeting, reversing, turning around, parking, intersection crossing, railway crossing, overpass and railway bridge bottom passing, emergency braking, avoiding pedestrians, goods scattering, etc.

3. Comprehensive cognitive ability test is for the vehicle's comprehensive ability of cognition, autonomous driving and intelligent navigation. The test can be divided into urban road test, intercity road test, rural road test and combined test. Within a certain distance, the vehicle has to identify obstacles, traffic signals, vehicle lights,

horn sounds, infrastructures, traffic behaviors, and complete automatic driving and intelligent navigation according to the requirements. The main performance indicators are the number of correct recognition results, the time consumed for recognition and the completion status of assigned tasks, including achievement, speed, location, time, behavior, etc.

9.3.3 Business Model

Based on the Driverless Town, with industrial investment funds, the automatic driving and Internet connected driving industry chain will be formed rapidly. The coordinated development of technology, capital and talent will create and export a huge value.

Through the upstream and downstream of industrial chain, supply chain, service chain, value chain and capital chain, the intelligent automobile industry cluster is constructed, the intelligent community is formed, and the information consumption trend is triggered. Intelligent automobile manufacturers are introduced to provide supporting facilities, boosting the output value of local parts manufacturers, which also has a positive effect on related industries.

The intelligent manufacturing industry chain of the autonomous vehicle social laboratory and its evolution mode are shown in Figs. 9.24 and 9.25.

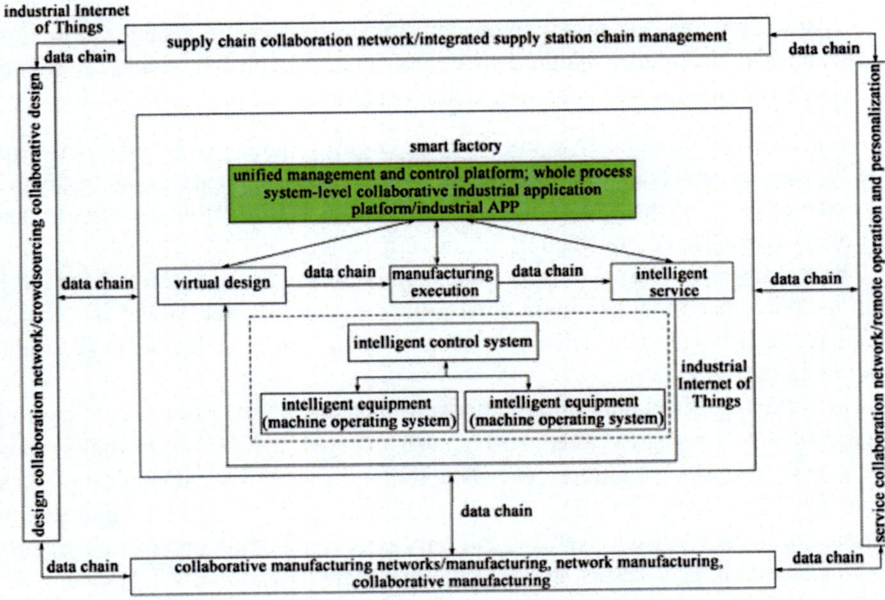

Fig. 9.24 Intelligent manufacturing industry chain of the autonomous vehicle social laboratory

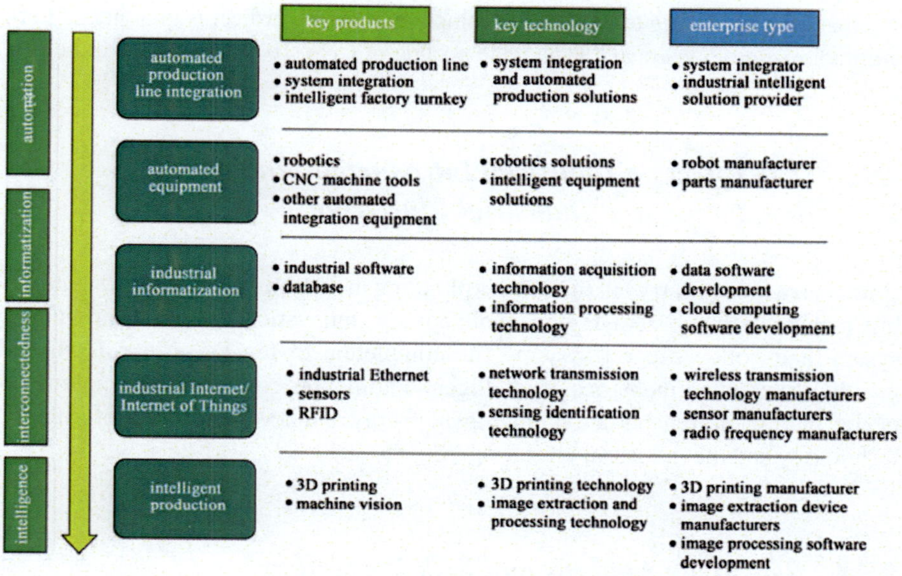

Fig. 9.25 Evolution mode of the intelligent manufacturing industry chain of the autonomous vehicle social laboratory

9.3.4 Resource Integration Ability and Source of Profit

1. Resource Integration

Government. The Driverless Town integrates relevant resources from the Ministry of Industry and Information Technology, the Ministry of Housing and Urban–Rural Development, the National Development and Reform Commission, the Ministry of Transport, the Ministry of Science and Technology, the Central Internet Information Office, the Ministry of Civil Affairs, and the Ministry of Commerce.

Enterprises. The Driverless Town introduces BAIC, SAIC, BYD, Volkswagen, Chang'an, Ford, GM, Honda, Nissan and other well-known automobile enterprises to settle in.

Universities. The Driverless Town introduces Tsinghua University, Peking University, Beijing Institute of Technology, Beijing University of Technology, as well as other well-known universities in the United States and Germany, to integrate their academic strengths.

Media. The Driverless Town introduces Xinhua, People.cn, China.com.cn, People's Daily, CCTV, CSTNET and other media to promote the project.

2. Source of profit

The project relies on national industrial funds, private capital, and the profitability of its own technology and service.

Operating according to the above business model, the project is guaranteed to be profitable within 3 years and can create a value of CNY 10 billion per year then.

9.3.5 The Effect on Industrial Introduction, Industrial Incubation and Industrial Clustering

Efforts are made to carry out in-depth exploration in the construction of new automobile manufacturing system, regional collaborative innovation system, talent training system and policy support system. The completion of the Driverless Town will greatly promote the industries of intelligent automobile, robot and artificial intelligence, intelligent transportation, intelligent energy and new generation information technology and the industrial clustering.

9.3.6 Sensitivity Analysis and Risk Prevention

This project plan can ensure the maximum protection of the interest of investors and enterprises in the zone. And it comprehensively guarantees the feasibility of the town planning from the perspectives of the national policy-making, the introduction of high-end academic resources, and the introduction of advanced technologies, so as to effectively avoid risks.

Chapter 10
From Autonomous Vehicle to AI City

10.1 Autonomous Vehicle Technology Motivating Intelligent Transportation

Data from the Ministry of Industry and Information Technology (MIIT) shows that market of traditional fuel vehicles in China has shown a downward trend in recent years, while the market of new energy vehicles has shown an upward trend.

During the first half of 2019, the production and sales volumes of automobiles reached 12.132 million units and 12.323 million units respectively, down 13.7 and 12.4% year-on-year. In June, the production and sales volumes reached 1.895 million and 2.056 million units, down 17.3% and 9.6%, respectively. The monthly automobile sales and year-on-year changes from 2016 to 2019 are shown in Fig. 10.1.

During the first half of 2019, domestic brand passenger vehicle sales in China reached 3.998 million units, down 21.7% year-on-year, accounting for 39.5% of total passenger vehicle sales. Among them, sedan sales are 1.001 million units, down 16.2% year-on-year, with a market share of 20.2%. SUV sales are 2.295 million units, down 23.3% year-on-year, with a market share of 53.4%. MPV sales are 509,000 units, down 25.9% year-on-year, with a market share of 75.9%. Monthly commercial vehicle sales and year-on-year changes from 2016 to 2019 are shown in Fig. 10.2.

During the first half of 2019, the production and sales of new energy vehicles reached 614,000 and 617,000 units, with year-on-year growth of 48.5% and 49.6% respectively. Among them, the production and sales of pure electric vehicles reached 493,000 and 490,000 respectively, with an increase of 57.3% and 56.6%. Plug-in hybrid vehicle production and sales reached 119,000 and 126,000 units, with an increase of 19.7% and 26.4% respectively. The monthly sales and year-on-year changes of new energy vehicles from 2016 to 2019 are shown in Fig. 10.3.

In the context of the changing and development of the automotive industry, the autonomous vehicle industry, with intelligence and networking as its main features, has become a strong driving force for the growth of the automotive economy. Automatic driving technology is expected to become the hotspot of competition in the transportation industry in the future.

M. Du, *Autonomous Vehicle Technology*, Unmanned System Technologies,
https://doi.org/10.1007/978-981-19-4143-6_10

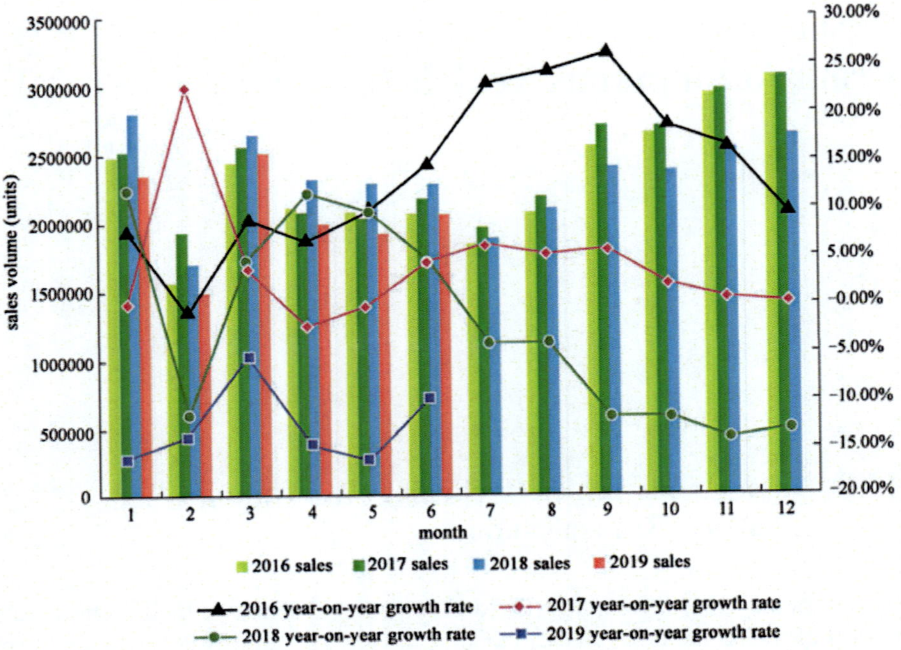

Fig. 10.1 Monthly automobile sales and year-on-year changes from 2016 to 2019. *Source* MIIT

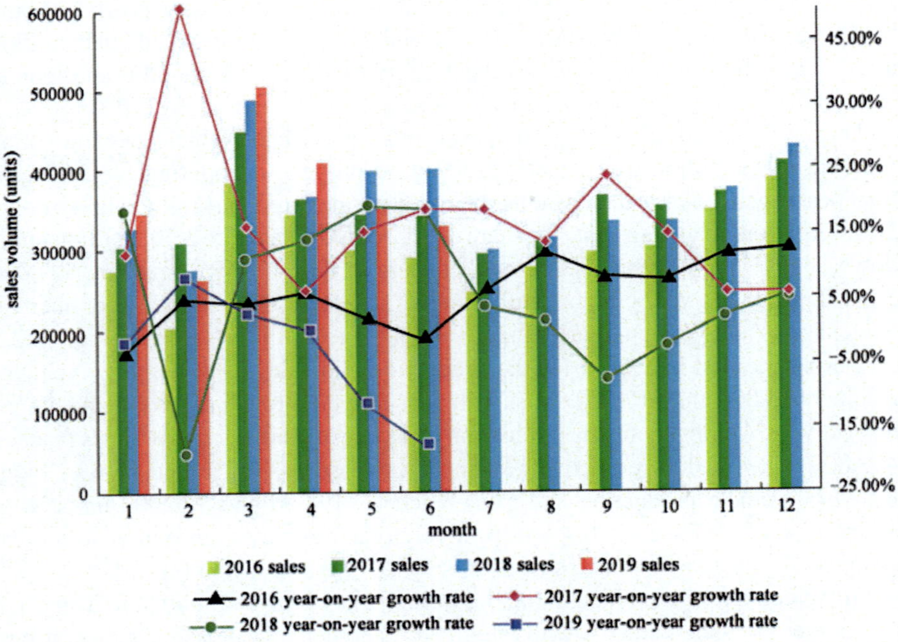

Fig. 10.2 Monthly commercial vehicle sales and year-on-year changes from 2016 to 2019. *Source* MIIT

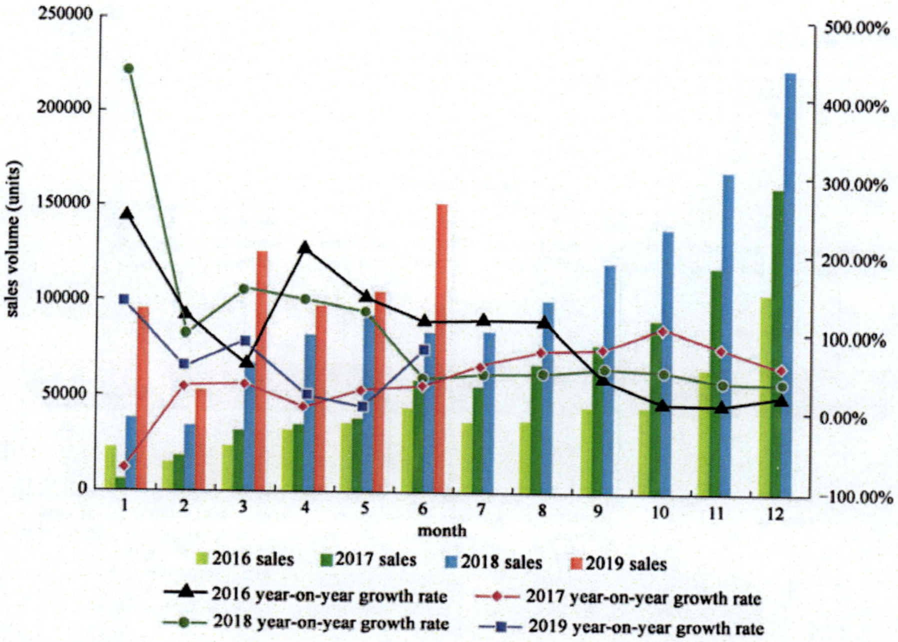

Fig. 10.3 Monthly sales and year-on-year changes of new energy vehicles from 2016 to 2019. *Source* MIIT

The development history and future trend of autonomous vehicle are shown in Fig. 10.4.

The period of 2016–2025 is the partially automatic stage, which will realize L4 automatic driving. L5 fully automatic driving will be achieved after 2025.

The development trend of intelligent connected vehicle is shown in Fig. 10.5.

By 2025, under the premise of mature fully automatic driving technology, urban automatic driving, urban vehicle networking and vehicle–road cooperative control will be realized.

Automotive electronics and vehicle networking technology brought by autonomous driving technology will trigger the rise of a new industry of intelligent transportation.

With the commercialization of autonomous vehicle technology and the in-depth implementation of the new smart city action plan, the integration and development of autonomous vehicle and smart city will gradually go deeper and lead the development of the smart transportation industry.

The autonomous vehicle industrial ecosystem has basically formed, and the new ecology of smart transportation is shown in Fig. 10.6.

The successful model of the high precision map giant Here is worth learning from, as Here has gradually evolved from the largest map provider to an integrated service provider of ADAS, cloud services, automatic driving technology, Internet of Vehicles, and transportation finance. The business model of Here is shown in Fig. 10.7.

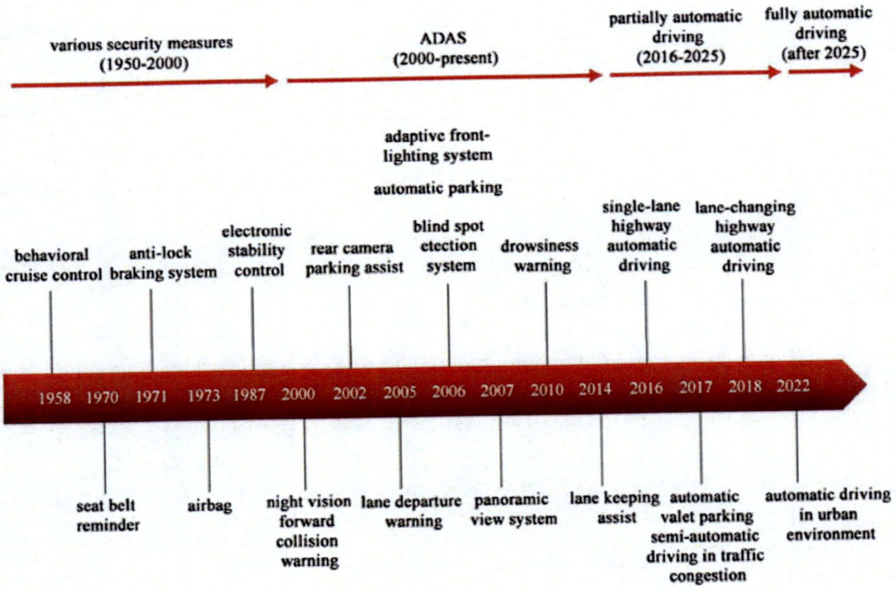

Fig. 10.4 Development history and future trend of autonomous vehicle

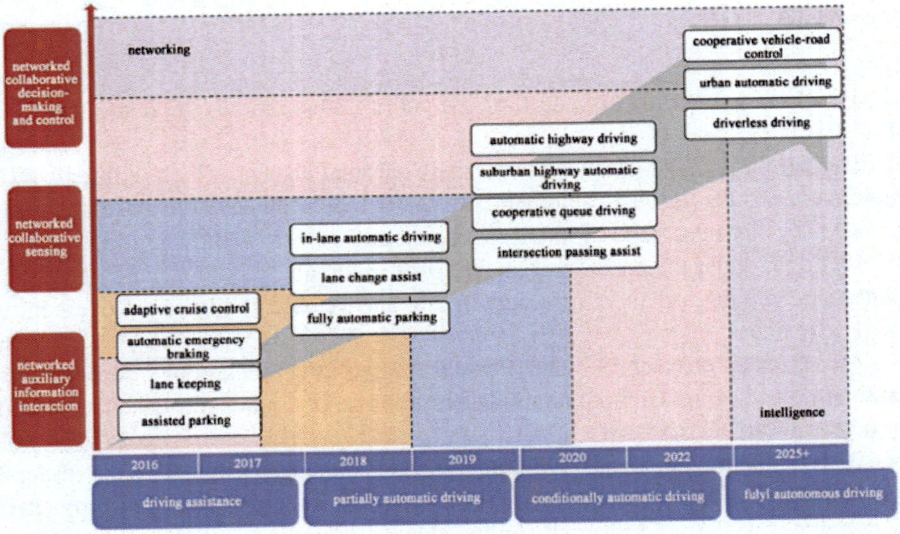

Fig. 10.5 Development trend of intelligent connected vehicle

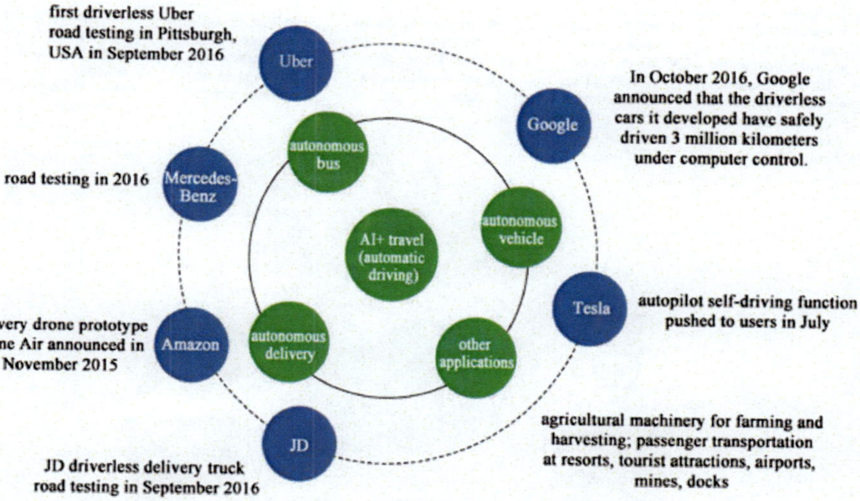

Fig. 10.6 New ecology of smart transportation

The future intelligent transportation system based on AI+ and 5G+ will become the mainstream trend. The future intelligent transportation system includes 7 elements, which are vehicle, road, people, event, information, environment, law. Among them, ring refers to the comprehensive environmental conditions of traffic. Information refers to the network and information communication. Law refers to laws, regulations and policies. The intelligent transportation system driven by the 7 elements will provide guarantee for the improvement of urban traffic condition and service improvement based on the spatio-temporal traffic data, thus contributing to the construction of a new generation of smart city. The typical analysis methods and results of traffic behavior based on spatio-temporal traffic big data are shown in Fig. 10.8.

Analyzing the relationships between vehicles, roads, people, events, information, environments and laws is the premise and foundation of extracting the knowledge from the city model. The community model has the advantages of rich user behavior, direct demand, wide coverage, timeliness and high accuracy, and is the key link connecting traffic and citizens. Urban smart planning, resource allocation, smart governance, population layout, and functional area layout all rely on the urban knowledge mining result, which is also the key entry point of improving people's living standards.

In the past two years, the development of autonomous vehicle, Internet of Vehicles, traffic big data and traffic cloud platform has greatly promoted the rapid advancement of intelligent transportation industry. At present, the integrated development of artificial intelligence and intelligent transportation system is mainly reflected in the following categories: autonomous vehicle and Internet of Vehicles, online ride

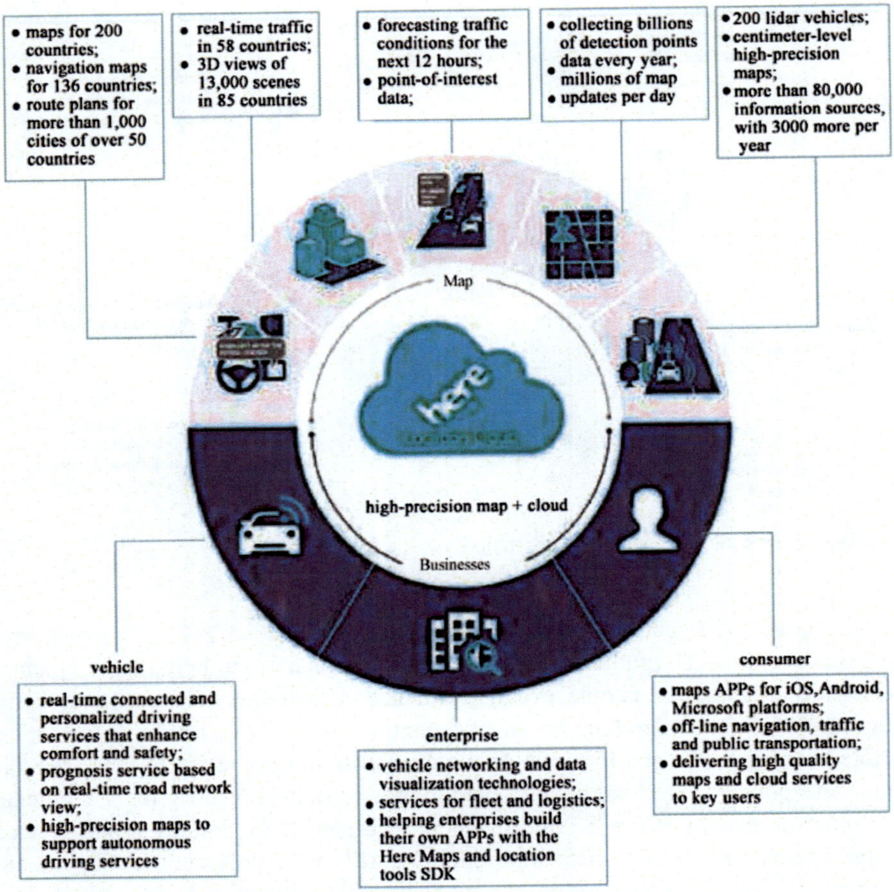

Fig. 10.7 Business model of Here

booking, shared transportation, automobile after-sales service and vehicle intelligent insurance service. The ecological map of the intelligent transportation industry is shown in Fig. 10.9.

Smart transportation is a relatively mature field of smart city. The development of autonomous vehicle, Internet of Vehicles, traffic big data and traffic cloud platform in recent two years has greatly promoted the rapid advancement of smart transportation. At present, the hotspots of intelligent transportation system mainly include autonomous vehicle and Internet of Vehicles, online ride booking, shared transportation, automobile after-sales service and vehicle intelligent insurance service, UAV, intelligent parking and traffic brain.

Ali ET City Brain is a typical case of the preliminary practice of artificial intelligence city. It uses real-time and abundant data resources to globally optimize public resources, timely correct urban operation defects, and achieve the breakthrough of urban governance mode, service mode and industrial development. In terms of urban

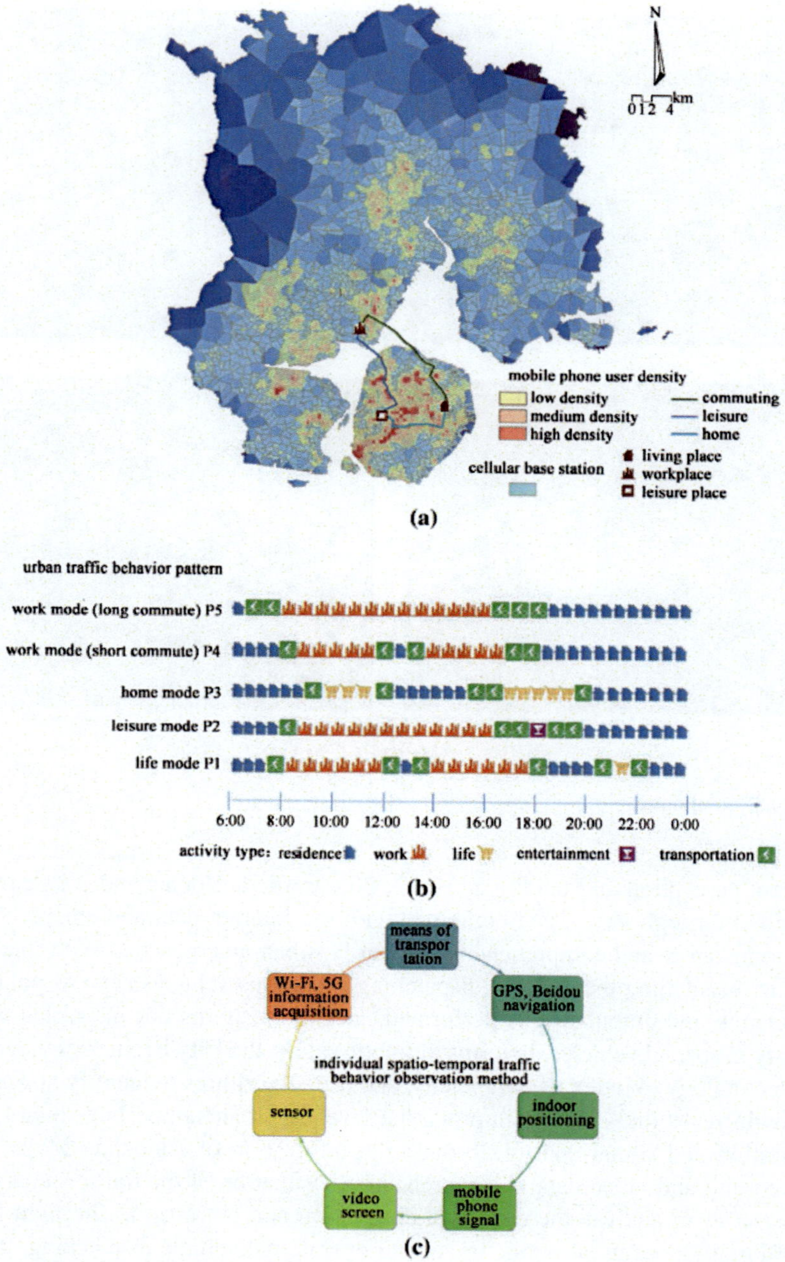

Fig. 10.8 Typical analysis methods and results of traffic behavior based on spatio-temporal traffic big data. **a** traffic behavior analysis based on cell phone user density. **b** classification of urban traffic behavior patterns. **c** individual spatio-temporal traffic behavior observation method. **d** heat map of travel direction

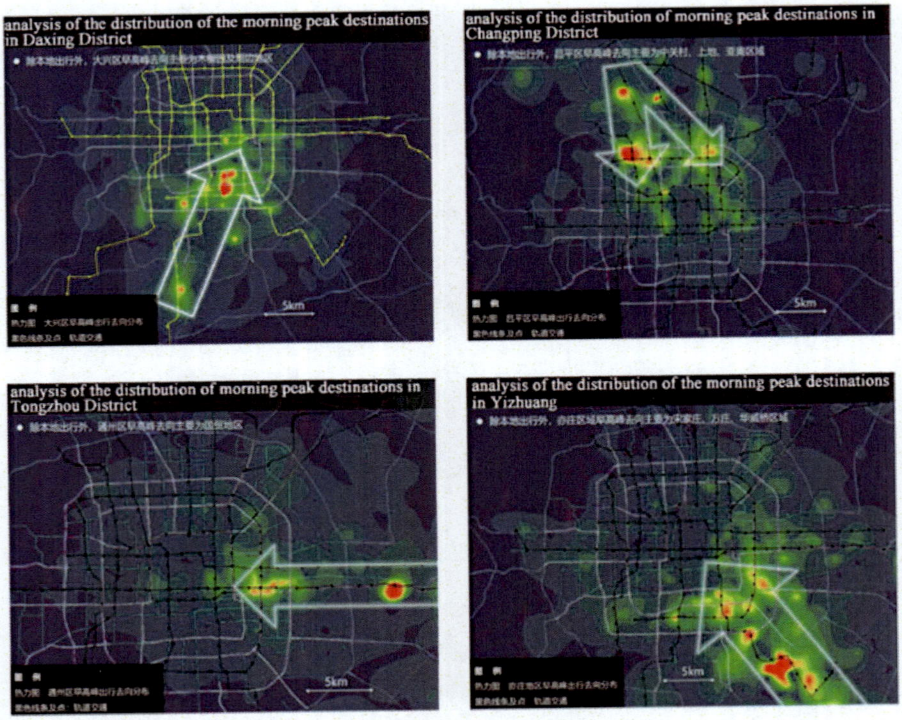

(d)

Fig. 10.8 (continued)

incident perception and intelligent processing, traffic accidents and congestion are identified through video, and combined with the Internet data and emergency call data to instantly and comprehensively identify urban emergencies. With intelligent vehicle dispatching technology, dispatching of various vehicles such as police car, fire engine and ambulance is performed, and traffic lights are instructed to give priority to special vehicles. In terms of public safety, the City Brain uses video analysis technology to index the whole city, and uses algorithms to identify and analyze videos in real time, so as to improve the level of urban safety. In terms of traffic congestion and signal control, through the integration of AutoNavi Maps, traffic police radio and video data, a comprehensive evaluation of the traffic status can be made, so as to analyze the causes of congestion and optimize traffic light signals accordingly. In terms of public travel and operational vehicle dispatching, through video, AutoNavi Maps, Wi-Fi probes, mobile operator data, the required transport capacity in crowded area can be calculated, in order to adjust bus schedules, shuttle routes and taxi dispatching according to ease the congestion.

The overall architecture of ET City Brain is shown in Fig. 10.10.

The actual operation of ET City Brain is shown in Fig. 10.11.

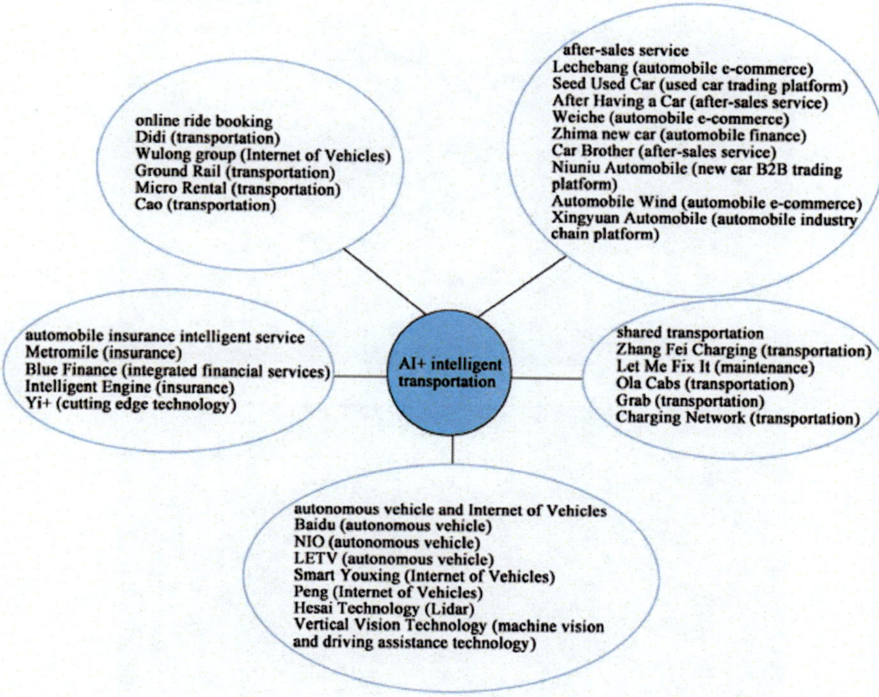

Fig. 10.9 Ecological map of the intelligent transportation industry

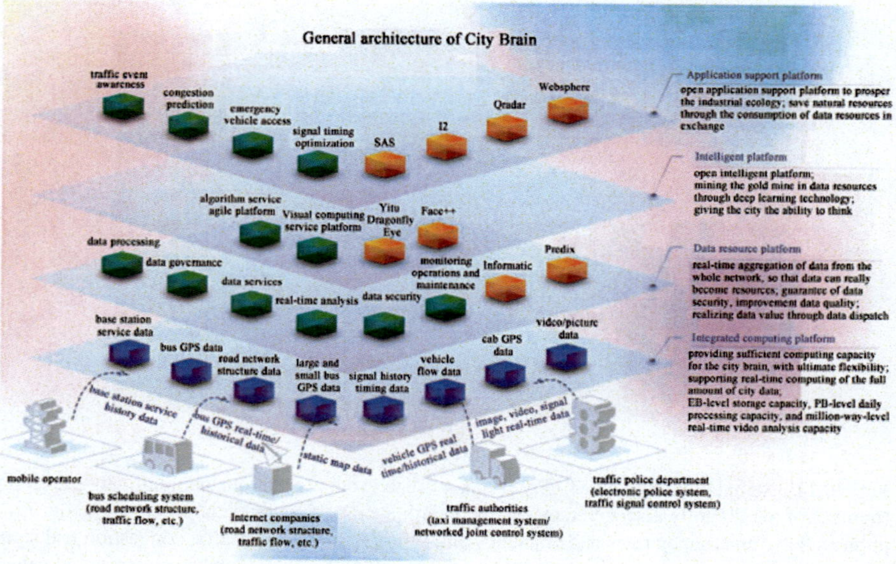

Fig. 10.10 Overall Architecture of ET City Brain

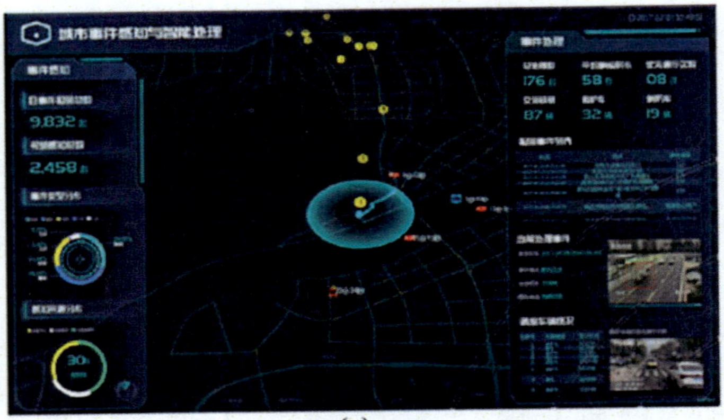

(a)

(b)

(c)

Fig. 10.11 Actual operation of ET City Brain. **a** urban event perception and intelligent processing (number of intelligently discovered events increased; average time of event processing reduced). **b** public safety (emergency response efficiency and safety improved). **c** traffic congestion and signal control (congestion level and pilot area travel time reduced). **d** public travel and operational vehicle scheduling (crowd reduced; public transportation utilization increased)

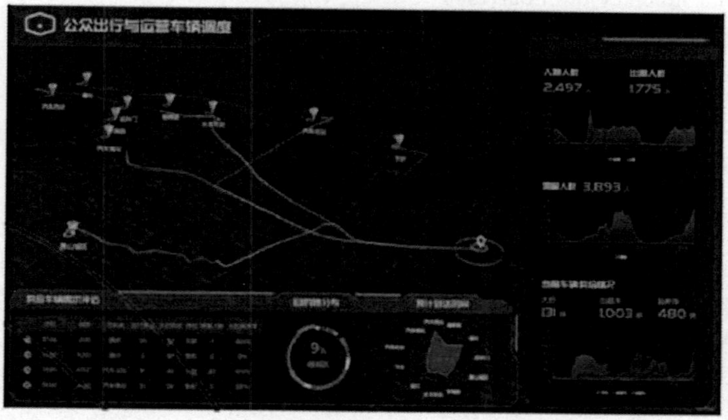

(d)

Fig. 10.11 (continued)

10.2 Evolutionary Stages of Smart City in China

At present, China is in a period of accelerated urbanization. The problem of "urban disease" in some areas is becoming increasingly serious. Many economic and resource challenges brought about by urbanization have led to the desire for smart cities, and smart city have become an important method of urban transformation around the world. With the continuous advancement of human society, cities will hold more and more people in the future. It is expected that by 2050, more than 70% of the world's population will be living in cities, which account for only 2% of the Earth's land area. The idea of smart city has attracted attention from all over the world. In the past 10 years, developed countries and regions have invested heavily in the construction of their own smart cities. In China, the construction of smart cities is an important way to implement the national policy of innovation-driven development and new urbanization. The evolution of smart city has been closely related to the development of digital economy, the fourth industrial revolution and new urbanization.

Since the mid-eighteenth century, human history has experienced a total of four industrial revolutions: the steam technology revolution (the first), the electric technology revolution (the second), the computer technology revolution (the third), and the green and intelligent industrial revolution (the fourth). Industry 4.0 is part of the high technology strategy proposed by the German government in *Ideas, Innovation, Growth—Germany's High Technology Strategy 2020*, which aims to enhance the intelligence of the manufacturing industry, establish intelligent factories with adaptability, resource efficiency and genetic engineering, and integrate customers and business partners in business processes and value streams, with the networking as the technological basis. The goal of Industry 4.0 is to establish a highly flexible, personalized, digital production model for products and services. In this model, traditional industry boundaries will disappear, giving rise to various new forms of cooperation. At present, countries around the world are actively taking actions to participate in the new round of industrial revolution. The United States proposed the ideas of Advanced Manufacturing Partnership, Re-industrialization and Industrial Internet. The United Kingdom proposed the idea of Industry 2050. France proposed the New Industrial France Plan, Japan proposed the idea of Society 5.0, and South Korea proposed Manufacturing Innovation Plan 3.0.

The essence of the new round of industrial revolution is the battle for the dominance of intelligent manufacturing ecosystem, which is a great concern for governments and enterprises. The revolution has become a powerful external driving force for urban development. Meanwhile, internal contradictions and demand-side changes in cities have become the internal driving force for urban development. Currently, urban development is facing many issues such as industrial transformation and upgrading, urban environment restoration, social governance, urban disease prevention and control, and smart city has become an inevitable path for urban transformation and upgrading. The application of Industry 4.0 concepts and technologies in the city provides new ideas and widens the space for the development of smart city. The large-scale networked collaboration, personalized customization and service extension advocated by Industry 4.0 are becoming power sources of smart city innovation. Industrial Internet, blockchain and artificial intelligence are redefining the urban industrial chain, giving birth to new economic forms in cities and becoming the foundations of smart cities.

The stages of smart city development in China are shown in Fig. 10.12.

Smart City 1.0 (2008–2012). In 2008, IBM put forward the concept of smart Earth, which introduced smart city into the budding period (or preliminary exploration period). Networking and informatization is the primary task of this stage of smart city construction, so it can also be called the digital city stage.

Smart City 2.0 (2012–2016). Smart City 2.0 mainly focuses on the construction of informationized infrastructure, e-government, and information services for the people. The construction plan is fragmented and lacks top-level design. It can be regarded as the starting period (or exploration period) of China's smart city development. In 2012, the Ministry of Industry and Information Technology issued a notice to solicit opinions on the evaluation system of smart cities. In the same year, the Ministry of Housing and Urban–Rural Development issued the *National Smart City*

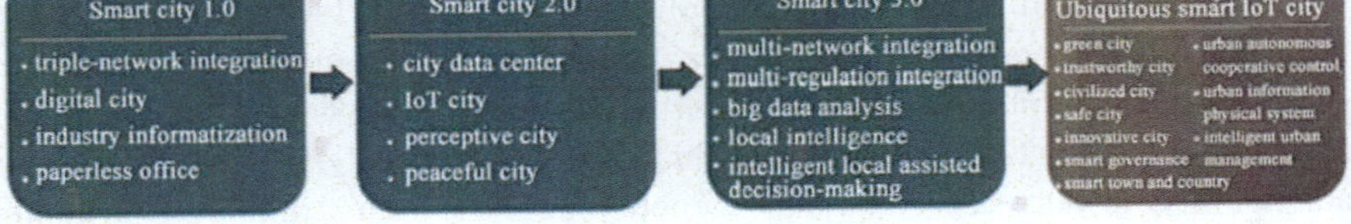

Fig. 10.12 Stages of smart city development in China

Pilot Plan and the *Smart City Index System*, and announced hundreds of national pilot smart cities successively. The *National New Urbanization Plan (2014–2020)* clearly proposes to promote the construction of new smart cities.

Smart City 3.0 (2016–2018). The new type of smart city discussed in the 13th Five-Year Plan period can be addressed as Smart City 3.0. With the full implementation of the concept of innovation, coordination, green and sharing, the idea of city is given a new connotation, and new requirements are put forward for the construction of smart cities. On the basis of a comprehensive survey of the national smart city construction, the Cyberspace Administration of China proposed the concept of new smart city in the face of new challenges, and took the lead in organizing 26 ministries and commissions, including the National Development and Reform Commission, to jointly promote the construction of new smart cities.

Smart City 4.0 (2018-present). The smart city in this stage places more emphasis on the importance of industry, especially strategic emerging industries, and on the role of the digital economy in urban transformation and sustainable development. The accelerated integration of new generation information technology and the real economy will find a solid footing in Smart City 4.0. The distributed and personalized high technology innovation with respect to the development needs of social components (individuals, enterprises, government institutions, etc.) is becoming the original driving force for urban development. With the deployment of rural revitalization and urban–rural integration strategies, the construction of smart city and smarty country is on the agenda. The smart governance of urban–rural integration will become an important trend in the future. Along with the issuance of State Council's *New Generation of Artificial Intelligence Development Plan* (No. 35, 2017, SC), the promotion and implementation of the Smart + strategy, and the arrival of the 5G era, China's smart city is advancing into the stage of AI city and ubiquitous intelligent IoT city. The year of 2018 is the first year in the AI city history, while the year of 2019 is the first year in the ubiquitous intelligent IoT city history.

10.3 New Generation AI Development

The book's understanding of the connotation of AI is shown in Figs. 10.13.

In terms of inspiration, AI includes anthropomorphic intelligence and bionic intelligence (imitation of dogs, antelopes, horses, fish, birds, ant colonies, bee colonies, etc.). In terms of existing form, AI includes tangible intelligence (robots, autonomous vehicles, drones, intelligent voice terminals, etc.) and invisible intelligence (used for intelligent reasoning, diagnosis, selection, prediction, classification, clustering, planning, analysis, decision-making, optimization, control, etc.).

In July 2017, the State Council released the *New Generation of Artificial Intelligence Development Plan* (No. 35, 2017, SC). It states that the development of AI has entered a new stage. After more than 60 years of evolution, with the joint drive of new theories and technologies such as mobile Internet, big data, supercomputing, sensor networks, brain science, and the strong demands for economic and social

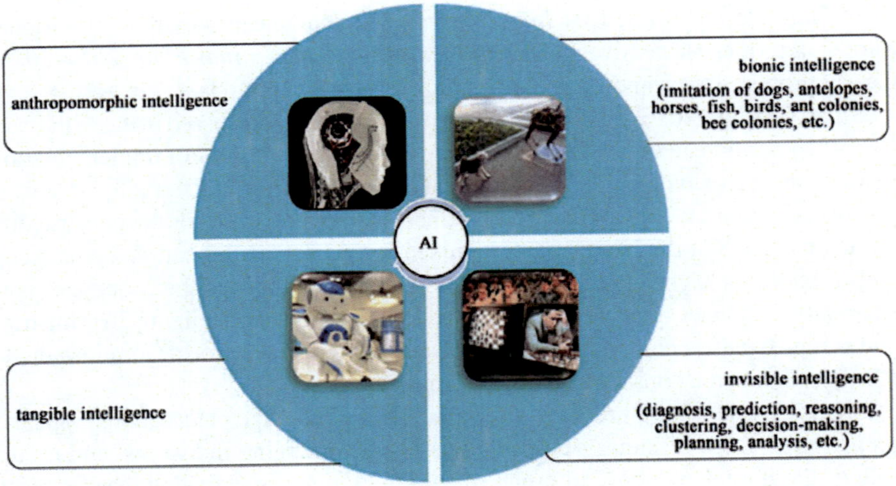

Fig. 10.13 Connotation of AI

development, AI technology has accelerated its development, showing new features such as deep learning, cross-border integration, human–machine collaboration, group intelligence, and autonomous control. Big data driven knowledge learning, cross-media collaborative processing, human–machine collaboration enhanced intelligence, group integrated intelligence, and autonomous intelligent systems have become the focuses of AI development. Brain-like intelligence inspired by brain science research is ready for rapid advancement. The trend of chip-based, hardware-based and platform-based technologies has become more obvious, and the development of AI has entered a new stage. In order to implement the concept of *Made in China 2025* and the *New Generation of Artificial Intelligence Development Plan*, accelerate the development of AI industry, and promote the deep integration of AI and the real economy, the Ministry of Industry and Information Technology issued the *Three-Year Action Plan for Promoting the Development of the New Generation Artificial Intelligence Industry (2018–2020)* in December 2017. In October 2018, the Political Bureau of the Central Committee of the Communist Party of China held the ninth collective study on the current situation and trends of AI development. Secretary General Xi Jinping, while leading the study, stressed that AI is a strategic technology leading this round of technological revolution and industrial change, with a strong spillover drive and a "head goose" effect. Accelerating the development of the new generation AI is an important strategic step to win the initiative in the global competition of science and technology. The new generation AI can also provide important strategic resources for the development of science and technology, industrial optimization and upgrading, and productivity jump.

Algorithms, computing power and data are the fundamentals of AI. Compared with the previous generation of AI, the new generation has a substantial improvement in terms of computing power and data.

In recent years, China's AI industry has entered a period of explosive growth. By the end of 2017, the scale of China's AI core industry exceeded CNY 18 billion, and the scale of related industries reached CNY 220 billion. China has accumulated 15,700 patents in the field of AI, ranking second in the world. Since 2018, with leading international companies competing to enter the Chinese market, capital investment has been increasing. The top five segments that have received the most investment in AI in China are computer vision (R&D), natural language processing, private virtual assistant, intelligent robotics, and voice recognition. In the future, the capital will mainly flock to the fields of machine learning and scene application. Driven by both policy and market, China has made significant progress in AI application in the field of smart city and smart society. The number of scientific and technical papers published, and invention patents granted of China ranks second in the world. The research and commercial application in the fields of speech recognition, machine vision, Chinese language information processing, autonomous vehicle and intelligent service robot is among the best in the world. In terms of industry and enterprises, the scale of the industry in China continues to grow, and a number of unicorn enterprises and leading listed companies have emerged. The scale of the core industry is expected to exceed CNY 150 billion by 2020, driving the scale of related industries to exceed CNY 1 trillion. In terms of the integration of AI and new urbanization, AI has been applied in manufacturing, transportation, logistics, construction, home life, medical, security, retail, finance and other fields in the city, and new business models are emerging in various fields. It is promoting the continuous upgrading of urban consumption and urban industry.

In general, automatic driving technology will bring about the rapid development of AI city (Fig. 10.14).

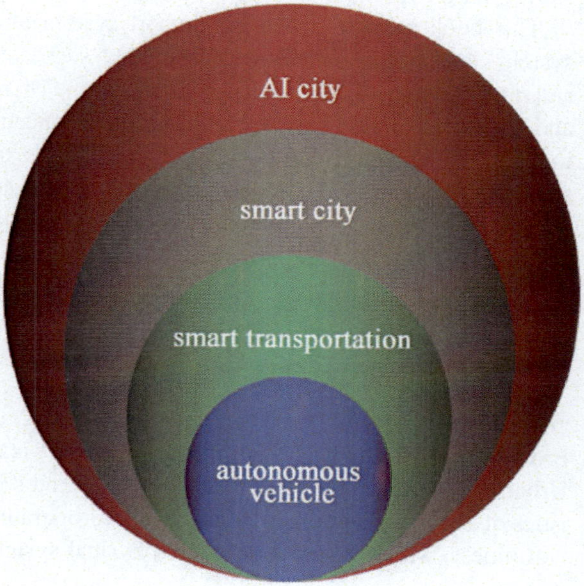

Fig. 10.14 From autonomous vehicle to AI city

The AI city discussed in this book means that, on the premise of following the law of urban development and meeting the needs of social and economic development, supported by urban science, AI, cyber-physical systems (CPS) and system engineering theory, under the unified management of urban brain, with the support of human intelligence space, information space and physical space, new generation information technologies such as big data, cloud computing, Internet of Things, mobile Internet, industrial Internet, modern communication, blockchain and quantum computing are comprehensively adopted to build a highly intelligent city capable of real-time perception, efficient transmission, autonomous control, autonomous learning, intelligent decision-making, self-organization and coordination, self-optimization and evolution. AI city refers to the combination of AI and city, while "AI with city" focuses on AI technology, and "city with AI" focuses on urban vertical application.

The essence of AI city is that the agent composed of data and algorithms flows in a closed loop among the human intelligence space, information space and physical space. The human intelligence space enables the space of city cyber-physical systems (CCPS), creating city anthropomorphic intelligence which surpasses human beings after self-learning. Problems of modern urban system in interconnection, resource sharing, energy consumption, service management and economic development can be solved by the intelligence above.

The goal of AI city is to perform automatic city operation and management, optimal matching between supply and demand, and intelligent governance, so as to create a better life for people in the city and promote the healthy and sustainable development of the city.

The panorama of an AI city is shown in Fig. 10.15.

The panorama of the reference framework model of AI city, AICITYC1.0, is shown in Fig. 10.16.

This book considers that the support system of AI city consists of AI theory system, AI adjoint technology system, AI segmentation application system, and smart city industry application scenario system. These components support, interact and collaborate with each other to form a concentric circle structure as shown in Fig. 10.16, which serves the 8 core features of AI city including real-time sensing, efficient transmission, autonomous control, autonomous learning, intelligent decision self-organizing collaboration, self-optimizing evolution, and personalized customization.

10.4 Artificial Intelligence City

From the perspective of system theory, smart city is a typical complex giant system. Urban complex giant system is an economic and social system with the characteristics of self-organization, self-learning, interoperability, reconfigurability and trustworthiness, which is based on cyber-physical system (CPS) theory and AI theory.

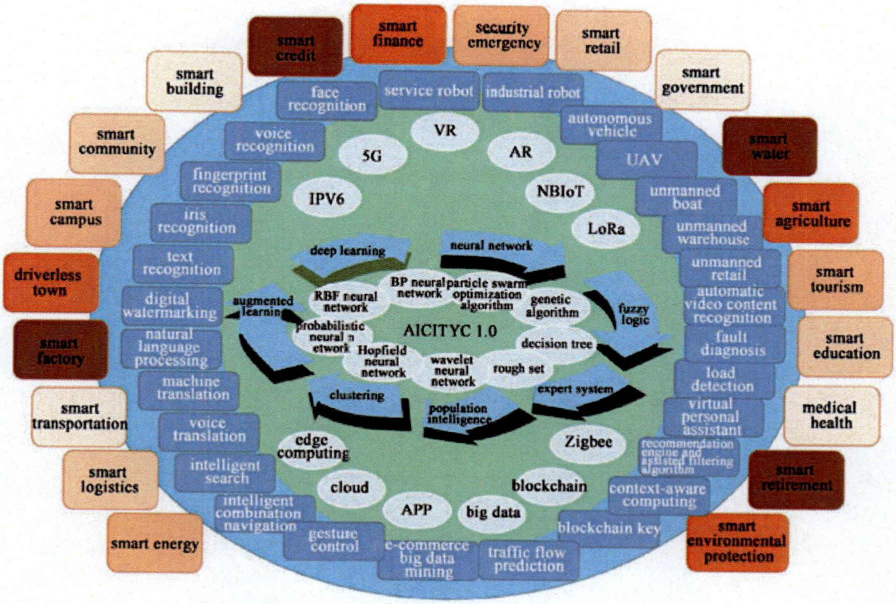

Fig. 10.15 Panorama of an AI city

It integrates urban domain knowledge and urban business system model, and serves urban production and life. Its hierarchical structure and components are shown in Fig. 10.17.

The urban complex giant system is divided into four levels:

Level 1: city level (1 brain + 4 sections);
Level 2: domain level (20 vertical domains);
Level 3: domain sub-system level ($20 \times N_i$);
Level 4: unit level ($20 \times N_i \times M_i$).

The figures in the brackets above are the benchmark numbers of components in each level (according to the current industry development status; may float up or down). N_i ($i = 1, 2\ldots$), M_i ($i = 1, 2\ldots$) are the numbers of internal components in each level (the third level corresponds to the sub-system, and the fourth level corresponds to the unit). The calculation of urban swarm systems can be based on the data of a single city.

The tree structure organization model of urban complex giant system is shown in Fig. 10.18.

The reference framework model of AI city, AICITYC1.0, is shown in Fig. 10.19.

There are two sets of flows in AICITYC1.0 connecting all the elements, namely the value flow and the information flow, as shown in Fig. 10.20.

There are three dimensions or threads AICITYC1.0, namely smart city dimension, artificial intelligence dimension, product and service dimension (or industry dimension).

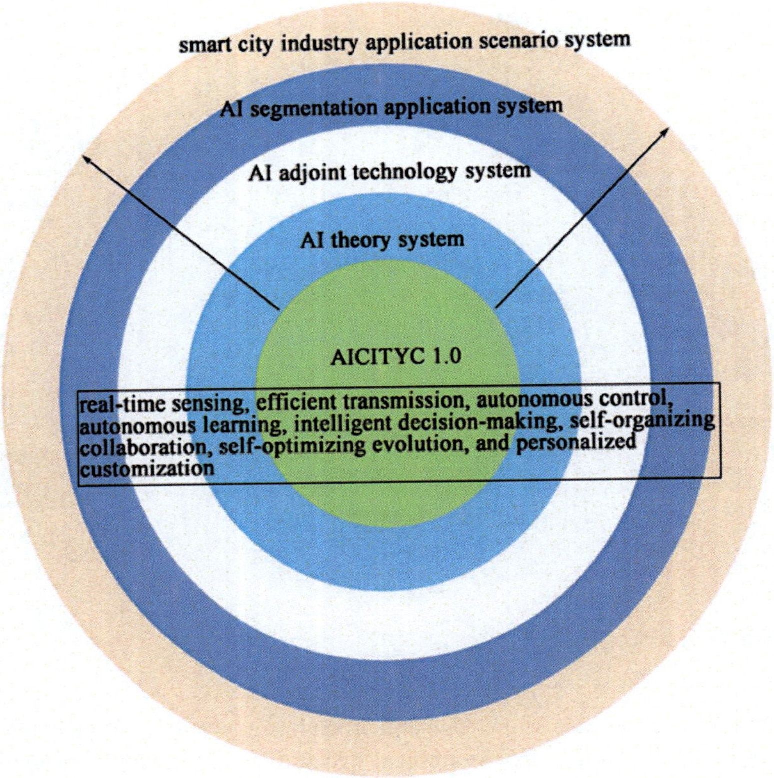

Fig. 10.16 Simplified diagram of AICITYC1.0 panorama

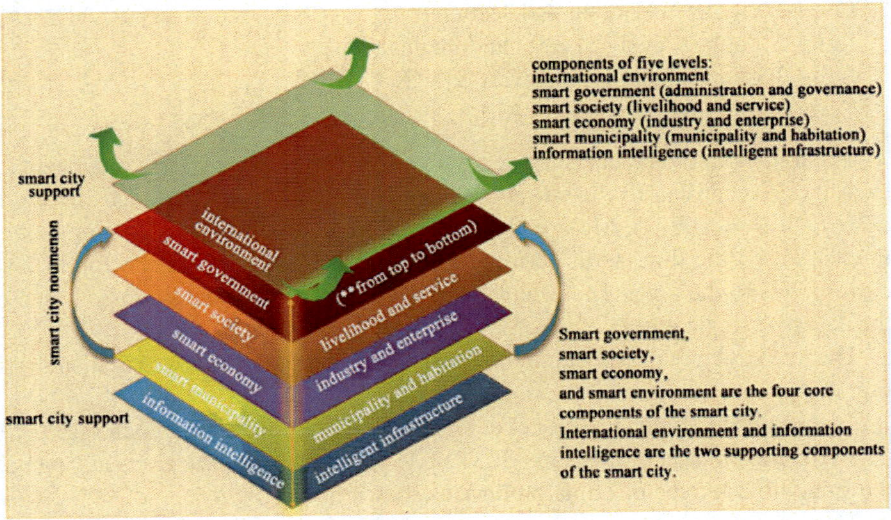

Fig. 10.17 Overall hierarchical structure and components of urban complex giant system

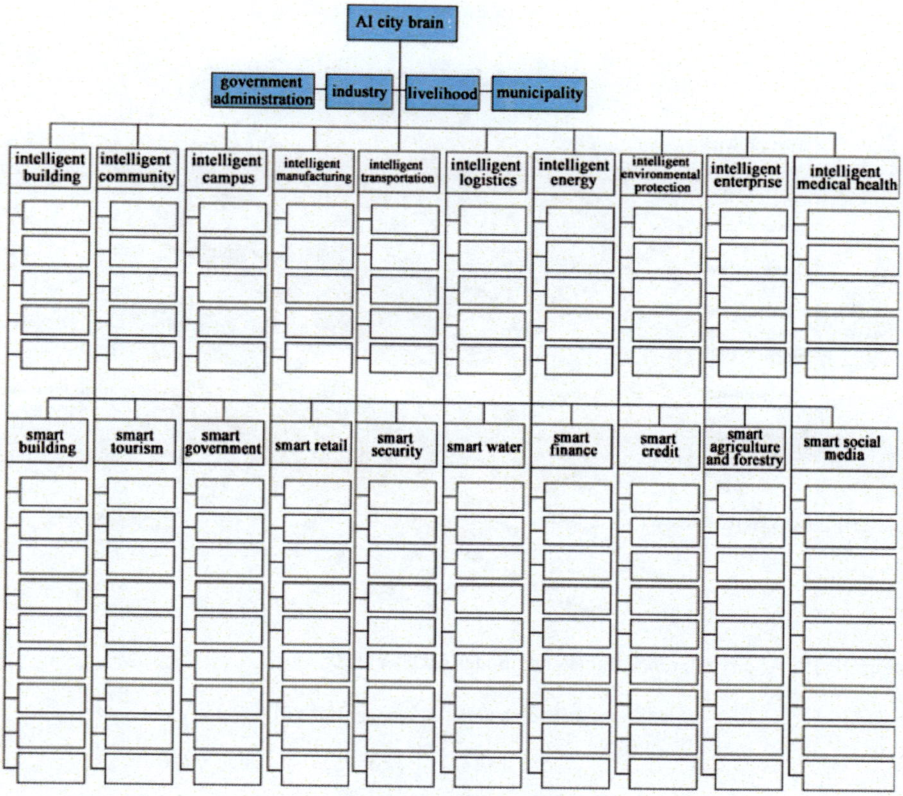

Fig. 10.18 Tree structure organization model of urban complex giant system

The framework of the smart city dimension in AICITYC1.0 is shown in Fig. 10.21.

The framework of the artificial intelligence dimension in AICITYC1.0 is shown in Fig. 10.22.

The dimension of artificial intelligence includes three layers:

Layer 1 is the basic resource layer, including hardware (reflecting computing power), algorithm model and data.

The algorithm model includes expert system, fuzzy system, neural network, deep learning, clustering, bionic system, swarm intelligence, knowledge mapping and other AI research branches.

Layer 2 is the AI technology layer, including common individual AI technology (image, voice semantics, text) and hybrid AI technology (robot, autonomous vehicle, UAV, etc.).

Layer 3 is the industry application of AI, which is the ramification of AI and application scenes.

Layer 1 provides core resources for the construction of layer 2, and produces various relatively universal AI technical directions for Layer 3. The development

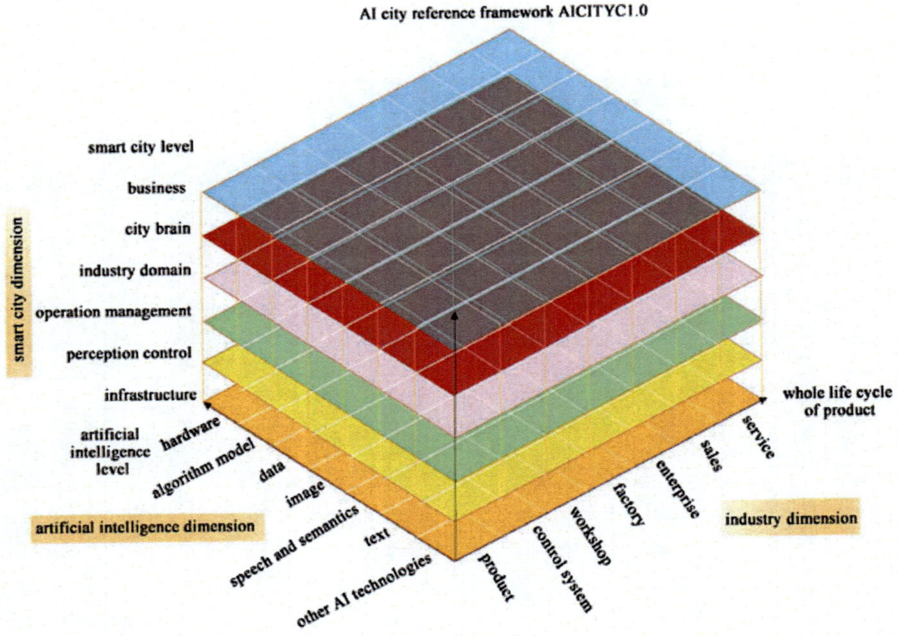

Fig. 10.19 AI city reference framework model AICITYC1.0

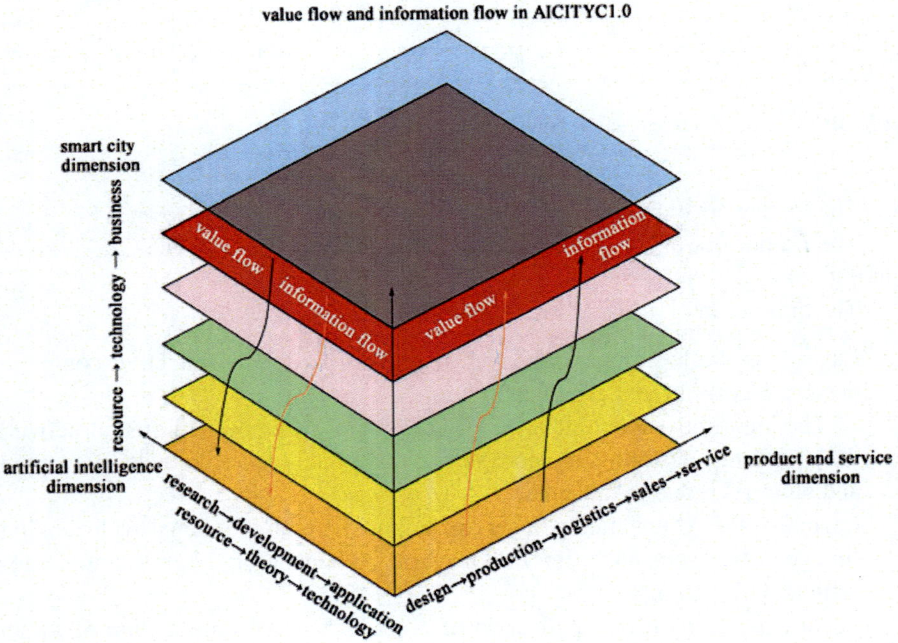

Fig. 10.20 Value flow and information flow in AICITYC1.0

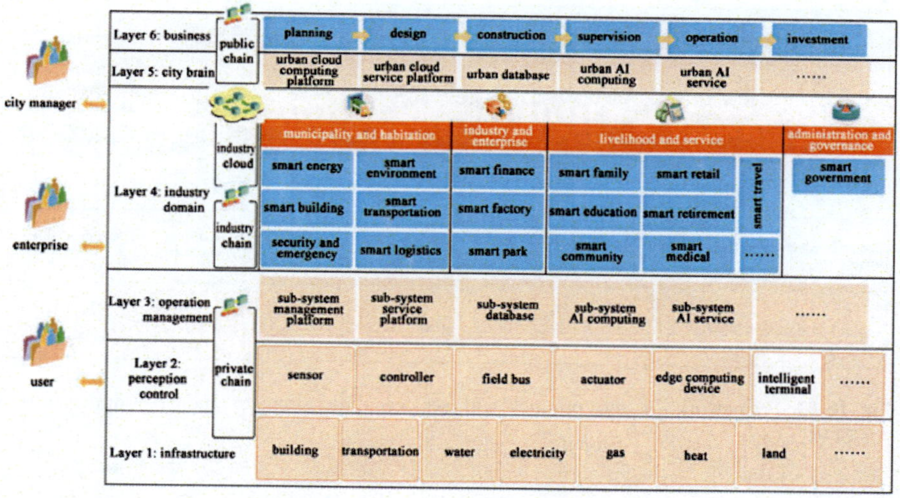

Fig. 10.21 Framework of the smart city dimension in AICITYC1.0

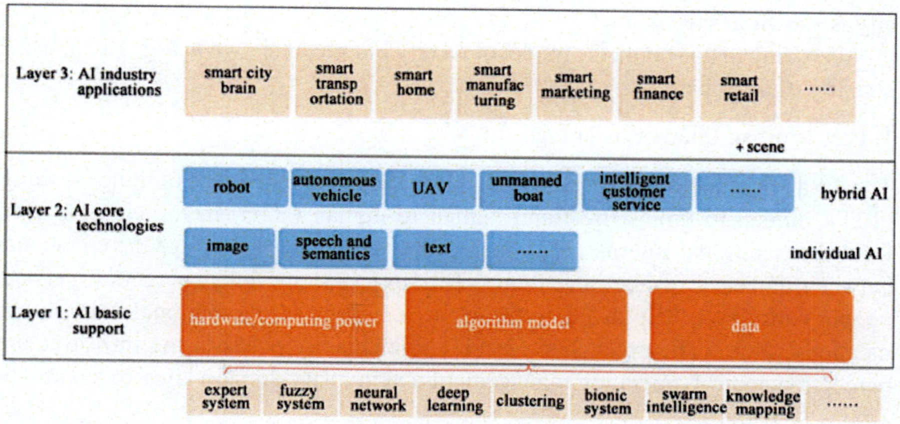

Fig. 10.22 Framework of the artificial intelligence dimension in AICITYC1.0

and implementation of layer 3 should be based on Layer 1 and Layer 2. Combining Layer 1 and Layer 2 with industrial scenarios, various applications forms can be created for Layer 3.

The framework of the product and service dimension (or industry dimension) in AICITYC1.0 is shown in Fig. 10.23.

The product and service dimension refers to the product and service industrial chain of smart city, which is the technical and economic relationship among various intelligent manufacturing industry departments, production factors, enterprises, consumers and users serving urban construction, consumption and application.

Based on the three-tier architecture of the artificial intelligence dimension, the "AI + city" industrial chain can be constructed. It consists of four coordinated and

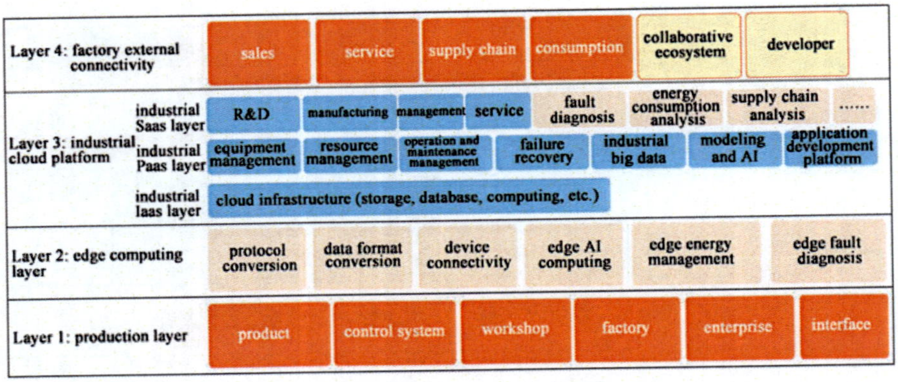

Fig. 10.23 Framework of the product and service dimension (or industry dimension) in AICITYC1.0

balanced sections, including value chain, enterprise chain, supply chain and space chain. AI industrial chain is formed among various industrial sectors according to specific logical and spatio-temporal relationship. The industrial chain model of AI city is shown in Fig. 10.24.

According to this book, the model of AI city Internet is shown in Fig. 10.25, which consists of three parts.

1. Part 1: urban Internet of Things

The urban Internet of Things mainly involves the interconnections of things to things (T2T), human to things (h2t) and human to human (H2H) in the city. Different from the traditional Internet, H2T means that human beings utilize the connection between universal devices and things. H2H refers to the interconnection between people without the help of PC (refer to the definition of International Telecommunication Union). The urban Internet of Things mainly exists in the infrastructure, perception control, operation management and industry domain layer of smart city.

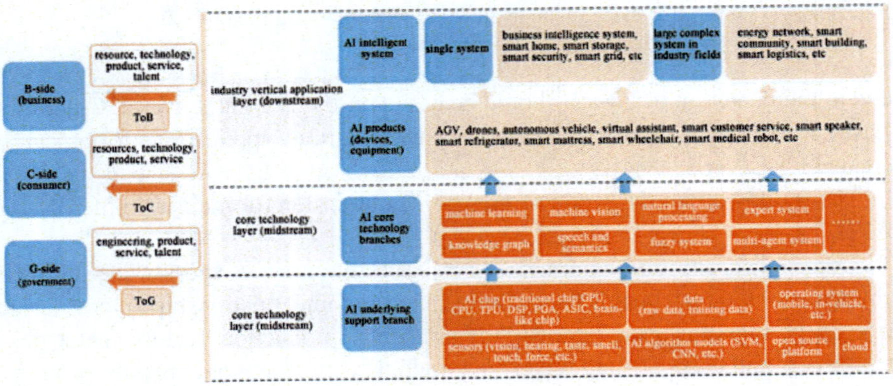

Fig. 10.24 Industrial chain model of AI City

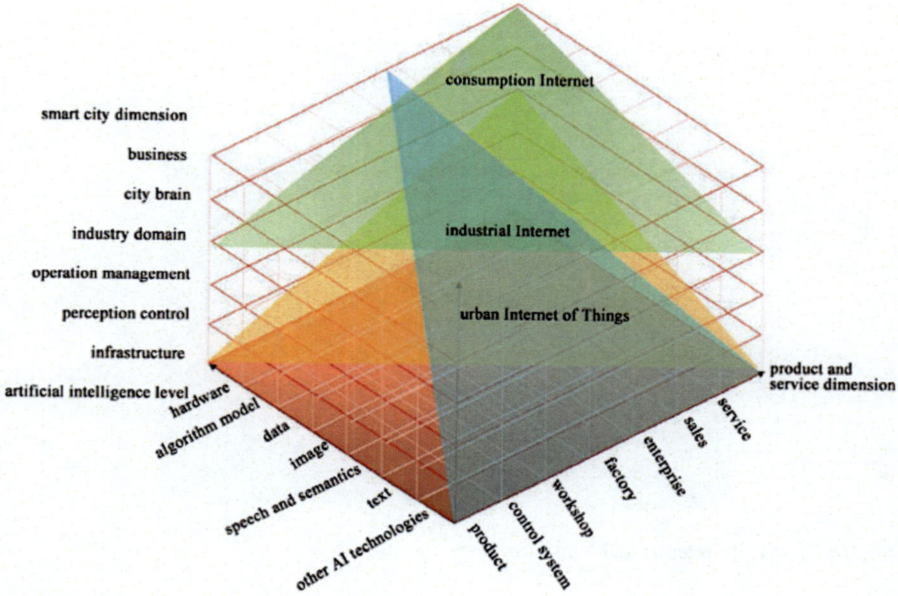

Fig. 10.25 Model of AI city Internet

2. Part 2: urban consumption Internet

The urban consumption Internet refers to the human to thing (H2T) and human to human (H2H) interconnections closely related to people's life in the city, which are formed with urban consumption as the service goal, and with PC and mobile phone as the interconnection tools. It mainly exists in the industry field, city brain and business layer of smart city.

3. Part 3: urban industrial Internet

The urban industrial Internet refers to the Internet which directly serves urban industrial development and indirectly serves urban life. It is closely related to factories and production.

The relationship among the three parts of AI city Internet, the primary, secondary and tertiary industry of the national economy, and the digital economy is shown in Fig. 10.26.

The core elements of smart city, including city, information, human and their relationship, can be described by the cyber physical space of smart city shown in Fig. 10.27.

AICITYC1.0 AI-CPS space model consists of four subspaces: city, information, human and AI brain. CPS space plus AI space equals AI-CPS space (corresponding to AI brain in the figure). The first generation smart city is basically in a semi-autonomous state. The second generation smart city (under the collaborative command of AI brain) is in a fully autonomous state.

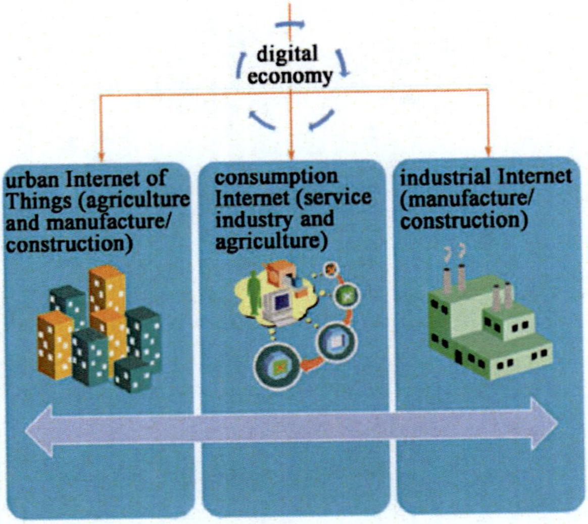

Fig. 10.26 AI city Internet and digital economy

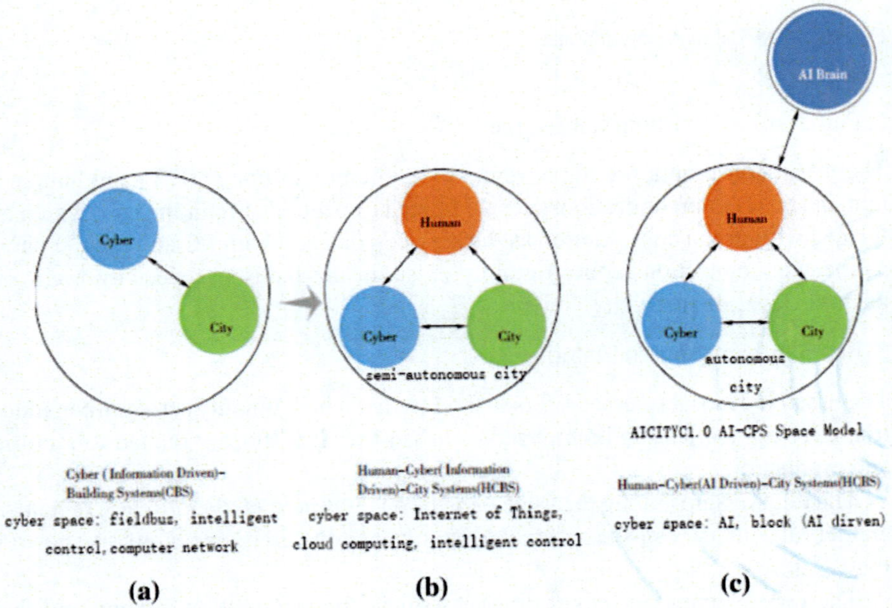

Fig. 10.27 Cyber physical space of smart city. **a** digital city; **b** first generation smart city; **c** second generation smart city

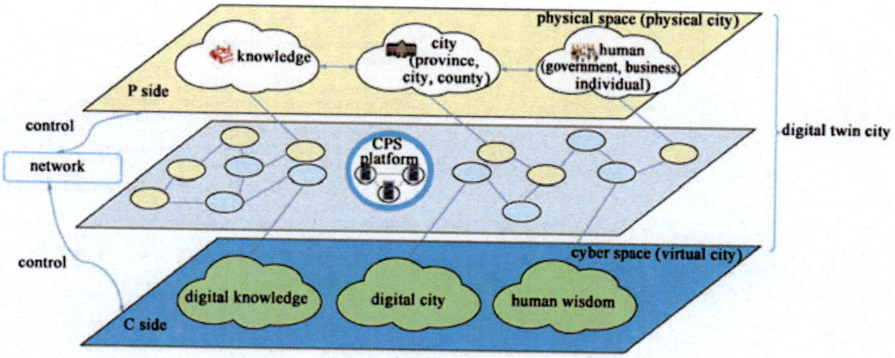

Fig. 10.28 AICITYC1.0 Digital Twin City conceptual model

The AI brain stores an image of physical city system, so it can be an intelligent twin of the physical city in reality.

The AI brain can also be regarded as an operating system, the AI-CPS City OS, integrated with technology and commerce. It can be used to form a digital and intelligent power to promote the self-organization, self-optimization, self-evolution and self-governance of the city.

Based on the AICITYC1.0 AI-CPS space model, this book proposes the AICITYC1.0 Digital Twin City conceptual model, as shown in Fig. 10.28. Between the twin cities, information is transmitted in two directions. When the information is transmitted from the physical city to the digital city, the data from the sensors can be used to observe the operation of the city. On the contrary, when information is transmitted from the digital city to the physical city, data and models can be used to monitor, study, predict, control, modify and improve the physical city with AI algorithm. That is namely the mutual control of cyber side and physical side.

The AICITY1.0 open ecosystem model proposed in this book is shown in Fig. 10.29.

The blockchain technology model consists of data layer, network layer, consensus layer, incentive layer, contract layer, application layer and evaluation layer. Data layer, network layer and consensus layer are the necessary elements of blockchain technology. Incentive layer, contract layer, application layer and evaluation layer are not necessary for each individual blockchain application. The data layer encapsulates the chain structure of the underlying data blocks, as well as the related asymmetric public private key data encryption technology and timestamp technology, which is the basic data structure in the whole blockchain technology. The network layer includes P2P networking mechanism, data dissemination mechanism and data verification mechanism. The consensus layer encapsulates all kinds of consensus mechanism algorithms of network nodes. Consensus mechanism algorithm is the core technology of blockchain, because it determines who is responsible for bookkeeping, which has an influence on the security and reliability of the whole system. At present, there

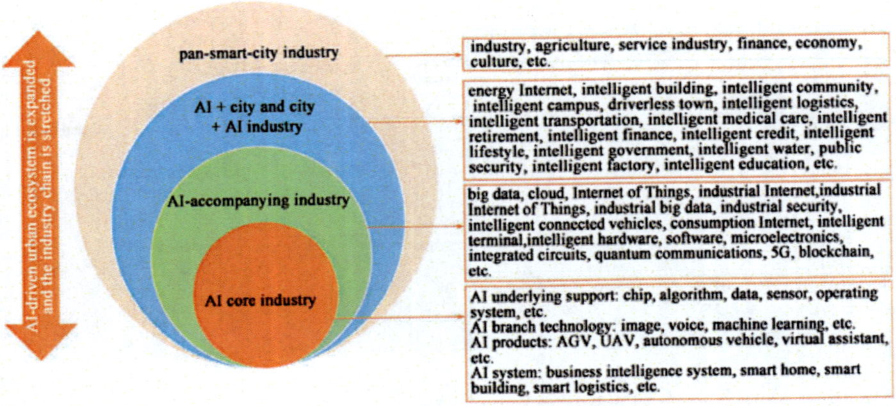

Fig. 10.29 AICITY1.0 open ecosystem model

are more than ten consensual mechanism algorithms, among which the most well-known are proof of work (PoW), proof of stake (PoS), delegated proof of stake (DPoS), etc. The incentive layer integrates economic factors into the blockchain technology system, including the issuance and distribution mechanisms of economic incentives, which mainly appear in the public chain. The contract layer encapsulates scripts, algorithms and smart contracts, which are the basis of the programmability of blockchain. The application layer encapsulates various application scenarios and cases of blockchain, and the future programmable city and programmable society will be built on the application layer. The evaluation layer encapsulates the evaluation and feedback content after the application, including the level of credibility, credit score, improvement measures and so on.

The AICITY1.0 urban blockchain model proposed in this book is shown in Fig. 10.30.

With the further development of AI city, urban intelligence will show the overall evolution trend as shown in Fig. 10.31.

With the advent of the 5G commercialization era, 5G and AI will jointly give birth to ubiquitous intelligent IoT city, which also represents the future of smart city (Fig. 10.32).

AI city is a new form of smart city with AI + as the symbolic driving force in the evolution of smart city during the fourth industrial revolution. It has three core elements: technology, economy and humanity. Each element has its own structure and interacts with each other. In the future, smart city will develop towards AI city. The industry of smart city will be more flexible, customized, personalized, real-time collaborative and diversified. The natural endowment, classic culture and modern industry of city will be integrated organically, and urban governance and social governance will evolve towards transparency and diversification.

The rapid development of China's new urbanization has provided enough space for the development of AI city. After years of development, the new generation information technology with the new generation AI as the main engine has gradually

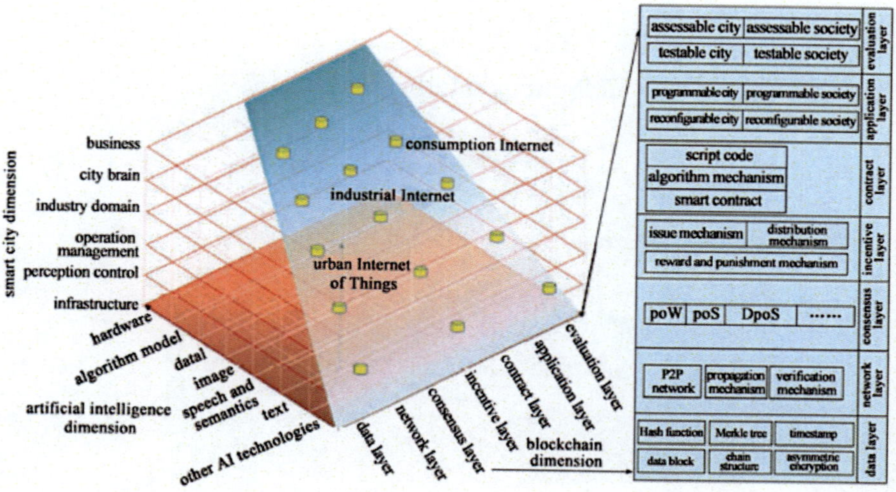

Fig. 10.30 AICITYC1.0 urban blockchain model

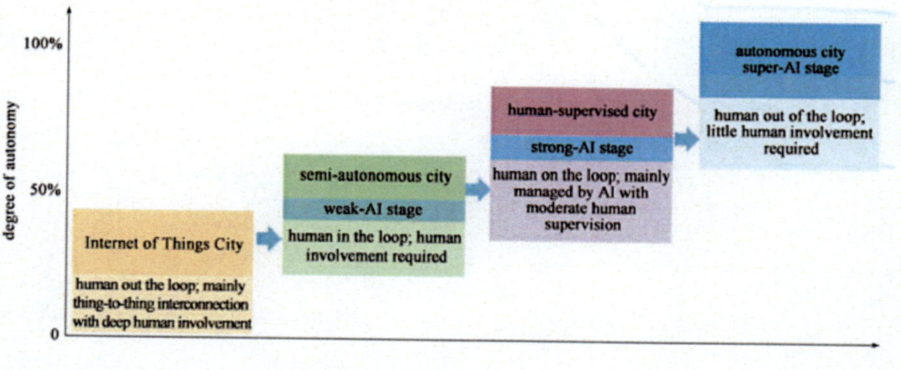

Fig. 10.31 Evolution trend of urban intelligence

become mature. Yet there is still a long way to go for the integration of AI and smart city. The construction of AI city in China is still in its infancy. Successful applications and business models in many subdivision fields, and the accelerated shaping of the Internet of Value have brought us enough hope and confidence. Taking autonomous vehicle and its industrial ecology as the entry point, we can accelerate the construction of AI city. In the near future, a digital country of China supported by AI city technology will rise.

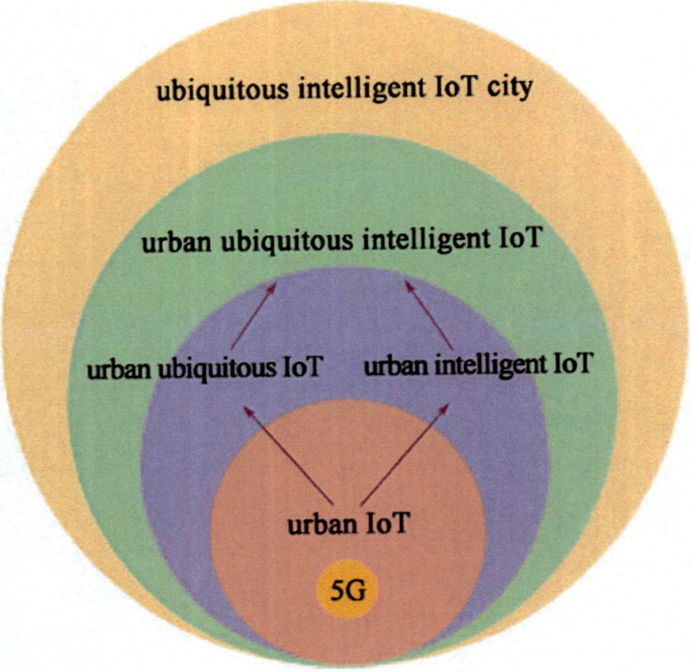

Fig. 10.32 Future of smart city: ubiquitous intelligent IoT city

Bibliography

1. Du M (2015) Research on vision-based technology for autonomous vehicle road environment understanding. Doctoral dissertation. Beijing University of Institute, Beijing, China. [in Chinese]
2. Liu L, Jiang J (2016) Analysis of the key points of automobile fuel system arrangement. Southern Agric Mach 47(12):98 and 107. [in Chinese]
3. Gan L (2017) Analysis of automobile fuel system structure principle and fault diagnosis. Southern Agric Mach 48(15):144–145. [in Chinese]
4. Yue C (2012) Micro pure electric vehicle power system design. Doctoral dissertation. Nanjing University of Aeronautics and Astronautics, Nanjing, China. [in Chinese]
5. Liao C, Li J, Ma Z (2015) Application of wireless energy transmission technology in wireless charging of electric vehicles. Modern Phys Knowl 27(02):51–53. [in Chinese]
6. Gao D, Wang S, Yang F (2015) Research progress of wireless charging technology for electric vehicles. J Automot Saf Energy Conserv 6(04):314–327. [in Chinese]
7. Li B, Liu C, Chen Q et al (2013) Wireless charging technology for electric vehicles. Jiangsu Electr Eng 32(01):81–84. [in Chinese]
8. Liu H (2017) Research on wireless charging technology and its application in electric vehicle charging. Doctoral dissertation. North China Electric Power University, Beijing, China. [in Chinese]
9. Zhao G (2017) Research on powertrain matching and overall optimization of pure electric vehicles. Doctoral dissertation. Jiangsu University. Zhenjiang, Jiangsu, China. [in Chinese]
10. Yao F (2007) Overview of oil-electric hybrid vehicles and their classification. Sci Technol Inf Dev Econ 17(09):173–174. [in Chinese]
11. Yuan Y, Xu S, Wan Y (2017) Analysis of fuel cell vehicle powertrain scheme. Power Technol 41(01):165–168. [in Chinese]
12. Wu B, Chen P, Leng H et al (2007) On-board hydrogen supply system. Shanghai Automot (09):9–11. [in Chinese]
13. Ni G, Ni P, Xiong S (2016) Modern electric vehicles, hybrid electric vehicles and fuel cell vehicles-fundamentals, theory and design, 2nd edn. Machinery Industry Press, Beijing. [in Chinese]
14. Duo S, Guo X, Li P et al (2015) Hydrogen and fuel cells-emerging technologies and their applications, 2nd edn. Machinery Industry Press, Beijing. [in Chinese]
15. China State Council (2016) Thirteenth five-year national informatization plan, (Guo Fa No. 73, 2016). [in Chinese]
16. Federal Government of Germany (2020) Ideas, innovation, growth-Germany's high technology strategy 2020
17. Government of Japan (2016) Fifth science and technology basic plan (2016–2020)
18. Federal Government of Germany (2015) Reference architecture model Industry 4.0

M. Du, *Autonomous Vehicle Technology*, Unmanned System Technologies,
https://doi.org/10.1007/978-981-19-4143-6

19. Government of Japan (2016) Japan intelligent manufacturing reference framework IVRA (Industrial Value Chain Reference Architecture)
20. United States Industrial Internet Consortium (IIC) (2017) Industrial Internet reference architecture Ver1.8
21. Central Committee of the Communist Party of China, China State Council (2014) National new urbanization plan (2014–2020). [in Chinese]
22. Pentland A (2015) Smart society: big data and social physics. Zhejiang People's Publishing House, Hangzhou. [in Chinese]
23. General Office of the CPC Central Committee, General Office of the China State Council (2016) Outline of the thirteenth five-year plan for national economic and social development of the People's Republic of China. [in Chinese]
24. China State Council (2016) Thirteenth five-year plan for the development of national strategic emerging industries (Guo Fa No. 67, 2016). [in Chinese]
25. China State Council (2015) Notice of the state council on the issuance of "Made in China 2025" (Guo Fa No. 28, 2015). [in Chinese]
26. China State Council (2016) The state council's guiding opinions on deepening the integration and development of manufacturing and the internet (Guo Fa No. 28, 2016). [in Chinese]
27. China State Council (2017) Notice of the state council on the issuance of a new generation of artificial intelligence development plan (Guo Fa No. 35, 2017). [in Chinese]
28. Ministry of Industry and Information Technology of China (2017) Three-year action plan to promote the development of a new generation of artificial intelligence industry (2018–2020). [in Chinese]
29. Du M (2017) Ai modeling and practice of new smart city application system. China Constr Inf (18):34–37. [in Chinese]
30. Du M, Wang J et al (2016) Multi-scale road perception for ground robots based on semantic tree Markov random field model. J Mil Eng 37(03):512–517. [in Chinese]
31. Du M, Wang J et al (2014) Application of single-layer wavelet packet approximation to compressed perception for road understanding. J Comput-Aided Des Gr 26(11):2007–2015. [in Chinese]
32. Du M et al (2014) Autonomous mobile robot adaptive outdoor road detection. Chin J Gr 19(07):1046–1053. [in Chinese]
33. Du M (2018) Intelligent internet boosts rapid development of unmanned vehicles. China Inf Commun (02):70–73. [in Chinese]
34. United States Congress (2018) Artificial intelligence security council act of 2018 (draft for comment)
35. Du M, Zhu Y et al (2018) A new form of future smart city-trustworthy AI city. Xiong'an Digital City, (01). [in Chinese]
36. Chen C, Liu S (2016) Application of high precision map in automatic driving. https://36kr.com/p/1721319325697. [in Chinese]
37. Du M (2018) How artificial intelligence empowers smart cities in the era of high cross-border integration. http://www.sohu.com/a/253857482_99947626. [in Chinese]
38. Federal Government of Germany (2015) Reference architecture model Industry 4.0.
39. Government of Japan (2016) Japan intelligent manufacturing reference framework IVRA (Industrial value chain reference architecture)
40. United States Industrial Internet Consortium (IIC) (2017) Industrial internet reference architecture Ver1.8
41. AdamShan (2018) Introduction to convolutional neural network-real-time vehicle detection based on deep learning. https://blog.csdn.net/AdamShan/article/details/79193775
42. AdamShan (2018) Hierarchical finite state machine and autonomous vehicle behavior planning. https://blog.csdn.net/AdamShan/article/details/80633099
43. Lillicrap TP et al (2015). Continuous control with deep reinforcement learning. https://arxiv.org/abs/1509.02971

44. Department of Science and Technology, Ministry of Industry and Information Technology of China (2019) Interpretation of the action plan for the development of the internet of vehicles (smart connected vehicle) industry. http://www.miit.gov.cn/n1146285/n1146352/n3054355/n3057497/n3057503/c6565736/content.html. [in Chinese]
45. Li L, Lin Y, Cao D et al (2017) Parallel learning-a novel theoretical framework for machine learning. J Autom 43(01):1–8. [in Chinese]
46. Wang F, Zhang J, Wei Q et al (2017) PDP: parallel dynamic programming. IEEE/CAA J Autom Sin 4(01):1–5
47. Ge B, Han J, Wei Z et al (2015) Dynamic adaptive ant colony optimization algorithm for the minimum-maximum vehicle path problem. Pattern Recognit Artif Intell 28(10):930–938. [in Chinese]
48. Xu J, Wang Y, Lin P (2015) Dynamic shortest path algorithm in big data environment. J South China Univ Technol (Natural Science Edition) 43(10):1–7. [in Chinese]
49. Mo Y (2016) Research on key technologies of traffic information distribution and processing in the IoV environment. Doctoral dissertation. Jilin University, Changchun, Jilin, China. [in Chinese]
50. Zhang W (2011) Research on key issues of driver fatigue recognition based on machine vision. Doctoral dissertation. Tsinghua University, Beijing, China. [in Chinese]
51. Hu M (2014) Research on seat belt recognition method based on in-vehicle machine vision. Doctoral dissertation. Jiangsu University, Zhenjiang, Jiangsu, China. [in Chinese]
52. Jiang R (2012) Multi-dimensional road modeling and vehicle localization optimization for vision-assisted driving system. Doctoral dissertation. Shanghai: Shanghai Jiaotong University, Shanghai, China. [in Chinese]
53. Shen H (2010) Research on visual environment perception technology for intelligent vehicles. Nanjing University of Aeronautics and Astronautics, Nanjing, Jiangsu, China. [in Chinese]
54. Yu G, Li Q, Wang D (2013) Research on intelligent lane line tracking method for vehicles. In: Proceedings of the eighth annual china intelligent transportation conference-intelligent transportation and safety. [in Chinese]
55. Wang J (2016) Research on key technologies of environment sensing system for autonomous vehicles. Doctoral dissertation. University of Science and Technology of China, Hefei, Hebei, China. [in Chinese]